D0204854

# STEEL–CONCRETE COMPOSITE STRUCTURES

## Stability and Strength

## *Related titles*

### SHELL STRUCTURES: STABILITY AND STRENGTH
*edited by R. Narayanan*

1. Stability and Collapse Analysis of Axially Compressed Cylindrical Shells A. CHAJES
2. Stiffened Cylindrical Shells under Axial and Pressure Loading J. G. A. CROLL
3. Ring Stiffened Cylinders under External Pressure S. KENDRICK
4. Composite, Double-Skin Sandwich Pressure Vessels P. MONTAGUE
5. Fabricated Tubular Columns used in Offshore Structures W. F. CHEN and H. SUGIMOTO
6. Collision Damage and Residual Strength of Tubular Members in Steel Offshore Structures T. H. SØREIDE and D. KAVLIE
7. Collapse Behaviour of Submarine Pipelines P. E. DE WINTER, J. W. B. STARK and J. WITTEVEEN
8. Cold-Formed Steel Shells G. ABDEL-SAYED
9. Torispherical Shells G. D. GALLETLY
10. Tensegric Shells O. VILNAY

### STEEL FRAMED STRUCTURES: STABILITY AND STRENGTH
*edited by R. Narayanan*

1. Frame Instability and the Plastic Design of Rigid Frames M. R. HORNE
2. Matrix Methods of Analysis of Multi-Storeyed Sway Frames T. M. ROBERTS
3. Design of Multi-Storey Steel Frames to Sway Deflection Limitations D. ANDERSON
4. Interbraced Columns and Beams I. C. MEDLAND and C. M. SEGEDIN
5. Elastic Stability of Rigidly and Semi-Rigidly Connected Unbraced Frames G. J. SIMITSES and A. S. VLAHINOS
6. Beam-to-Column Moment-Resisting Connections W. F. CHEN and E. M. LUI
7. Flexibly Connected Steel Frames K. H. GERSTLE
8. Portal Frames Composed of Cold-Formed Channel- and Z-Sections G. J. HANCOCK
9. Braced Steel Arches S. KOMATSU
10. Member Stability in Portal Frames L. J. MORRIS and K. NAKANE

### CONCRETE FRAMED STRUCTURES: STABILITY AND STRENGTH
*edited by R. Narayanan*

1. Stability of Reinforced Concrete Building Frames J. G. MACGREGOR
2. Stability of Compression Members U. QUAST
3. Nonlinear Analysis and Optimal Design of Concrete Framed Structures C. S. KRISHNAMOORTHY
4. Reinforced Concrete Columns in Biaxial Bending I. E. HARIK and H. GESUND
5. Design of Concrete Structures for Torsion R. NARAYANAN
6. Reinforced Concrete Deep Beams F. K. KONG
7. Column-supported Shear Walls L. CERNY and R. LEON
8. Design of Reinforced Concrete Flat Slabs P. E. REGAN
9. Progressive Collapse of Slab Structures D. MITCHELL and W. D. COOK

# STEEL–CONCRETE COMPOSITE STRUCTURES

## Stability and Strength

*Edited by*

R. NARAYANAN

M.Sc.(Eng), Ph.D., D.I.C., C.Eng., F.I.C.E., F.I.Struct.E., F.I.E.
*Manager (Education and Publications), The Steel Construction Institute, Ascot, United Kingdom*

ELSEVIER APPLIED SCIENCE
LONDON and NEW YORK

ELSEVIER APPLIED SCIENCE PUBLISHERS LTD
Crown House, Linton Road, Barking, Essex IG11 8JU, England

*Sole Distributor in the USA and Canada*
ELSEVIER SCIENCE PUBLISHING CO., INC.
52 Vanderbilt Avenue, New York, NY 10017, USA

WITH 17 TABLES AND 186 ILLUSTRATIONS



**British Library Cataloguing in Publication Data**

Steel–concrete composite structures:
stability and strength.
1. Composite construction 2. Structural
stability
I. Narayanan, R.
624.1′8341 TA664

**Library of Congress Cataloging-in-Publication Data**

Steel–concrete composite structures.

Includes bibliographies and index.
1. Composite construction. I. Narayanan, R.
TA664.S74 1988 624.1′821 87-24471

ISBN 1-85166-134-4



Phototypesetting by Keyset Composition, Colchester, Essex
Printed in Great Britain at the University Press, Cambridge

# PREFACE

It is my privilege to write a short preface to this book on Steel–Concrete Composite Structures, the seventh in the planned set of volumes on *Stability and Strength of Structures*.

Recent years have seen a substantial increase in the use of composite construction and an enhanced research effort aimed at developing techniques for combining steel and concrete effectively. Composite construction has several advantages over the traditional reinforced concrete or steel structures: these include high strength-to-weight ratios, structural integrity, durable finishes, dimensional stability and sound absorption. From the point of view of structural performance, composite construction results in an efficient design, achieved by the most suitable arrangement of the highly stressed elements in a structure; significant economies of cost and construction time have been observed especially in building floors and bridges by employing this form of construction. The associated technology is in a continuing state of development on account of the potential for improvements in the cost-effectiveness of this form of building. Codes of Practices on Composite Construction are being revised in the UK and in Europe, based on the substantial amount of new knowledge gained as a result of studies in recent years.

This book presents the principles and methods of analysis and design of steel–concrete composite construction and serves to describe the background research carried out on both sides of the Atlantic. We have continued the policy of inviting several expert contributors to write a chapter each, so that the reader is presented with the 'state-of-the-art' on a number of related topics with sufficient introductory material. The topics chosen are not exhaustive, but reflect the diverse fields in which composite construction has a demonstrated potential.

I am indebted to Professor R. P. Johnson who has provided an overview of the established types of steel–concrete composite structures (viz. columns, steel decking slabs, beams, steel frames with concrete in-fill, bridges etc.) discussed in the book. The final chapter describes the current state-of-the-art on Ferrocement, a valid form of steel–concrete composite, which is increasingly popular in countries where labour costs are not high when compared with those prevailing in the Western World.

As Editor, I wish to express my gratitude to all the contributors for the willing co-operation they extended in producing this volume. I sincerely hope that readers find the book informative and stimulating.

R. NARAYANAN

## CONTENTS

*Preface* . . . . . . . . . . . . v

*List of Contributors* . . . . . . . . . ix

1. Composite Structures: A Preview . . . . . . . 1
   R. P. JOHNSON

2. Steel–Concrete Composite Flooring Deck Structures . . 21
   H. R. EVANS and H. D. WRIGHT

3. Behavior and Design of Composite Beams with Web Openings 53
   DAVID DARWIN

4. Composite Girders with Deformable Connection between Steel and Concrete . . . . . . . . . . 79
   V. KŘÍSTEK and J. STUDNIČKA

5. Steel Frames with Concrete Infills . . . . . . . 115
   T. C. LIAUW

6. Steel–Concrete Composite Columns—I . . . . . . 163
   H. SHAKIR-KHALIL

7. Steel–Concrete Composite Columns—II . . . . . 195
   R. W. FURLONG

8. Design of Composite Bridges (Consisting of Reinforced Concrete Decking on Steel Girders) . . . . . 221
   J. B. KENNEDY and N. F. GRACE

9. Composite Box Girder Bridges . . . . . . . 249
R. GREEN and F. A. BRANCO

10. Ferrocement Structures and Structural Elements . . . . 289
P. PARAMASIVAM, K. C. G. ONG and S. L. LEE

*Index* . . . . . . . . . . . . . 339

# LIST OF CONTRIBUTORS

F. A. BRANCO
*Professor, Centro de Mecânica e Engenharia Estruturales, Universidade Tecnica de Lisboa CMEST, Instituto Superior Técnico, 1096 Lisboa Codex, Portugal*

D. DARWIN
*Professor, Department of Civil Engineering and Director, Structural Engineering and Materials Laboratory, University of Kansas, Lawrence, Kansas 66045, USA*

H. R. EVANS
*Professor and Head, Department of Civil and Structural Engineering, University College, Cardiff CF2 1TA, UK*

R. W. FURLONG
*Donald J. Douglass Centennial Professor of Engineering, University of Texas at Austin, Austin, Texas 78712, USA*

N. F. GRACE
*Department of Advanced Technology, Giffels Associates, Southfield, Michigan, USA*

R. GREEN
*Professor, Department of Civil Engineering, University of Waterloo, Waterloo, Ontario N2L 3G1, Canada*

R. P. JOHNSON
*Professor, Department of Engineering, University of Warwick, Coventry CV4 7AL, UK*

J. B. KENNEDY

*Professor, Department of Civil Engineering, University of Windsor, Windsor, Ontario N9B 3P4, Canada*

V. KŘÍSTEK

*Professor, Faculty of Civil Engineering, Czech Technical University, Thákurova 7, Dejvice, 166 29 Prague 6, Czechoslovakia*

S. L. LEE

*Professor and Head, Department of Civil Engineering, National University of Singapore, Kent Ridge, Singapore 0511, Republic of Singapore*

T. C. LIAUW

*Reader, Department of Civil and Structural Engineering, University of Hong Kong, Pokfulam Road, Hong Kong*

K. C. G. ONG

*Senior Lecturer, Department of Civil Engineering, National University of Singapore, Kent Ridge, Singapore 0511, Republic of Singapore*

P. PARAMASIVAM

*Associate Professor, Department of Civil Engineering, National University of Singapore, Kent Ridge, Singapore 0511, Republic of Singapore*

H. SHAKIR-KHALIL

*Lecturer, Department of Civil Engineering, University of Manchester, Oxford Road, Manchester M13 9PL, UK*

J. STUDNIČKA

*Associate Professor, Faculty of Civil Engineering, Czech Technical University, Thákurova 7, Dejvice, 166 29 Prague 6, Czechoslovakia*

H. D. WRIGHT

*Lecturer, Department of Civil and Structural Engineering, University College, Cardiff CF2 1TA, UK*

*Chapter 1*

# COMPOSITE STRUCTURES: A PREVIEW

R. P. JOHNSON

*Department of Engineering, University of Warwick, Coventry, UK*

*SUMMARY*

*The material in each of chapters 2–9 is part of an international 'State of the art' that has many inconsistencies, as do the national design methods for composite structures in current use. Attention is drawn in this Introduction to some of these inconsistencies, and reasons for them are discussed. The contents of chapters 2–9 are related to the writer's experience and to the multinational body of knowledge that was used in drafting Eurocode 4,* Composite Structures of Steel and Concrete *(1985). Reference is also made to gaps in recent design Codes, and to the information needed to fill them.*

## 1.1 GENERAL

### 1.1.1 Introduction

The authors of the various chapters of this book have been drawn from many different countries. The material presented is more comprehensive, but less coherent, than it would be in a book based on practice in one region, such as North America, Western Europe or the USSR.

This could be confusing to readers who are practising engineers rather than specialists in composite structures. An attempt is made in this chapter to place the material in a wider context. Some reflections are given on the nature of the information presented and then, for each of the other chapters, comments are made from a viewpoint different from that of its author or authors, or supplementary information is given. Sections 1.2 to

1.9 of this chapter refer to Chapters 2 to 9, respectively, and each may conveniently be read after the relevant chapter.

The use of limit state design philosophy for new codes is now so general that no reference is made in this chapter to permissible-stress design methods. The emphasis is on recent developments in Western Europe and North America, regions whose design codes continue to influence practice in other continents such as South America and Africa. The USSR was one of the earliest nations to develop limit state design philosophy, but so little of this work or of recent Russian research is readily available in English, that it is not referred to here.

### 1.1.2 Types of Information

Most of the material published on a technical subject such as composite structures falls into one or more of the following categories:

(1) An account of original research, sufficiently detailed for other research workers to evaluate and check its conclusions.
(2) A critical review of knowledge of a subject, either as it developed over a period of time, or in the form of a 'state of the art' at the time of writing.
(3) A review of developments in the conception and use of a certain type of structure or structural form.
(4) A review of one or more design procedures, and of their origins in research and practice (e.g., a commentary on a code of practice).
(5) A statement of a design procedure, without explanation or comment (e.g., a code of practice).

Each of Chapters 2 to 9 includes material of types (2) and (4). In addition, Chapters 3 (openings in webs), 4 (shear connection) and 5 (infilled frames) give detailed accounts of research or theory; and Chapters 6 and 7 (columns) include material of type (3).

### 1.1.3 Inconsistency of Information

In pure science, if knowledgeable authors in ten countries were all invited to review the same subject, one might expect the results to be similar. There would be agreement on what were established facts and accepted theories; and in controversial areas the authors would refer to the same leading hypotheses.

Engineering is different. Facts of interest to engineers often relate to components (e.g., profiled steel sheeting) made to specifications that differ from country to country, and to properties of components determined by tests (e.g. a push test) that also differ between countries.

Engineers everywhere are interested in design and construction of objects, such as bridges, useful to man; but the information sought by any one engineer may depend on the location of his work. For example, the influence of de-icing salt on the durability of concrete will be of interest in the UK, and resistance to earthquakes is important in California; but the converse is not true.

These are examples of questions to which the accepted answers are becoming independent of location, as knowledge spreads; but for design concepts and procedures, world-wide convergence is much slower. These are still strongly regional for reasons, now discussed, that are rooted in history and geography. There were, for example, marked differences between the conceptions in the 1970s of composite highway bridges in Switzerland and in the UK. This was found by Johnson and Buckby (1979) to be mainly due to differences in the imposed loadings specified in those countries, which in turn resulted from different types of traffic and different historical developments of their systems of transportation.

Research on the response of structures to their loading and environment continues to reveal detail of great complexity which, for design purposes, must be simplified. If all the necessary simplifications are made 'on the safe side', the result is ultra-conservative and uneconomic design; so we use conservatism in one simplification to offset optimism in another, in ways influenced by experience and even by tradition. Codes of practice for composite structures differ from country to country both for this reason and because no one drafting committee can possibly know, or even know of, all of the relevant experience and research literature. It will know best, and will tend to use, the work done in its own country.

### 1.1.4 Examples of Differences between Design Rules for Composite Structures

#### *1.1.4.1 Factors $\varphi$ and $\gamma_m$*

The resistance or capacity reduction factors $\varphi$ used in the USA are broadly equivalent to the reciprocals of the $\gamma_m$ factors used in Eurocodes and the British codes; but the concepts represented are not identical so that, in general, we find that $\varphi \neq 1/\gamma_m$.

#### *1.1.4.2 Levels of Loading for Highway Bridge Decks*

The Autostress design method for composite bridges (*Standard Specifications for Highway Bridges* (1983) and Haaijer, Carskaddan and Grubb (1983)) uses three levels of imposed loading: service load, overload, and maximum design load. There is no simple relationship between these concepts and those represented by the serviceability and ultimate levels of HA and HB loading in the British bridge code BS 5400: Part 2 (1978).

*1.1.4.3 Classification of Compressed Steel Plates*
The behaviour of plate elements of steelwork stressed in compression depends mainly on their breadth/thickness ratios. Codes classify such elements according to slenderness, but in ways that vary from code to code. Some of the differences are in presentation only, but others are technical and may be related to the design methods given for the various classes; for example, to the permitted amounts of redistribution of moments in continuous beams.

*1.1.4.4 Partial Shear Connection*
Section 1.1.4.3 illustrates the fact that design rules are devised in related packages. Another example of this is found in the limitations to the use of partial shear connection in composite beams. The main objective is to ensure that the probability of sudden collapse due to the failure of all of the connectors in a shear span is lower than the probability of at least one less abrupt mode of failure.

It was deduced by Johnson (1986) from the limited test data that the shear resistance of a beam depends on the length of its shear spans, the degree of partial shear connection and the load-slip properties of the connectors used. These in turn are influenced by the strength and density of the concrete and the method used for determining the effective breadth of the concrete flange, because that affects the number of connectors needed for full shear connection.

In the draft Eurocode 4 (1985), account is taken of this behaviour by defining 'flexible' connectors, by giving a less conservative design method for such connectors than for 'stiff' connectors, and by allowing partial shear connection only down to 50% of full connection, and only in beams of span not exceeding 20 m.

In another code, one may find no distinction between flexible and non-flexible connectors, or a different limit on span. Such differences may arise from different simplifications of test data of limited scope, but usually they are related to other differences between the two design packages. The penalty for simplifying a package, or extending its scope, is always that it becomes more conservative in some of the situations covered.

### 1.1.5 Use of More than one Design Code
The preceding examples show why designers are warned of the danger of using the 'best' bits of different codes in a single design. Usually, they are

required to follow one specified national code. No such code should be supplemented by the use of material from a foreign code, without a good understanding of the backgrounds of both documents.

## 1.2 COMPOSITE FLOORS USING STEEL SHEETING

The pioneering work on the use of profiled steel sheeting was done in North America, as is evident from the papers cited in Chapter 2. It became better known in Western Europe after Professor J. Fisher spent several months at the Federal Polytechnic of Lausanne. There was thus an input of American practice to a Committee of the European Convention for Constructional Steelwork chaired by Professor Reinitshuber, and this in turn influenced British Standard 5950: Part 4 (1982) and draft Eurocode 4 (1985), so that there is good agreement between these codes and practice in North America. The Eurocode includes the design of composite beams supporting sheeting, whereas in BS 5950 this is not in Part 4 but in Part 3 (in preparation).

### 1.2.1 Determination of $m_r$ and $k_r$

In Sections 2.3.1 and 2.3.2 of Chapter 2, reference is made to two types of testing programme for finding the shear transference characteristics $m_r$ and $k_r$ (defined in Fig. 2.14) of a particular type of profiled steel sheet.

The second type of testing, the more widely used, is specified in almost identical terms in BS 5950: Part 4 (1982) and in draft Eurocode 4; but for Eurocode 4 it was decided that the procedure for calculating $m_r$ and $k_r$ should be consistent with that for the evaluation of push tests on shear connectors, and with the clauses on testing in draft Eurocode 3 (1984). The differences between this procedure and that given in BS 5950 are now explained.

The essential features of Fig. 2.14 are reproduced in Fig. 1.1. A minimum of three tests is done in each of two regions, A and B. For each region the variation of each result from the mean should not exceed $\pm 10\%$ (EC4) or $\pm 7{\cdot}5\%$ (BS 5950). The values of $k_r$ and $m_r$ are determined from a line (CD in Fig. 1.2) which:

—in EC4, is 10% below the lowest result in each region, and
—in BS 5950, is 15% below the regression line (EF) through the two groups of results (or 10% below if eight or more tests are done).

If the limits on variation are not satisfied, then at least three further tests are to be done in the region concerned. In EC4, the line CD then passes through the characteristic value for that region, determined following Eurocode 3 and using all the results; whereas in BS 5950, the regression line is found using all the results, and line CD is found from it as before.

In both codes, the concrete strength is taken as the mean of the results of tests on four specimens, cured under the same conditions as the test slabs. This cannot, of course, be the curing under water normally used for standard concrete cubes and cylinders.

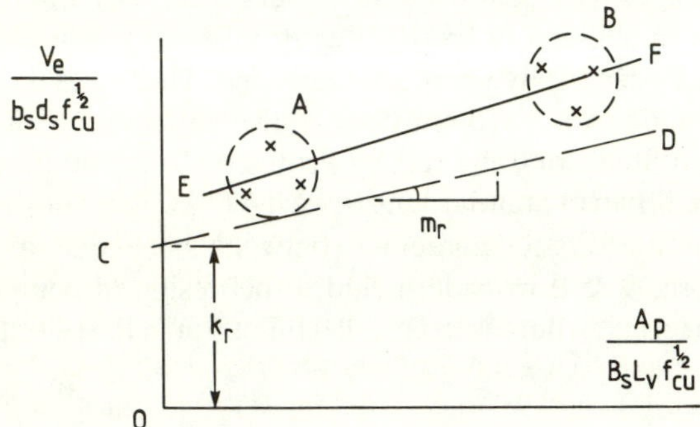


FIG. 1.1. Determination of shear bond capacity.

Neither of these methods is inherently 'better' than the other. They simply give two different approximations to unknown values that could only be found more accurately if many more tests were done.

### 1.2.2 Strength and Stiffness of Stud Shear Connectors

Reference is made in Section 2.4.1 to the reduction factor for the strength of stud connectors in ribbed slabs that was published by Grant, Fisher and Slutter in 1977. The formula is for decks with corrugations at right angles to the steel beam, as in Fig. 2.3. It gives $P_{dr}$, the design shear strength of a stud of overall height $h$ in a rib of breadth $b_r$ and height $h_r$, as

$$P_{dr}/P_d = 0{\cdot}85n^{-0{\cdot}5}(b_r/h_r)[(h/h_r) - 1]$$

where $P_d$ is the design shear strength for the stud in a solid concrete slab and $n$ is the number of studs per rib. The ratio $P_{dr}/P_d$ may not be taken as greater than 1·0, and $n$ may not be taken as greater than 3.

This same formula is given by Lawson (1983) and in draft Eurocode 4 (1985). The results referred to in Chapter 2 suggest that it is unreliable for the particular decking, studs and welding techniques used in the tests—but the formula cannot be studied in isolation because, in design, $P_{dr}$ depends on the method used to calculate $P_d$, and several different methods are in use (Oehlers and Johnson (1987)).

It is remarkable that in the push tests reported in Fig. 2.19, specimen 6 is almost 50% stronger than specimen 4 at 3 mm slip, and that at low slips the ratio of their stiffnesses is as great. It would be interesting to know whether the equation above predicts this difference. The significant difference between curves 2 and 3 results from a variable, the position of the stud in the trough, not included in the formula (because this variable obviously cannot affect $P_d$). In regions where the direction of shear can reverse, studs should be placed near the centre of the troughs, and design formulae should relate to that location.

The modified push test used by the authors of Chapter 2 is much more realistic (in relation to beams) than is the standard test. The effect of providing studs at two cross-sections, rather than one (Fig. 2.18), is to increase the amount of redistribution of load from one slab to the other that can occur, and so to increase the failure load per stud found in the test (Oehlers and Johnson (1987)). The extent of the increase for this shape and size of slab is important only if data are also used from tests with only one stud per slab. The more important question is whether either type of test reliably predicts the behaviour of studs in beams. An analysis of the results of the eight beam tests from this point of view could be useful.

In summary, the test results referred to in Chapter 2 appear to raise new questions, and they suggest that much more testing is needed—which is surprising in view of the long and extensive use of through-deck stud welding in North America. For longitudinal shear in beams, current design methods appear to be conservative, even though the push tests gave lower shear strengths than predicted; but for stiffness, draft Eurocode 4 allows the increase in flexibility due to slip to be neglected altogether (Clause 5.4.2), whereas even the increase in flexibility given by the linear interaction theory is reported in Chapter 2 to be insufficient.

## 1.3 COMPOSITE BEAMS WITH WEB OPENINGS

Chapter 3 is primarily a review of research, most of which originates from North America. It deals mainly with rectangular holes in the webs of rolled steel I-section beams that are composite with a concrete slab, where the edges of the holes are not stiffened.

It has limited applicability to plate girders (where holes through the thinner webs would often need local stiffening) and does not refer to the related problem that occurs at the holes in castellated beams. This has been the subject of extensive research, to which Knowles (1985) lists 40 references.

The subject of Chapter 3 is not treated in any British code of practice for composite structures, nor in draft Eurocodes 3 or 4, perhaps because in Europe the use of composite construction has not advanced as far in buildings as in bridges. Holes of the type considered occur in diaphragms in box-girder bridges (and are covered by codes), but not in webs of main girders.

The design method given in Section 3.3.2 is appropriate for design for the ultimate limit state of sagging moment regions of beams for buildings. In beams larger and more slender than those tested, it seems likely that behaviour would be modified by web buckling near a corner of the hole, so it would be useful if the limits of applicability of the method could be more precisely defined.

Attention is drawn in Section 3.2.3 to the occurrence of cracking at loads as low as 21% of the ultimate. Cracking is not necessarily a form of unserviceability; but this result suggests that serviceability criteria, including extra deflection due to yielding of steel, could in some circumstances govern design, or limit the scope of the method.

## 1.4 GIRDERS WITH DEFORMABLE SHEAR CONNECTION

Two distinct types of structure are treated in Chapter 4: conventional uncased composite beams, and members consisting of coaxial hollow cylinders of steel and of concrete.

### 1.4.1 Composite Beams for Buildings and Bridges

A summary is given of available methods of partial-interaction elastic analysis for beams of uniform section, assuming a linear relation between load and slip for the shear connection, and neglecting uplift. These are the appropriate assumptions for analyses for stresses and deflections in service. Even so, the accuracy of the results will be limited, as the authors emphasize, by uncertainty in the creep properties of the concrete, the flexibility ($K$) of the shear connection, and the effects of residual stresses in the steelwork. When a structure is being designed, it may not be possible to determine any of these data accurately.

The property $K$ is equal to the spacing of the shear connectors along the beam, $p$, divided by the relevant slope of the load/slip curve for a connector, $k$. When connectors other than studs are included, $k$ can range from about 100 to 400 kN/mm (Johnson and May, 1975), and although the stiffer connectors tend to be used at wider spacing, $p/k$ can be less than $0{\cdot}5 \times 10^{-3}$ mm$^2$/N in long-span bridges, and greater than $3 \times 10^{-3}$ mm$^2$/N in beams for buildings with partial shear connection at the 50% level. Fortunately, even a tenfold change in the flexibility $K$ does not greatly influence the stresses calculated by partial-interaction analysis, as is shown by Fig. 4.9.

All the methods of analysis given in Chapter 4 become fairly complicated when $K$ varies along the span, or when the beam is continuous or has a non-uniform cross-section or irregular loading. This is one reason why no current design code requires the use of partial-interaction analysis. Another reason is that stresses calculated by elastic theory rarely enter into the limit-state design of beams for buildings. They are important in the design of bridge girders of slender section, but in these the shear connection is so stiff (to provide appropriate resistance to the effects of travelling and repeated loading) that full-interaction theory is accurate enough.

The main use of partial-interaction analysis is for the derivation and checking of certain simplified design methods. Examples of these are the methods given in codes for:

(a) the use of partial shear connection;
(b) the additional shear connection required near end supports for the primary effects of temperature difference;
(c) the spacing of shear connectors across the breadth of closed-top box girders; and
(d) estimating the influence of slip on deflections.

The method for (d) given in Chapter 4 should be useful in the common situation when design of a simply-supported composite beam is governed by its deflection. The extension of the method to continuous beams is probably not necessary, because deflections of such beams are rarely critical.

*1.4.1.1 Effect of span of beam on loss of interaction due to slip*

In geometrically similar beams of different span, loss of interaction due to slip is greatest at very short spans. A study of this subject (Johnson, 1981) has shown that even in composite plates spanning only 3 m, the resulting increases in bending stresses are acceptable.

### 1.4.2 Hollow Composite Tubes

The conclusions of Chapter 4 on the effect of shear connection on bending stresses in hollow composite tubes are consistent with current practice in the design of composite columns, where shear connection (other than by natural bond) is not normally provided. The important exception to this practice is where a concentrated longitudinal force is applied to either the steel or the concrete component of the column, and local shear connection may be needed to transfer part of the force to the other material. Guidance on the design of the connection is given in draft Eurocode 4 (1985).

## 1.5 INFILLED FRAMES

The comprehensive review given in Chapter 5 of the literature on infilled frames includes masonry infill panels to steel, concrete or composite frames, whereas the rest of the chapter is concerned solely with steel frames and concrete infills.

Designers have hitherto made little use of this extensive body of research, partly because few codes give relevant design methods. There is, for example, no mention of infilled frames in the draft Eurocodes or in any current British code of practice.

### 1.5.1 Masonry Panels

A Draft for Public Comment of a British Code (Draft standard for the use of structural steel in buildings (now BS 5950), Part 3 (1976)), was issued in 1976. This included a chapter 'Composite action between masonry panels, beams and columns' and a commentary. The design methods, given for the ultimate limit state, could be assumed to satisfy also the serviceability limit state in respect of excessive cracking of the masonry, and of deflections of supporting beams and of masonry-braced frames up to four storeys in height.

The clauses 'Masonry panels on beams' made provision for arching action by reducing the bending moment at midspan of the beam (of span $L$), due to the weight $W_w$ of the panel, from $W_w L/8$ to a value between $W_w L/30$ and $W_w L/140$ that depended on the weight per unit length of the panel, its thickness, and the stiffness of the steel or composite beam. The relevant research was by Riddington and Stafford Smith (1978).

The clauses 'Masonry infilled frames subjected to horizontal loading', meaning loading in the plane of the frame, gave a design method based mainly on the work by Stafford Smith and Riddington cited in Chapter 5, with contributions from work by Wood and by Mainstone, also referenced in Chapter 5.

In the draft code it was necessary to define the scope of these methods by geometrical limits, and to consider problems of the type that most research workers conveniently overlook, such as frames and panels that are not precisely co-planar, the presence of damp-proof courses and of coexisting out-of-plane loading, the effect of propping (or of not propping) the beam while the panel is built on it, and the behaviour of panels of cavity construction.

The most difficult problems were to make appropriate provision for the effects of any holes through the panel, whose location is known to the designer, and to specify appropriate restrictions on the making of further holes and on the removal of panels while the structure is in service. This last subject was treated as follows.

'If reliance is placed upon framed panels to ensure the lateral stability of any storey in a frame, a sufficient number of such panels should be placed in that storey to allow alternative paths for horizontal loads in the event of the simultaneous removal of all the panels from alternate internal bents. When such removal could serve no useful purpose, it is sufficient to consider the simultaneous removal of one panel from each bent in that storey.'

Despite such safeguards, designers of composite structures continue to be reluctant to take advantage of the in-plane stiffness and strength of masonry panels, probably because of the impossibility of predicting future alterations—yet designers of all-masonry structures do not have this inhibition. BS 5950: Part 3 was still in preparation in 1987. It then seemed unlikely that any reference would be made in it to infilled frames.

## 1.6 STEEL–CONCRETE COMPOSITE COLUMNS—I

### 1.6.1 Scope of Design Methods

It is generally agreed by research workers that, using numerical analysis, it is now possible to predict the behaviour up to failure of a composite column, even when it is a component of a frame, and when biaxial bending and end restraints are present. Results of such predictions for columns of uniform and doubly symmetric cross-section have been verified by tests (references in Chapters 6 and 7, and also May and Johnson (1978) and Johnson and May (1978)) and used in design methods (e.g., by Dowling, Chu and Virdi (1977) and Roik, Bode and Bergmann (1977)). The claim is believed to be valid for the types of cross-section illustrated in Chapter 6, all of which are doubly symmetric, provided that the following conditions are satisfied:

(1) The detailing is such that premature local buckling of steel and spalling of concrete cover do not occur.
(2) The method of application of loading is such that local bond stress between structural steel and concrete is not excessive near the ends of the column.

Rules relating to these conditions are given in codes, and those for item (2) may occasionally lead to the provision of shear connection, as noted in Section 1.4.2.

It is shown in Chapters 6 and 7 that the results of the numerical analyses have enabled simple design methods to be derived from columns of uniform and symmetric section; but where asymmetry and/or longitudinal taper are present the designer is unlikely to find either a simplified method or a relevant computer program, and must cautiously interpret and use the existing methods.

**1.6.2 Influence of Sequence of Construction on Resistance of a Column**

The comments in Section 1.6.1 relate to the common situation where the concrete of a composite column is cast at a time when the stresses in the steel component are low. In certain types of multi-storey construction, referred to in Section 7.2, they may not be low. The designer should then ensure that yield of the steel in compression does not occur at service loads.

At ultimate load it is conservative to use an interaction condition of the type

$$N_s/N_{us} + N_c/N_{uc} \ngtr 1 \tag{1.1}$$

where $N_{us}$ and $N_{uc}$ are the design resistances to axial load (allowing for associated bending moments) of the steel and composite members, and $N_s$ and $N_c$ are the design loads for those members, which depend on the sequence of construction.

By analogy with design methods for beams of compact section built unpropped, this problem can be ignored in short columns (i.e., those whose resistance is not influenced by second-order effects), and the normal condition $N_s + N_c \ngtr N_{uc}$ can be used.

For columns so slender that account is taken in design of the influence of creep on buckling, use of eqn 1.1 is recommended. Little information on intermediate situations is available. Often, the steel section is so small a component of the final section that its presence can be ignored when $N_{uc}$ is calculated, and the conditions

$$N_s/N_{us} \not> 1 \text{ and } N_c/N_{uc} \not> 1$$

can be used.

### 1.6.3 Concrete Contribution Factor

It is explained in Chapter 6 that the squash load of a column, $N_u$, is the sum of contributions from the structural steel, the reinforcement and the concrete, written here as $N_s$, $N_r$ and $N_c$ respectively. The concrete contribution factor, used in several codes, is defined by $\alpha = N_c/N_u$.

Ideally, a design method for composite columns should be consistent with the design of steel columns when $N_r = N_c = 0$, and with the design of reinforced concrete columns when $N_s = 0$. These limits correspond to values 1 and 0, respectively, for a *steel contribution ratio*, $\delta$, defined by

$$\delta = N_s/N_u$$

The corresponding values for $\alpha$ are 0 and $1 - N_r/N_u$, and the second of these is less convenient than the limit $\delta = 0$. This is why the design methods of draft Eurocode 4 (1985) are presented in terms of $\delta$ rather than $\alpha$. It will be noted that, in general,

$$\delta \neq 1 - \alpha.$$

### 1.6.4 Vectorial Action Effects

This term is used in draft Eurocodes 1 to 7 to refer to a stress resultant that has more than one component; for example, a combination of axial force $N$ and bending moment $M$ acting on a cross-section of a column. It is stated that when one component is favourable, this should be allowed for by reducing the relevant partial safety coefficient $\gamma_F$ applied to that component.

As explained in Section 6.3.3, this situation arises in composite columns subjected to a small axial load. Its effect on the bending resistance is beneficial, as shown in Fig. 1.2, which is a typical interaction diagram for the resistance of the cross-section of a concrete-filled rectangular hollow section. The use of the Roik–Bergmann method for the design of a slender column in this situation is now discussed.

Let $N_{d1}$ and $M_d$ be the unfactored maximum design axial load and bending moment, respectively, for a cross-section of the column, and let $\gamma_F$

be the partial safety factor for loading at the ultimate limit state. Then, if $N_a$ is the resistance of the column to axial load, found from a strut curve, the first design check is to ensure that the bending resistance $M_r/M_u$ given by line AB on Fig. 1.2 exceeds $\gamma_F M_d/M_u$.

Because of the shape of portion CD of the curve, the designer should also determine the minimum design axial load that can coexist with $M_d$, $N_{d2}$ say, multiply that by an appropriate partial safety factor $\gamma_{F2}$ less than 1·0, and check that $M_r/M_u$ given by line EF on Fig. 1.2 also exceeds $\gamma_F M_d/M_u$. With this lower axial load, the column is of course less susceptible to buckling. In the example shown in Fig. 1.2, this more than offsets the

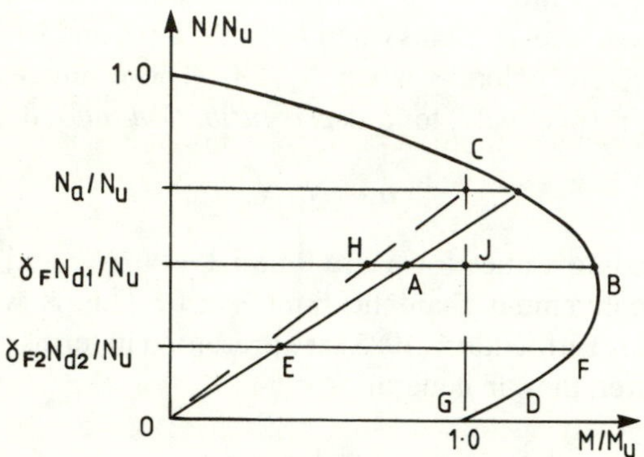


FIG. 1.2. Design of composite columns.

reduction in the bending resistance of the cross-section, so that line EF is longer than AB. It can be very conservative to use the alternative simple method proposed in Section 3.3 of Chapter 6. This is to replace curve CD by the line CG. The available bending resistance $M_r/M_u$ is then given by line HJ, and no minimum design axial load need be considered.

## 1.7 STEEL–CONCRETE COMPOSITE COLUMNS—II

### 1.7.1 Design Compressive Strengths of Concrete in Columns

In Chapter 6, the squash load $N_u$ is presented in terms of a strength $f_{cd}$, and the ultimate moment of resistance $M_u$ in terms of a different strength, $\bar{f}_{cd}$. The reason is that account has to be taken in $\bar{f}_{cd}$, but not in $f_{cd}$, of the unsafe error that results from assuming that the rectangular stress block for concrete in compression extends to the neutral axis. (Design of composite sections becomes very complex if a more realistic stress block, stopping short of the neutral axis, is used.) For example, the values used in the British code of practice CP 110 (1972) were

$$f_{cd} = 0{\cdot}45 f_{cu}, \quad \bar{f}_{cd} = 0{\cdot}40 f_{cu}$$

where $f_{cu}$ is the characteristic cube strength.

Parametric studies by Stark, reported by Smith and Johnson (1986), have shown that the unsafe error in using the higher value for uncased composite beams is negligible, and that in the design of columns it can be allowed for by taking $M_u$ as 90% of the calculated value. This has enabled a single value for the design compressive strength of concrete, $0{\cdot}85\, f_c/\gamma_m$, to be used throughout draft Eurocode 4. In this expression, $f_c$ is the cylinder strength, and the partial safety factor $\gamma_m$ is usually taken as 1·5.

The Roik–Bergmann method of column design described in Section 6.3 is used in draft Eurocode 4, and the simplified procedure for calculating points on the interaction diagram for $N$ and $M_x$ (or $N$ and $M_y$) is explained in an Appendix. It is not possible to comment on the effect of using the assumptions in the preceding paragraph rather than taking the design compressive strength of concrete as $0{\cdot}75 f_c$ (Chapter 6), without considering also the values of $\gamma_F$ and $\gamma_m$ (or the $\varphi$-factors) envisaged in Chapter 6, because, as noted earlier, each design procedure is a package.

### 1.7.2 Resistance of Slender Columns to Axial Load

This subject is discussed in Section 6.6, and recommendations are given in the context of design practice in the United States. It is instructive to compare them with the proposals in Eurocode 4.

In both methods, the influences of creep and cracking on buckling are allowed for by using an effective elastic flexural stiffness of the composite section, $(EI)_e$ say. In the notation of Chapter 6, the recommendation there is

$$(EI)_e = kE_s(I_s + I_r) + (0{\cdot}5E_c)(0{\cdot}67I_c) \tag{1.2}$$

where $k$ is between 0·5 and 0·75. It is now taken as 0·65.

The relationship between the mean short-term elastic modulus $E_c$ and the cylinder strength of the concrete, given in draft Eurocode 2, is represented to within 1% (Smith and Johnson (1986)) by

$$E_c = 9{\cdot}5(f_c + 8)^{0{\cdot}33} \tag{1.3}$$

with $f_c$ in N/mm$^2$ and $E_c$ in kN/mm$^2$. Then, from eqn 1.2,

$$(EI)_e = 0{\cdot}65E_s(I_s + I_r) + 3{\cdot}17I_c(f_c + 8)^{0{\cdot}33} \tag{1.4}$$

The corresponding expression from draft Eurocode 4 is applicable when the eccentricity of loading is less than twice the overall depth of the cross-section. It is

$$(EI)_e = E_s(I_s + I_r) + 0{\cdot}6I_c f_c(1 - N_{Gd}/2N_d) \tag{1.5}$$

where $N_d$ is the design axial load, and $N_{Gd}$ is the part of that load that is permanent.

TABLE 1.1
EFFECTIVE VALUES OF $E_c I_c$, IN kN/mm$^2$ UNITS

| | | $f_c$ (N/mm$^2$) | |
|---|---|---|---|
| | | 20 | 40 |
| (1) | $E_c I_c$, eqn 1.4 | $9{\cdot}52I_c$ | $11{\cdot}37I_c$ |
| (2) | $E_c I_c$, eqn 1.5, $N_{Gd} = 0{\cdot}4N_d$ | $9{\cdot}60I_c$ | $19{\cdot}20I_c$ |
| | Ratio, (1) ÷ (2) | 0·99 | 0·59 |
| (3) | $E_c I_c$, eqn 1.5, $N_{Gd} = 0{\cdot}8N_d$ | $7{\cdot}2I_c$ | $14{\cdot}4I_c$ |
| | Ratio, (1) ÷ (3) | 1·32 | 0·79 |

The ratio between the 'steel' terms in eqns 1.4 and 1.5 is constant at 0·65. The ratios of the 'concrete' terms are given in Table 1.1 for two values each of $N_{Gd}/N_d$ and of $f_c$. The overall comparison depends on the ratio of the steel term to the concrete term. If, for example, $E_s(I_s + I_r) = E_c I_c$ and $E_s/E_c = 10$, then $E_s(I_s + I_r) = 20I_c$ in kN/mm$^2$. For $N_{Gd} = 0{\cdot}8N_d$ and $f_c = 20$ N/mm$^2$, from eqn 1.4,

$$(EI)_e = 13I_c + 9{\cdot}52I_c = 22{\cdot}5I_c,$$

and, from eqn 1.5,

$$(EI)_e = 20I_c + 7{\cdot}2I_c = 27{\cdot}2I_c$$

The ratio is 22·5/27·2, or 0·83. The results for the other three cases in Table 1.1 are lower. It appears that the method of Chapter 6 normally gives lower effective stiffnesses than the method of Eurocode 4, particularly when the 'steel' contribution is large compared with the 'concrete' contribution.

It does not follow that the former method is more conservative, because eqn 1.5 is based on a fictitious value of $E_c$, $600f_{cu}$, chosen to optimise agreement between test data and the results of the method as a whole when used with the European strut curves. The results of a true comparison would therefore depend on which strut curves are used with the method of Chapter 3, and on other details of the methods. Complete designs would have to be done, by both methods, for a wide range of situations, and the results compared with the very limited test data on long-term strength of slender composite columns.

This comparison again illustrates that design methods are packages, not easily compared with each other.

## 1.8 COMPOSITE BEAM-AND-SLAB BRIDGE DECKS

The material presented in Chapter 8 relates to North American design practice and philosophy. This has influenced developments in Japan and in Australasia more than those in Western or Eastern Europe. Although engineers from the USA and Canada actively support IABSE and have long participated in the work of the Euro-International Committee for Concrete (CEB), the membership of the IABSE/CEB/FIP/CECM Joint Committee on Composite Structures (1971–1981) was exclusively European. It produced a model code, Composite Structures (1981), which became the main source document for draft Eurocode 4 (1985). This code also made little use of North American experience and practice, mainly because the task of reconciling the practices of twelve European nations had to take priority.

Despite the lack of formal collaboration in code drafting, it is clear that the differences between the emerging 'European' practice and that presented in Chapter 8 are fewer than those that existed within Europe in the 1950s. A few of the current differences are now summarised.

### 1.8.1 Loading and Global Analysis

Work on a European loading code for bridge decks began only in 1986. At present, there is much inconsistency between practices in different countries.

For example, the use of impact factors by AASHTO and of dynamic load allowances in Ontario is described in Section 8.2.2. These terms are not used in the British Standard BS 5400 (1978), where account is taken of impact in the specified highway loadings, and vibration is considered only for footbridges.

The lateral distribution of vehicle loading between the beams of a multi-beam deck, allowed for by the distribution factors $D_d$ of AASHTO and the orthotropic-plate idealisation of Ontario, is in some other countries determined by computer analysis in which the deck is idealised as a grillage and the contribution of transverse bracing or diaphragms is included. Each is a valid simplification, within its range of application, for the relevant national loads; but no two of the three will give identical results for any particular bridge deck.

### 1.8.2 Analysis of Cross-sections for Flexure

The use of performance factors $\varphi$ in Section 8.4.1 does not correspond in detail with the use of factors $\gamma_M$ or $\gamma_m$ in either the draft European or the British code, because neither the sets of uncertainties allowed for by the factors nor the assessments of those uncertainties are identical in the three situations.

The plastic distributions of bending stress shown in Fig. 8.3 are, in European practice, only applicable to cross-sections defined as 'plastic' or 'compact' by limiting slenderness ratios that exclude most hogging moment regions of continuous main girders.

To devise simple but rational design procedures for these regions is perhaps the most difficult problem in the whole field of composite bridge design. Account has to be taken of local buckling of both webs and flanges, distortional lateral buckling, the behaviour of stiffeners, and uncertainty in the level of the neutral axis for flexure, due to shear lag and cracking in the concrete slab. National codes have not converged on any one solution, and it is not yet known to what extent the proposals in draft Eurocodes 3 and 4 will be supported. In research, the problem is being tackled mainly by numerical analyses (e.g., Bradford and Johnson, 1987). Testing bridge girders of realistic scale to failure is so expensive that money for it tends to become available only if a series of disasters occurs (e.g., the box-girder failures in the early 1970s).

## 1.9 BOX GIRDER BRIDGES

The subject of Chapter 9 is a particular problem that arises during the erection of steel open-top box girders of trapezoidal section, before the box is closed by construction of the concrete deck. This is that unbraced open-top boxes have very low stiffness in longitudinal torsion and transverse distortion. A related problem is the low resistance of the steel top flanges to lateral buckling at midspan.

These structures are unusual in that their St Venant torsional stiffness is so low that torsional behaviour is governed by the extent to which longitudinal warping can be prevented. This fact is exploited in the detailed qualitative guidance given in Chapter 9 on the effects of various types of bracing. These effects are illustrated by the results of tests on a pair of quarter-scale girders, of span 12.2 m.

It is usually cheaper to leave erection bracing in place than to remove it. The effects of this bracing on the behaviour of the completed girder are illustrated in Section 9.6. This account, like the rest of Chapter 9, relates to behaviour in the elastic range. Recent design codes (e.g. BS 5400: Part 3: 1982 in the UK) allow certain types of inelastic behaviour to be exploited in design for the ultimate limit state, even in box girders. The forces that have to be resisted by any cross-bracing between girders then become impossible to predict accurately, because they depend on differences between the inelastic behaviours of the two girders joined. This was discovered during the testing of a curved bifurcated model viaduct with open-top boxes by Owens *et al.* (1982). Even in the elastic range, such bracing is prone to fatigue damage, because the stress ranges in it are difficult to predict.

The comment in Chapter 9 that 'the torsionally closed (steel) section is seldom used for composite bridge structures' relates to North American practice. In Europe, closed-top boxes are at least as common as open-top boxes, and have been used for spans ranging from 21 to 174 m, often with integral composite cross-members (Johnson and Buckby (1986)). Cross-members have also been used in decks with open-top boxes, and they then provide much of the bracing needed during erection.

## REFERENCES

AMERICAN ASSOCIATION OF STATE HIGHWAY AND TRANSPORTATION OFFICIALS (1983) *Standard Specifications for Highway Bridges, Division 1—Design*. 13th edn.

BRADFORD, M. A. and JOHNSON, R. P. (1987) Inelastic buckling of composite bridge girders near internal supports. *Proc. Instn Civ. Engrs*, **83** (Part 2), 143–59.

BRITISH STANDARDS INSTITUTION (1972) *The structural use of concrete*. Code of Practice CP 110.

BRITISH STANDARDS INSTITUTION (1976) *Draft standard for the use of structural steel in buildings. Part 3: Composite construction*. Document 76/11877DC.

BRITISH STANDARDS INSTITUTION (1978) *Steel, concrete and composite bridges. Part 2: Loads*. BS 5400: Part 2.

BRITISH STANDARDS INSTITUTION (1982) *Structural use of steelwork in building. Code of practice for design of floors with profiled steel sheeting*. BS 5950: Part 4.

COMMISSION OF THE EUROPEAN COMMUNITIES (1984) *Eurocode 3, Steel structures, draft for National comment*. Report EUR 8849 EN.

COMMISSION OF THE EUROPEAN COMMUNITIES (1985) *Eurocode 4, Composite steel and concrete structures, draft for National comment*. Report EUR 9886 EN.

DOWLING, P. J., CHU, H. F. and VIRDI, K. S. (1977) The design of composite columns for biaxial bending. *Prelim. Report, 2nd Int. Colloq., Stability of Steel Structures*, Liège, IABSE, 165–74.

GRANT, J. A., FISHER, J. W. and SLUTTER, R. G. (1977) Composite beams with formed metal deck. *Eng. J. Amer. Instn Steel Constr.*, **14** (1), 24–42.

HAAIJER, G., CARSKADDAN, P. S. and GRUBB, M. A. (1983) Autostress design of steel bridges, *J. Struct. Eng., Amer. Soc. Civil Engrs*, **109** (1), Jan., 188–99.

JOHNSON, R. P. (1981) Loss of interaction in short-span composite beams and plates. *J. Constr. Steel Res.*, **1** (2), Jan., 11–16.

JOHNSON, R. P. (1986) Limitations to the use of partial shear connection in composite beams. *Research Report CE21*, Engineering Department, University of Warwick.

JOHNSON, R. P. and BUCKBY, R. J. (1979 and 1986) *Composite Structures of Steel and Concrete. Vol. 2, Bridges*. 1st edn, Granada, 1979; 2nd edn, Collins, 1986.

JOHNSON, R. P. and MAY, I. M. (1975) Partial-interaction design of composite beams. *Struct. Engr*, **53**, Aug., 305–11.

JOHNSON, R. P. and MAY, I. M. (1978) Tests on restrained composite columns. *Struct. Engr*, **56B**, June, 21–8.

JOINT COMMITTEE ON COMPOSITE STRUCTURES (1981) *Composite Structures*. Construction Press, London.

KNOWLES, P. R. (1985) *Design of Castellated Beams for use with BS 5950 and BS 449*. Constrado, Croydon, 54 pp.

LAWSON, R. M. (1983) Composite beams and slabs with profiled steel sheeting. *Report 99*, CIRIA, London.

MAY, I. M. and JOHNSON, R. P. (1978) Inelastic analysis of biaxially restrained columns. *Proc. Instn Civ. Engrs*, **65** (Part 2), June, 323–37.

OEHLERS, D. J. and JOHNSON, R. P. (1987) The strength of stud shear connections in composite beams. *Struct. Engr*, **65B** (Part B), 44–8.

OWENS, G. W. *et al.* (1982) Experimental behaviour of a composite bifurcated box girder bridge. *Proc. Int. Conf. on Short and Medium Span Bridges*, Vol. 1, Canad. Soc. for Civ. Engg, Toronto, 357–74.

RIDDINGTON, J. R. and STAFFORD SMITH, B. (1978) Composite method of design for heavily loaded wall–beam structures. *Proc. Instn Civ. Engrs*, **64** (Part 1), Feb. 137–51.

ROIK, K., BODE, H. and BERGMANN, R. (1977) Composite column design. *Prelim. Report, 2nd Int. Colloq., Stability of Steel Structures*, Liège, IABSE, 155–64.

SMITH, D. G. E. and JOHNSON, R. P. (1986) *Commentary on the 1985 draft of Eurocode 4, 'Composite steel and concrete structures'*. Building Research Establishment, Garston, Watford.

*Chapter 2*

# STEEL–CONCRETE COMPOSITE FLOORING DECK STRUCTURES

H. R. EVANS and H. D. WRIGHT

*Department of Civil and Structural Engineering, University College, Cardiff, UK*

*SUMMARY*

*Recent research has resulted in the development of an effective composite flooring deck system which has enhanced the competitiveness of steel-framed construction for high rise building. This chapter reviews this research, describes the structural behaviour of the flooring deck, outlines its advantages and those factors that influence its design. The three phases of the structural action that must be considered by the designer, identified as the construction stage behaviour, the composite slab behaviour and the composite beam behaviour, are discussed. Present design procedures, which are partly empirical, are described and, by comparisons with test results, are shown to be conservative in some instances.*

## 2.1 INTRODUCTION

In recent years, the development of an effective composite flooring deck system has greatly enhanced the competitiveness and effectiveness of steel-framed construction for high-rise buildings. This chapter will be devoted to a consideration of this new flooring system and will outline its advantages, its structural behaviour, factors that influence its design and related research and development. Some of these aspects have been summarised in a recent paper by Wright *et al.* (1987) and have been considered in detail in a definitive report by Harding (1986).

Although the flooring deck system has been in use since the early 1950s in the USA, its adoption by the British construction industry is much more recent. Indeed, it is only during the last five years that its use has become widespread in the UK. At the present time, used in conjunction with a steel frame, it is winning an increasing share of the high-rise construction market; its potential for refurbishment projects is also now being realised.

The composite flooring system consists of a cold-formed, profiled steel sheet which acts, not only as the permanent formwork for an in-situ cast concrete slab, but also as the tensile reinforcement, as shown in Fig. 2.1. The use of such sheeting as permanent formwork is not new, but its additional function as tensile reinforcement has been fully developed only in recent years.

The essential composite action between the steel deck and the concrete slab is provided by some form of interlocking device, capable of resisting horizontal shear and preventing vertical separation at the steel/concrete interface. Some profiled sheets rely mainly on their geometry to ensure composite action, see Fig. 2.2(a); others, although less common, may make use of transverse wires welded to the upper flanges of the profile, see Fig. 2.2(b). However, the most common and satisfactory method of

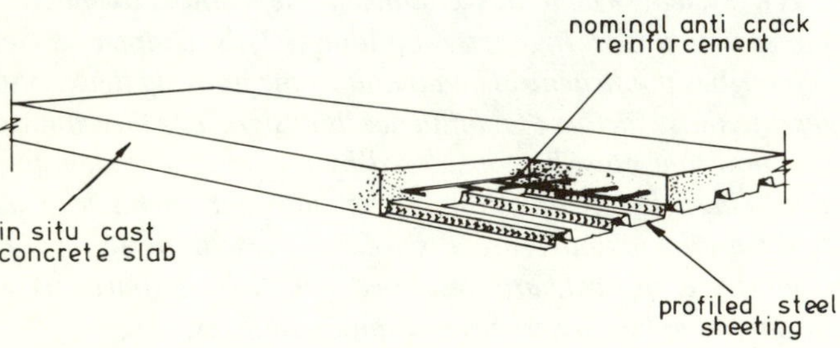


FIG. 2.1. Composite steel deck slab.

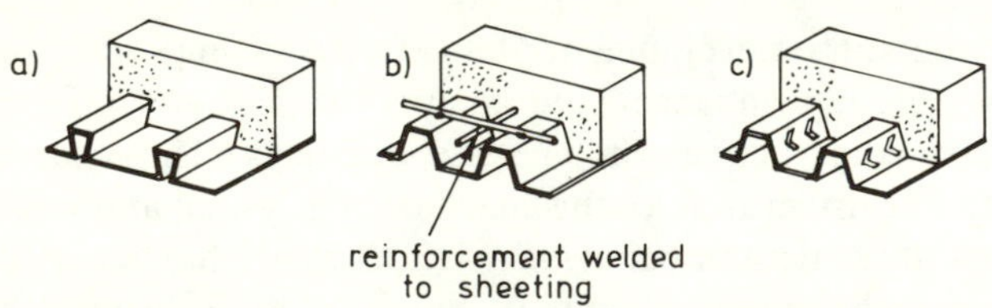


FIG. 2.2. Types of profile.

achieving the composite action is by rolling a pattern of 'embossments' into the surface of the steel sheeting; Fig. 2.2(c) shows an effective chevron type of embossing pattern.

The structural action thereby developed is termed the 'composite slab action' and it gives the flooring deck a spanning capacity in one direction, as in Fig. 2.1. A spanning capacity in the transverse direction, resulting in an effective two-way spanning flooring deck, can be achieved by making use of recent advances in on-site welding of stud shear connectors. Such

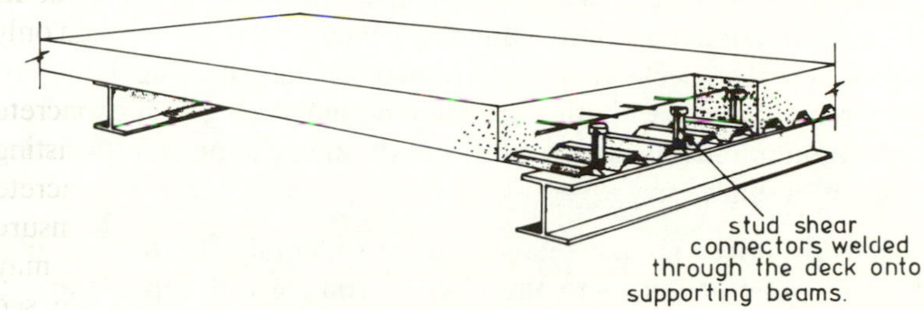


FIG. 2.3. Composite steel deck beam.

connectors can now be welded conveniently and reliably through the galvanised steel decking onto the supporting steel beams, as in Fig. 2.3. The concrete floor slab can then be made to act compositely with the supporting beams to provide the required transverse spanning capability; this action is termed the 'composite beam action' of the deck.

### 2.1.1 Advantages

The recent rapid expansion in the use of the system in the UK may be attributed to the significant advantages that it offers. These have been clearly identified by Harding (1986) who listed the following six advantages as being the most important:

(1) The steel deck acts as permanent shuttering for the in-situ cast concrete slab. Thus there is no need to erect and remove forms and falsework, with a consequent saving in time and labour.

(2) The steel deck, once in position, immediately provides a platform to support construction loads and a safe, sturdy working surface. Since supporting falsework is not required, finishing trades can operate on the floor immediately below the one being constructed; this obviously facilitates the construction programme.

(3) The steel decking acts as the tensile reinforcement, thereby eliminating the time-consuming placing and fixing of reinforcing bars for the slab.
(4) The steel deck geometry can result in a reduction of about 30% in the amount of concrete fill required for the floor. The consequent significant reduction in dead weight leads to lighter superstructures and reduced foundation loads.
(5) The cellular geometry of the deck permits the formation of ducting cells within the floor so that services can be incorporated and distributed within the floor depth. This gives the possibility of increased headroom or a reduction in building height.
(6) Since the steel decks are formed from thin gauge sheet steel, they are extremely light, facilitating the handling and placing by site workers. Also, hundreds of square metres of decking can be transported to site by a single lorry.

Collectively, these six advantages can considerably shorten the time involved in construction, with the obvious commercial benefits arising from the curtailment of interest charges payable on borrowed project capital and from early sale or let of the building.

### 2.1.2 Disadvantages

The system also has some inherent disadvantages which must be noted. Generally, these are minor, occurring at the construction stage, and should be readily overcome by good construction practice. Harding (1986) has again noted the following disadvantages:

(1) In areas of concentrated traffic or storage, the upper surface of the steel decking must be protected against damage from high local loads.
(2) Prior to concreting, the surface of the decking must be cleaned of all dirt, debris, water or any other foreign matter, to ensure proper bonding between steel and concrete.
(3) Steel decks serving as working platforms tend to be slippery to walk on.
(4) High winds during site construction may disrupt the laying and fixing of the light decking.
(5) The most important disadvantage arises from the difficulty in achieving an adequate fire rating. However, this is now largely being overcome as further fire test information becomes available.

### 2.1.3 Structural Action

The overall behaviour of the flooring deck system divides naturally into three phases. The structural action involved in each phase is different so that each individual phase must be carefully considered during design. It is, therefore, possible to identify three major aspects for consideration during the design of a composite flooring deck:

(1) The steel sheeting itself must be sufficiently strong and rigid to support the weight of wet concrete during casting—construction stage behaviour.
(2) The steel sheeting acting compositely with the hardened concrete, and spanning between the supporting steel beams, must support the imposed live loading—composite slab action.
(3) The steel beams, acting compositely with the hardened concrete through the stud shear connectors, and spanning in the transverse direction, must support the imposed live loading—composite beam action.

The authors' research and development work has covered each of these different aspects in considerable detail. With the sponsorship of seven different organisations, full-scale tests have been conducted to study each phase of the structural action. The remainder of this chapter will be devoted to a discussion of these tests and the allied theoretical developments for each phase in turn.

## 2.2 CONSTRUCTION STAGE

During the construction of a composite floor slab the profiled steel sheeting alone carries the weight of the wet concrete, workmen and tools. This loading is often the critical case for the determination of span, and accurate methods of analysis are therefore required.

Installation of the sheeting is carried out by laying one or more span lengths over supporting beams. The sheeting is then fixed to the beams with shot-fired pins and connected along the longitudinal edges with rivets or self-drilling screws. For the purpose of analysis, these site conditions are often approximated to either single or continuous simply supported spans.

The load imposed by wet concrete and by workmen is also approximated in the analysis to a uniformly distributed load, although the actual loading due to wet concrete is quite different. The concrete acts as a liquid at the

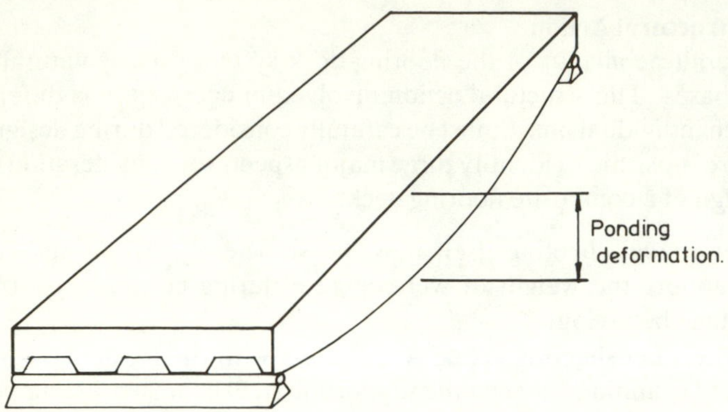


FIG. 2.4. Ponding deformation.

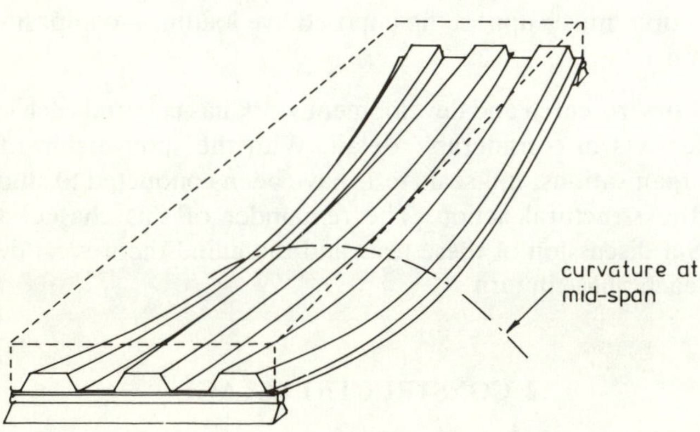


FIG. 2.5. Lateral distortion.

casting stage and therefore exerts a pressure normal to each of the plates. In addition, the deflection of the sheet causes the concrete to 'pond', giving rise to additional loading (Fig. 2.4).

The sheeting behaves as a folded plate structure and, for low load levels, the behaviour is similar to simple beam behaviour. Unrestrained edges will, however, deform more than the centre, giving rise to transverse deflection as shown in Fig. 2.5. At higher load levels, buckling of the component plates may occur. Such buckling does not give rise to immediate collapse as the buckle is contained at the fold lines, Fig. 2.6, but some loss of stiffness of the system does occur. Final collapse may occur either due to yield or when the compression buckle folds, Fig. 2.7. Final

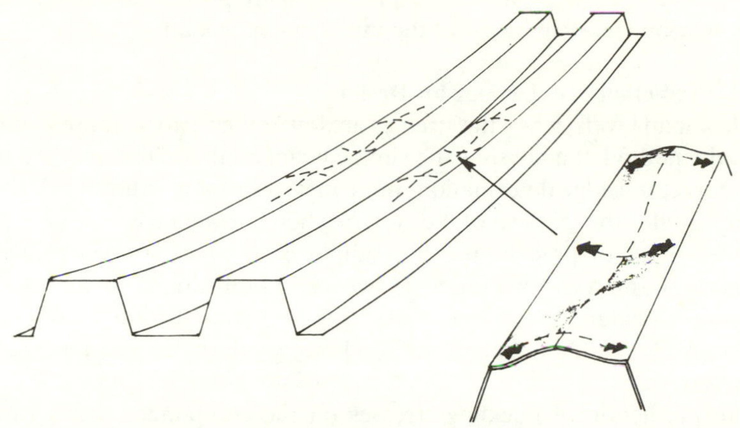

FIG. 2.6. Buckles contained within fold lines.

FIG. 2.7. View of decking after collapse.

collapse may occur at a load of three to four times the load required to cause initial buckling and, consequently, many profiles are designed to carry working loads in excess of the initial buckling load.

### 2.2.1 Prediction of Behaviour for Design

For low load levels, when the stresses are lower than critical for buckling, a suitable method of analysis is the simple beam method. This method does not predict the edge deformation, nor can it cater for ponding load. It also overestimates the stiffness of the system when buckling commences.

Most codes of practice use the simple beam theory when providing methods of analysis for design. However, modifications to the method are used to cater for the loss of stiffness and strength due to buckling, and equivalent additional uniform loads are given to account for ponding deformation.

The prediction of buckling stresses on the component plates can be achieved using classical or energy methods and is well documented by Timoshenko and Gere (1982). The behaviour of the plate once buckling has commenced is more complex and the empirical effective width method is used by most codes. This method was first proposed by Von Karman (1932) and subsequently modified in the light of extensive experimental work by Winter (1947). The method substitutes an equivalent width of plate, subjected to the uniform longitudinal stress calculated from a simple beam analysis, for the complete plate width which is under the action of the more complex stress pattern due to buckling, Fig. 2.8.

In the British code BS 5950: Part 4 (1982) the effective breadth of plate is equated to the original breadth of plate, thickness of plate and stress in the plate by the expression:

$$b_{ed} = \frac{857t}{\sqrt{\sigma_c}}\left(1 - \frac{187t}{b\sqrt{\sigma_c}}\right)$$

where $b_{ed}$ = the effective width of plate, $\sigma_c$ = the compressive stress in the plate, $t$ = the thickness of the plate, and $b$ = the original width of plate. In this equation $\sigma_c$ is unknown and has to be evaluated using simple beam analysis, assuming the full breadth of plate.

The method provides a simple design solution of the complex post-buckling behaviour but involves iteration and is therefore tedious. Further complications arise when the plate is stiffened, as is often the case, by longitudinal grooves running along the plate.

These stiffeners may be of such proportion that they do not bend with

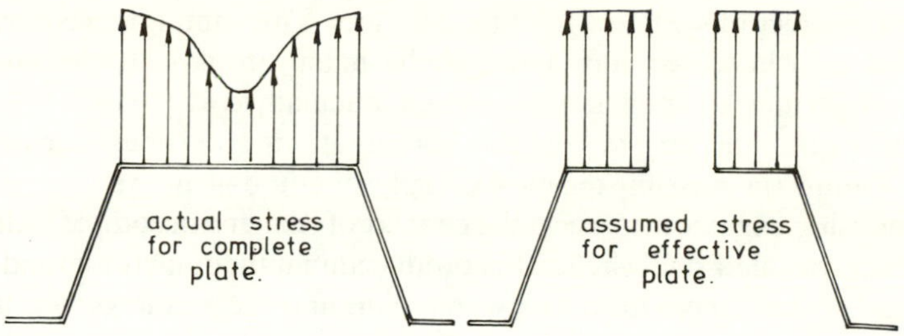


FIG. 2.8. Effective width concept.

the plate once buckling commences. Each individual part of the plate on either side of the stiffener may then buckle independently and the stiffener is said to be fully effective. Winter (1947) gives empirical values of stiffener second moment of area that may be used to determine whether or not the stiffener is effective. These factors are used in both the American AISI Specification for the design of cold formed steel structures (1968) and the British code BS 5950: Part 4 (1982).

In the British code the $I$ value of the stiffener must be greater than $I_{\min}$ where:

$$I_{\min} = 3{\cdot}66t^4 \sqrt{\left(\frac{b}{t}\right)^2 - \frac{25\,600}{P_y}}$$

or

$$I_{\min} = 18{\cdot}4t^4$$

where $b$ = the breadth of adjoining plates, $t$ = the thickness of the plate, and $P_y$ = the yield stress of the steel.

These empirical relationships are for stiffeners with adjoining plates with a breadth to thickness ratio less than 80. For plates with higher breadth to thickness ratios the $I$ value must be greater than $I_{\min}$ where:

$$I_{\min} = 1{\cdot}83t^3 b$$

The size of fully effective stiffeners leads to profiles of heavy section which are often uneconomical. Most commercial profiles use relatively small stiffeners and the European code of practice ECCS–TC7 (CONSTRADO, 1983b) provides a method of analysis to evaluate the stiffening effect of such low rigidity stiffeners. This method considers the stiffener as a beam–column and introduces an effective thickness technique which may be applied to the stiffener and adjoining plates.

This last method involves lengthy computations and is best carried out by computer. It is possibly the most complex method of analysis acceptable for practising engineers to use in the context of design to a code of practice. However, profiled steel sheeting is produced in a manufacturing industry situation. The competition between manufacturers allows for more exacting analysis and even prototype testing to be economically viable.

A method of analysis that is ideally suited to this type of structural member is the folded plate analysis developed by Goldberg and Leve (1957). This method offers a solution based on the small deflection theory of plate bending and the classical elasticity theory. Each plate is assumed simply supported at each end; in the case of the profiled steel sheet system, this gives rise to the approximation that there is a diaphragm at each support which is infinitely stiff in its own plane but infinitely flexible perpendicular to its plane.

The load characteristics of each individual plate may be expressed as the summation of the harmonics of a Fourier series. Consequently, for each plate and for each harmonic, the forces in the plate may be defined as a function of the forces at a single point (generally the midpoint). The deformations of the plate under such loading may also be expressed as the summation of the first harmonics of a sine wave:

$$u_{(x)} = u_{(\text{midpoint})} \sin \left[\frac{m\pi x}{\text{span}}\right]$$

where $u$ = deformation, $x$ = distance along span, and $m$ = number of harmonic, 1, 3, 5.

The plate may be returned to a state of zero stress by applying forces at the edges of the plate equal and opposite to the reactions caused by a load. Again, these forces may be expressed as a function of the forces at the midpoint of each longitudinal edge:

$$u_{\text{eH}} = u_{\text{eH}(x)} \sin \left[\frac{m\pi x}{\text{span}}\right]$$

where eH = edge holding suffix.

When two or more plates are connected together these 'holding forces' must combine, taking into account the relative stiffnesses of the plates.

This may be achieved using a matrix stiffness method of analysis with the stiffness matrix formed from the equations of bending and inplane action of the plate. The matrix is normally formed in two parts separating the inplane and plate bending action, as shown below:

$$K_{ss} = \begin{vmatrix} K_{(\text{plate bending})} & 0 \\ 0 & K_{(\text{inplane action})} \end{vmatrix}$$

This analysis is based on classical elasticity methods and is therefore not a numerical approximation as are the finite strip and finite element methods. The computing requirements are such that the matrix must be solved for each harmonic, and this makes the solution quite long. This is countered, however, by the small size of the matrix. Each plate constitutes an element with four degrees of freedom at each of two nodes. A profiled steel sheet of, say, nine plates is therefore solved with a matrix of the order of $40 \times 40$.

Such an analysis may accurately predict the deformation of profiled steel sheeting under the actual load of wet concrete rather than the approximated uniform loading used in simpler analyses. As each individual plate may be loaded with a single sinusoidal load, both ponding, see Fig. 2.5, and the additional lateral loads arising therefrom may be calculated.

Effective width methods described earlier can also be incorporated in the folded plate method to allow for post-buckling behaviour. The iteration involved lengthens the computing time considerably and a conversion of effective width to effective thickness is required.

### 2.2.2 Experimental Study

Despite the accuracy of the folded plate analysis, it is still worthwhile for manufacturers to test prototype sheeting to obtain the best possible spans for specified loads. The British Standard code of practice, BS 5950: Part 4, specifies the test procedure recommended in the ECCS document 20 (1977). Loading of sample sheets supported on roller supports carried out by an air bag acting against a restraining frame, evacuation of air behind the sheet in a suction chamber or loading by four or more line loads is permissible.

It may be noted that none of these test methods accurately simulate the true load of wet concrete. Although both air bag and evacuation methods do provide pressure loading and allow lateral deformation, they cannot simulate ponding loads.

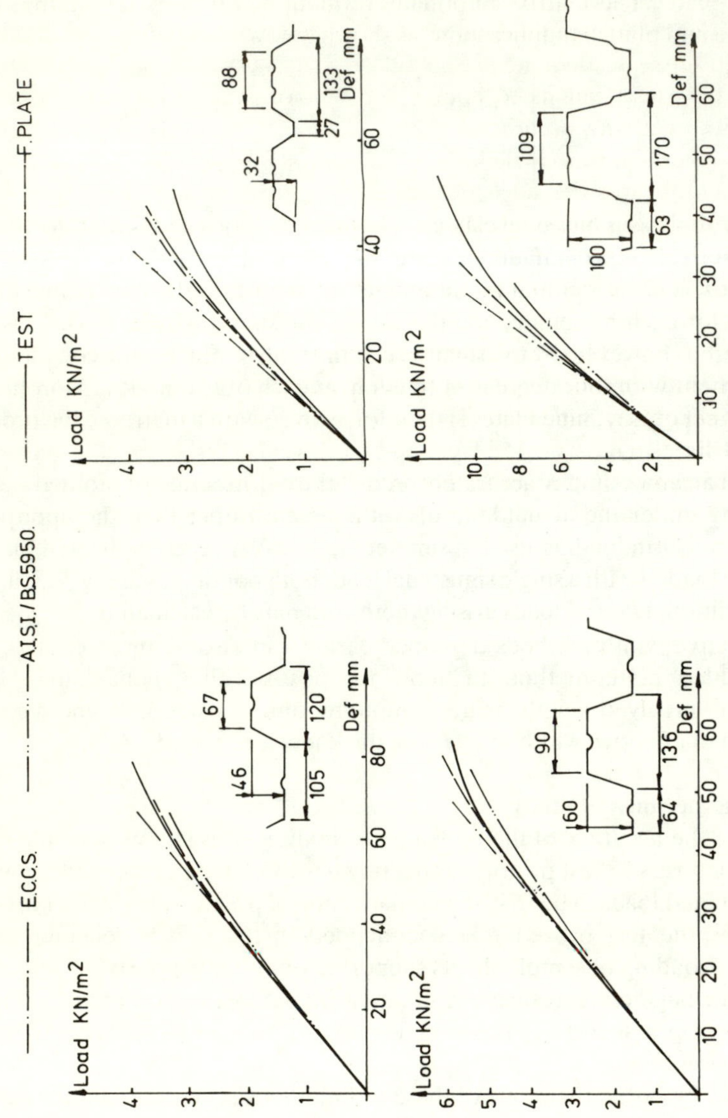


FIG. 2.9. Load–deflection plots for four profiles tested under wet concrete loading.

Twenty-four tests to the specifications of the European Recommendations have been carried out on the four profiles shown in Fig. 2.9 and, from these tests, it has been possible to compare the various methods of analysis with the experimental results. It can be seen from Fig. 2.9 that the methods give an adequate prediction in the elastic range but differ slightly when the sheet begins to buckle and cannot predict the elasto-plastic behaviour with any accuracy.

A further test was carried out by the authors to determine the ponding deformation and the edge deformation likely to occur under the wet concrete loading of a single sheet. The test was set up as though the sheet was to be concreted with timber formwork around the specimen. Sand was used as a loading medium as it was easier to handle than wet concrete. The results are shown in Fig. 2.10 and are compared to the folded plate analysis. The additional deformation due to ponding is considerable, especially at the sheet edges where the lateral distortion is compounded by the extra depth of sand load. The steel strains at these edges were found to be sufficient to have caused yield in the steel. This test shows the value of the folded plate analysis in the prediction of deformation and the considerable deformation that would result if the sheet edges were unsupported.

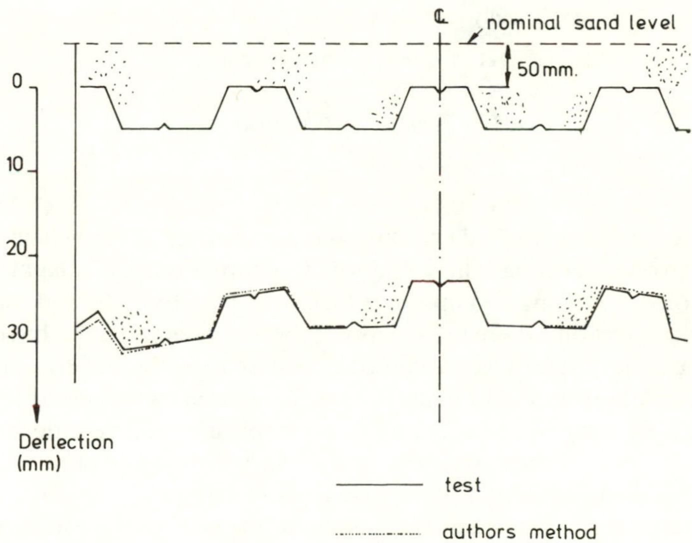


FIG. 2.10. Sand-loaded specimen.

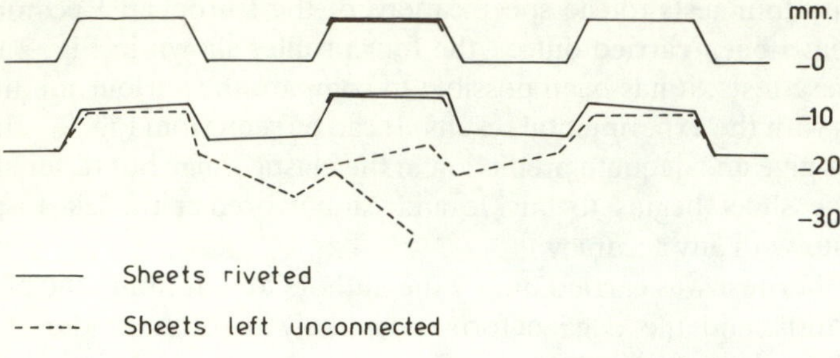


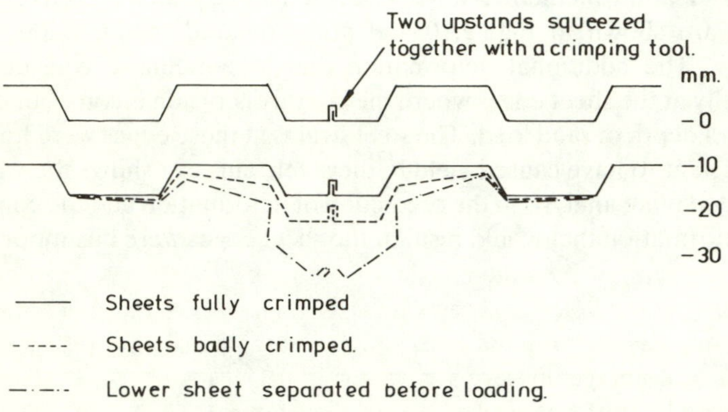


FIG. 2.11. Lapped and crimped joints.

The folded plate method has also been used to predict the behaviour of joints between sheeting. These joints may be formed either by lapping the sheets or by crimping two upstands together. The study showed that for perfect connection, where the sheets were well connected either by screws or rivets, the resulting connection was stiffer than the general profile. However, where imperfect fixing or some separation had occurred prior to loading, the joint would open, allowing seepage of concrete during the casting process. This is shown in Fig. 2.11 which again illustrates the adaptability and usefulness of the folded plate method.

As stated at the beginning of this section, the wet concrete load on the sheeting alone is often the critical condition for span design. The analytical and testing methods developed and described herein are helping in the optimum design of profile geometry. The new generation of profile shapes that are appearing on the market will utilise the post-buckling capacity of the constituent plates to the full and will necessitate careful stiffener design.

## 2.3 COMPOSITE SLAB ACTION

In this phase of the structural behaviour, the hardened concrete slab, acting compositely with the profiled steel sheet, spans between the supporting beams and carries the imposed live loads, as in Fig. 2.1. The steel deck acts as the tensile reinforcement to the slab and the development of composite action depends entirely upon adequate transference of horizontal shear forces at the steel/concrete interface. The imposed bending action also results in vertical separation between the steel and the concrete, and the profiled deck must be designed to resist this in addition to transmitting the horizontal shear.

The introduction to this chapter has already outlined ways in which various deck configurations attempt to satisfy these design requirements. The re-entrant, or dovetail, profile geometry of Fig. 2.2(a) is obviously ideal for resisting the vertical separation, but it has the disadvantage of being uneconomical in terms of cover width. The trapezoidal profile has an obvious advantage in terms of cover and, with proper design of an embossment pattern, is capable of developing composite action; the chevron pattern illustrated in Fig. 2.2(c) has been shown to be effective.

### 2.3.1 Evaluation of Shear Bond Capacity

The analysis of the composite slab action is complex since the conventional modular ratio or plastic methods do not take into account the possibility of a failure mode involving a loss of shear bond. However, many experimental studies, for example the series of full-scale tests reported by Schuster and Ekberg (1970), have identified such a loss of shear bond as the primary mode of failure for most composite slabs.

A shear bond failure mode arises from loss of composite action because of inadequate shear transference at the interface between the concrete slab and the steel deck. It is characterised by the formation of a major diagonal crack in the slab at approximately one-quarter to one-third span and by horizontal slip and vertical separation between the concrete and the steel.

The extent of the shear bond is difficult to predict theoretically since it is dependent upon several inter-related parameters. For example, the height, shape, orientation and frequency of the embossing pattern have a significant effect upon shear bond, and the geometry and flexibility of the profiled sheet itself are also important. Currently, it is not possible to predict shear bond capacity accurately and resort must be made to some form of performance testing to obtain a true indication of the capacity of any particular sheet. A very recent attempt by Prasannan and Luttrell (1984) to provide an empirical method that does not require further testing will be discussed later.

The British code BS 5950: Part 4 (1982) recommends such performance testing and specifies that one of two types of testing programme may be adopted. In the first option, a minimum of three prototype tests may be carried out for a particular construction situation. These tests will give the composite slab capacity pertaining to that particular situation only; the information obtained is thus limited and the expense involved in such tests could well not be considered worthwhile by the deck manufacturer.

The second type of testing programme would normally be more appropriate. This is based on the work of Porter, Ekberg and their colleagues (Porter and Ekberg, 1976; Porter *et al.*, 1976) where a linear regression analysis for the prediction of shear bond failure was established. This programme requires a minimum of six tests, and preferably eight tests, to be carried out to allow the determination of two factors ($m_r$ and $k_r$) which express the shear transference characteristics of a particular profiled sheet. Once established, these factors may be used subsequently to determine the capacity whenever the same sheet is used in different structures with various spans, slab thicknesses and concrete strengths. This second type of testing programme thus requires the investment in only a single testing series by the manufacturer of a particular sheet and is, in general, preferable to the first programme allowed by the design code.

### 2.3.2 Experimental Study

In addition to specifying a minimum number of tests to be carried out in the second type of testing programme, BS 5950 specifies testing procedure and the method of interpretation of test results; this method is based on the work of Porter and Ekberg (1976). Twelve tests have been conducted by the authors in strict accordance with BS 5950, and these tests will now be described to illustrate the required procedures. The particular profiled steel sheeting used in this test series was the CF46 deck produced by Precision Metal Forming Ltd; this deck has a typical trapezoidal profile with chevron embossments and is similar to that illustrated in Fig. 2.2(c).

As shown in Fig. 2.12, the slab was simply supported and subjected to a four-point loading system to represent a uniformly distributed load. The code of practice specifies the application of 10 000 loading cycles, varying from 0·5 to 1·5 times the design load, before the test to failure under static loading is commenced. This preliminary cycling ensures that any chemical bond developed between the steel and concrete is destroyed so that the subsequent test to failure gives a true indication of the mechanical bond developed by the embossment pattern. Such a preliminary cyclic loading regime was applied in all but two of the twelve tests conducted by the authors.

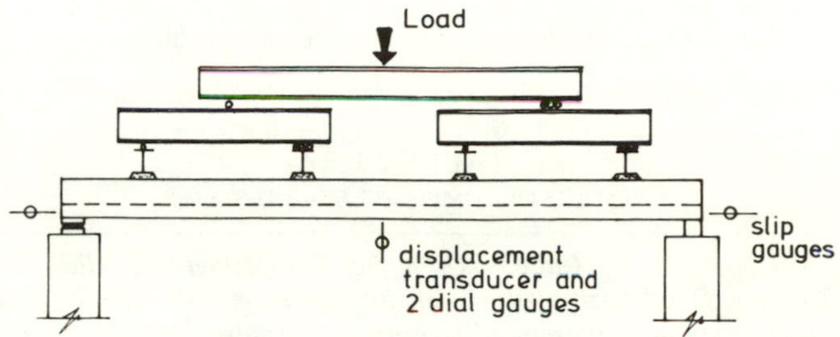


FIG. 2.12. Testing arrangement for composite slab.

Several parameters were varied during the testing programme, as outlined in Table 2.1. A lightweight concrete mix was used throughout with concrete strengths varying from 27·0 to 57·5 N/mm$^2$. Lytag aggregate was used in a mix representative of a pumped concrete and an average concrete density of 1900 kg/m$^3$ was achieved. Slab spans of 1·4 m, 2·1 m and 2·4 m were tested and slab depths varying from 95 mm to 131 mm were considered. The height of the chevron embossment was 3 mm for all decks, other than for the last two, where it was reduced to 2 mm.

During the tests, readings were taken of the vertical deflection at the midspan of the deck and of the differential movement (or slip) between the steel sheeting and the concrete slab. Also, electrical resistance strain gauges were positioned at midspan to monitor the longitudinal strains developed in the upper and lower flanges of the steel sheeting.

A similar response was observed in each test and, as expected from the previous work of Schuster and Ekberg (1970), failure occurred in each case as a result of the loss of shear bond. Typical load/deflection and load/slip curves are plotted in Fig. 2.13. These show that the initial linear load/deflection response terminates with the onset of slip and that the ultimate

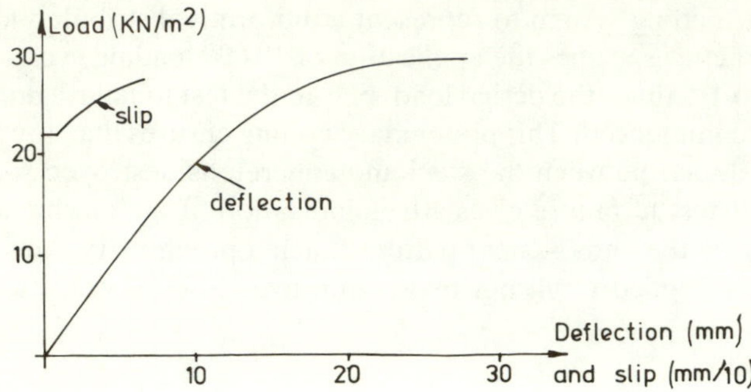


FIG. 2.13. Load–deflection relation for composite slab.

TABLE 2.1
RESULTS OF COMPOSITE SLAB TESTS

| | *Span (m)* | *Depth (mm)* | *Cube strength* ($N/mm^2$) | *Concrete density* ($Kg/m^3$) | *Embossment height (mm)* | *Cyclic load* | *Ultimate load* ($kN/m^2$) |
|---|---|---|---|---|---|---|---|
| SP 1 | 2·4 | 98 | 28·0 | 1 840 | 3 | yes | 31·14 |
| SP 2 | 2·4 | 100 | 52·0 | 1 900 | 3 | yes | 34·62 |
| SP 3 | 2·4 | 98 | 35·0 | 1 852 | 3 | yes | 29·75 |
| SP 4 | 2·1 | 131 | 48·0 | 1 896 | 3 | yes | 55·73 |
| SP 5 | 2·1 | 130 | 57·5 | 1 910 | 3 | yes | 57·12 |
| SP 6 | 1·4 | 95 | 49·0 | 1 904 | 3 | yes | 68·26 |
| SP 7 | 1·4 | 95 | 41·5 | 1 865 | 3 | yes | 67·13 |
| SP 8 | 1·4 | 130 | 43·5 | 1 870 | 3 | yes | 91·55 |
| SP 9 | 1·4 | 129 | 46·5 | 1 865 | 3 | yes | 90·43 |
| SP10 | 2·4 | 98 | 55·0 | 1 910 | 3 | no | 30·69 |
| SP11 | 2·4 | 98 | 27·0 | 1 850 | 2 | no | 12·75 |
| SP12 | 2·4 | 98 | 30·0 | 1 900 | 2 | yes | 12·50 |

load is then approached rapidly during the non-linear phase as the slip at the steel/concrete interface develops. The measured ultimate load for each deck is recorded in Table 2.1.

As stated earlier, the main objective of such a testing programme is to determine the $m_r$ and $k_r$ factors defining the shear transference characteristics of the sheet. Unfortunately, these two factors do not have a direct physical significance; they are simply empirical constants to be used in the determination of capacity. As specified by BS 5950, using the work of

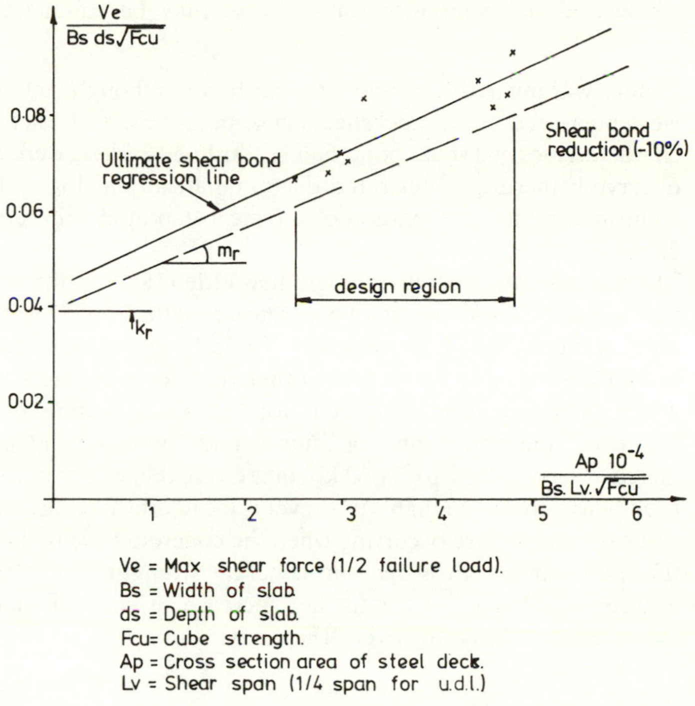


FIG. 2.14. Shear bond regression.

Porter and Ekberg (1976), they may be determined as shown in Fig. 2.14. This shows the experimental points obtained from the authors' study, plotted in accordance with the recommendations of the design code; the notation used is taken from the code and is fully defined in Fig. 2.14. The straight line drawn through the experimental points allows the $m_r$ factor to be obtained as the slope of the line and the $k_r$ factor to be obtained as the intercept, as shown.

### 2.3.3 Observations on Test Results

In addition to establishing the $m_r$ and $k_r$ factors for the particular profiled sheet, the tests clearly showed the effectiveness of the 3 mm high chevron embossing pattern. Failure loads considerably in excess of the required design loads were invariably achieved in the tests.

Also, results of more general interest were obtained, relating to the overall behaviour of composite slabs. These may be summarized as follows:

(a) Failure was invariably by loss of shear bond, although yield strains were measured in the steel sheets in some instances. The expected characteristics of a shear bond failure, outlined earlier, were clearly observed; these are illustrated diagrammatically in Fig. 2.15 and confirmed by the photograph of a typical slab after failure in Fig. 2.16.

(b) The variation in concrete strength had little effect on the ultimate load capacity; a wide range of concrete strength, from 27 to 57·5 N/mm$^2$, was considered. In particular, Table 2.1 shows that decks SP1 and SP2 were virtually identical other than for concrete strength, which increased from 28 N/mm$^2$ for SP1 to 52 N/mm$^2$ for SP2. However, the corresponding failure loads only showed a slight increase from 31·14 to 34·62 kN/m$^2$. Even the weakest concrete used was strong enough to prevent local crushing around the embossment, failure occurring when the concrete rode up and over the embossment. Provided the concrete strength is sufficient to ensure such behaviour, a further increase in strength does not significantly increase the overall capacity.

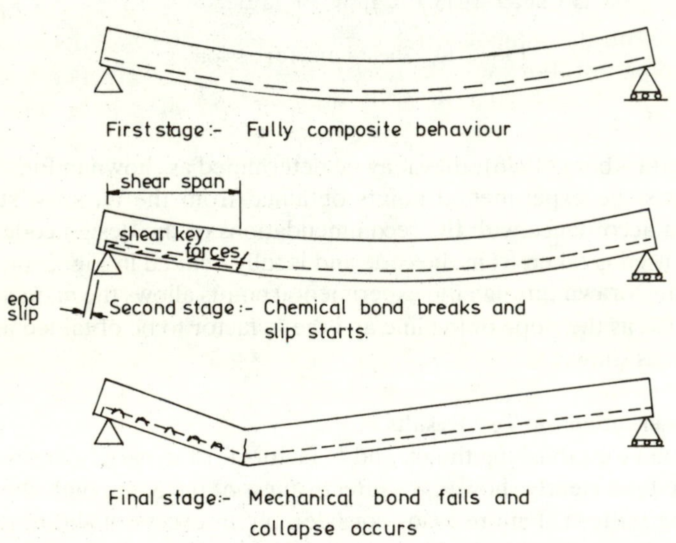


FIG. 2.15. Shear bond failure mechanism.

FIG. 2.16. The failed slab.

(c) The application of the preliminary cyclic loading regime had little effect on the ultimate strength measured in the subsequent static loading test. For example, decks SP11 and SP12 were identical but SP12 was subjected to cyclic loading and SP11 was not; the measured failure loads for the two decks were almost identical. A comparison of the results for decks SP2 and SP10 confirms this observation.

(d) The height of the embossment had a very significant effect upon the ultimate strength. Decks SP11 and SP12 were virtually identical to SP1 other than for a reduction in embossment height from 3 mm for SP1 to 2 mm in SP11 and SP12; the corresponding reduction in ultimate load level from 31·14 to 12·5 kN/m$^2$ shows the embossment height to be a crucial parameter.

### 2.3.4 Recent Empirical Approach

Whilst such performance tests are capable of establishing the capacity of a particular profiled sheet they can, of course, only be conducted once the sheet has been produced. They cannot, therefore, be used in the optimum design of a new profile without the considerable expense of repeated prototype production. A method of estimating composite slab capacity, without resorting to performance testing, would therefore be invaluable.

In 1984, Prasannan and Luttrell proposed such a method. They established an empirical formula to predict the capacity of any composite slab incorporating a trapezoidal sheet profile. The formula was based on the results of 79 full-scale tests and was expected to be applicable to any new decks, thereby removing the need for performance testing.

However, the authors have found that this empirical formula can give misleading results in some instances. This is believed to be because, although the formula does incorporate a number of important parameters, these are calibrated on previous test information. It presumably assumes an embossment calibration typical of American practice. The results obtained were in notably better agreement with the tests on 2 mm high embossments which are common in America at the present time than with the tests on decks with 3 mm high embossments which are common in Britain. The empirical formula does, however, represent a significant advance and, with further minor modifications, will provide a most useful design tool.

## 2.4 COMPOSITE BEAM BEHAVIOUR

Composite slab construction has become one of the most economical forms of flooring in modern office and commercial building. This is in spite of the fact that the fully composite strength and stiffness are often greater than those required for the imposed loading. The most economic of these floors also utilise composite behaviour in the transverse, or beam, direction by connecting the concrete slab to the steel support beam. For this behaviour also it is possible that the fully composite beam may be stronger than required. With composite beams it is possible to limit the degree of interaction between the slab and beam to that necessary to support the required live load whilst ensuring that the steel beam alone is adequate to support the dead load of the wet concrete.

### 2.4.1 Methods of Analysis

Composite beams are not new and have been used successfully for many years in bridge construction. A simple beam analysis, modified by modular ratio techniques, has generally been used, assuming that there is full interaction between the steel and concrete. Such interaction is almost impossible to achieve as the connection is rarely stiff enough to provide complete interaction and some slip between concrete and steel is inevitable. An elastic analysis taking this slip into account was developed by Newmark *et al.* (1951) but involves the solution of a fresh set of differential equations for each load or span condition. The determination of the shear force between concrete and steel assuming full interaction is also possible, and the simple expression for shear flow is quoted in many elementary texts. However, the accuracy of this last method is questionable as the connectors are often discrete and rarely act in a linearly elastic manner. The shearing forces at ultimate load are, however, readily calculable and therefore load factor design at the ultimate limit state is the most common method of analysis used today.

Johnson (1970, 1975) developed a simple analysis method based on tests and more exact analysis for the design of composite beams with incomplete interaction. This method is known as 'Linear partial interaction design' and assumes that the additional strength and stiffness of the beam caused by connection of concrete to the steel varies linearly with the degree or 'partial interaction' of that connection. The expression for ultimate strength of the composite beam with partial interaction is:

$$M_{\mathrm{comp}} = M_{\mathrm{plast}} + \frac{N}{N_{\mathrm{f}}}(M_{\mathrm{ult}} - M_{\mathrm{plast}})$$

where $M_{\mathrm{comp}}$ = the moment capacity of the composite beam, $M_{\mathrm{plast}}$ = the moment capacity of the steel beam alone, $M_{\mathrm{ult}}$ = the moment capacity of the fully composite section, $N$ = the number of connectors required, and $N_{\mathrm{f}}$ = the number of connectors required for full interaction.

The expression for mid-span deflection of a composite beam is:

$$\delta_{\mathrm{e}} = \delta_{\mathrm{c}} + 0{\cdot}5\left(1 - \frac{N}{N_{\mathrm{f}}}\right)(\delta_{\mathrm{o}} - \delta_{\mathrm{c}})$$

where $\delta_{\mathrm{e}}$ = the deflection of the composite beam, $\delta_{\mathrm{c}}$ = the deflection of the beam with full connection, $\delta_{\mathrm{o}}$ = the deflection of the steel beam alone, $N$ = the number of connectors required, and $N_{\mathrm{f}}$ = the number of connectors required for full interaction.

Figure 2.17 shows this relationship for both ultimate strength and stiffness in the elastic range. This analysis has proved satisfactory for traditional composite beams where the slab is solid.

When composite slabs are used, several aspects of this design may vary due to the presence of the sheeting and the corrugations in the profile.

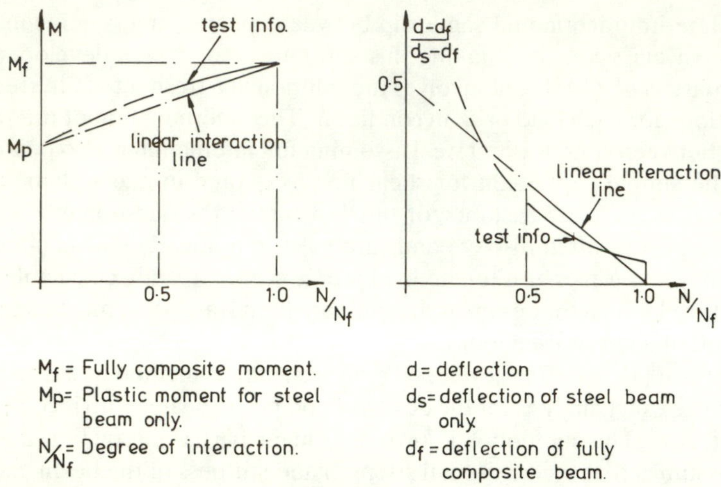


FIG. 2.17. Linear interaction concept.

Of major concern is the effect of welding shear connectors through the sheeting. All sheets used in Britain are galvanised and this may weaken the weld considerably. The authors have carried out over 80 tests on T. R. W. Nelson connector studs (currently the most popular type of connection) to determine the quality of the weld. These tests are reported in a paper by the Wright *et al.* (1984) and show that the single contactor system in current use gives consistently good performance, despite the presence of the galvanising.

The shape of the profiled sheet reduces the amount of concrete around the stud and this can lead to a reduction in shear strength of a stud welded in a sheeted slab, as compared to a stud in a solid slab. To test the amount of reduction for profiled steel sheets produced in Britain, a series of 24 push-out tests were carried out by the authors. The results of these tests are quoted in T. R. W. Nelson's brochure (1983). The push-out test is documented in CP 117 (1965) and has become the industrial standard for stud strength calculation. The authors have modified the standard test to include the profiled steel sheet; the extent of modification is shown in Fig. 2.18. The reduction in strength and stiffness of the 'through deck' tests as compared to the standard tests is shown in Fig. 2.19. These results were compared with a formula derived empirically by Grant, Fisher and Slutter (1977) for American profiles, and there was considerable discrepancy. In fact, only one of the profiles manufactured in Britain was calculated to

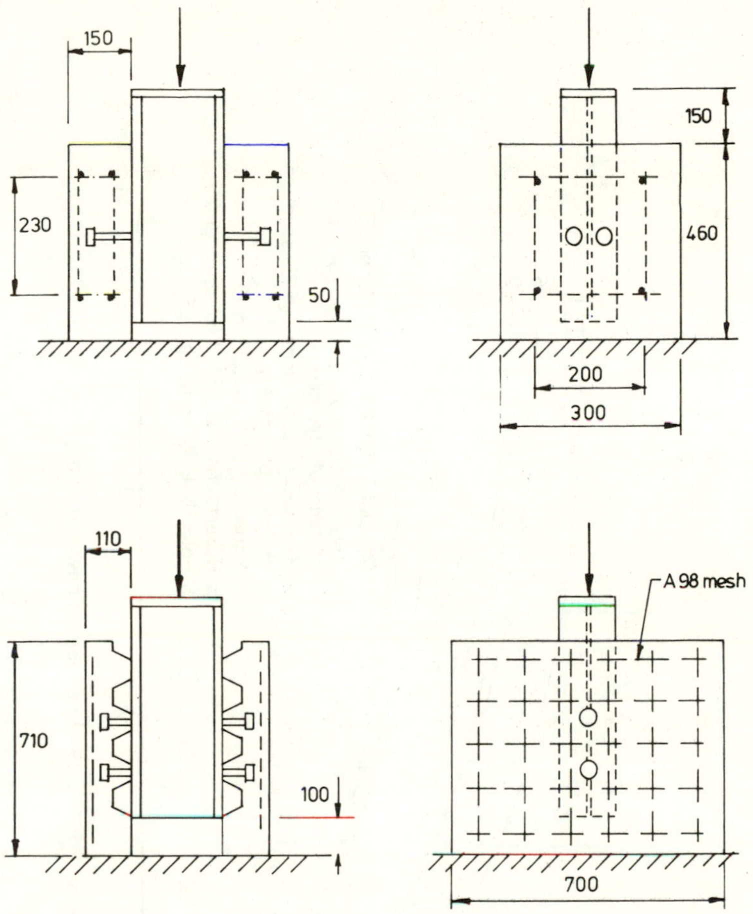


FIG. 2.18. Standard and through-deck push-out tests.

require a reduction in stud strength according to the formula, whereas all the profiles tested gave rise to a reduction in push-out strength. The authors conclude that the accurate determination of stud strength when welded through a profiled steel sheet is best carried out by 'through deck' push-out tests.

The final difference between a slab containing the profiled steel sheet and a solid slab is the presence of voids in the slab. The estimated strength and stiffness of the solid slab may utilise the complete depth of concrete,

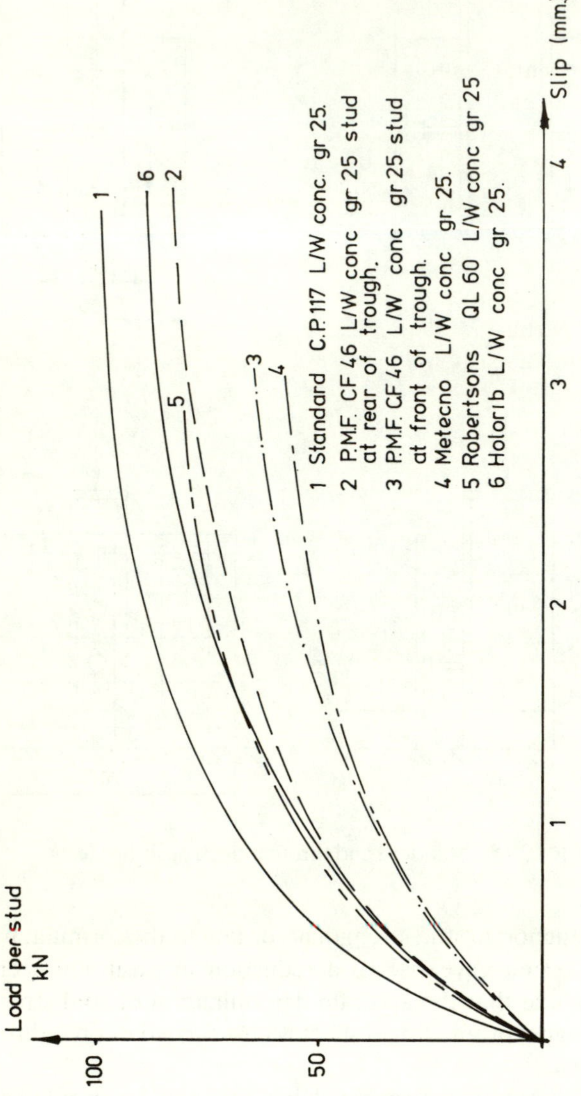


FIG. 2.19. Load slip for push-out tests.

but when this depth is interrupted by the sheet profile the stress distribution is unknown. The assumption is generally made that the area of concrete in the troughs of the sheet may be ignored in calculation if these troughs run perpendicular to the line of the beam.

### 2.4.2 Experimental Study

To study the overall effect of these variations, a series of eight full-scale beam tests have been carried out by the authors. The aim of the tests was to investigate the capacity of composite beams using profiled steel sheeting and T. R. W. Nelson through-deck welded studs and to verify that the proposed BS 5950 method of analysis could be used for their design. The initial test beams were designed to a draft version of the CONSTRADO design guide for composite beams (1983a) which is almost entirely based on the linear partial interaction theory of Johnson. The test set-up is shown in Fig. 2.20 and a list of the parameters varied is given in Table 2.2. Stud

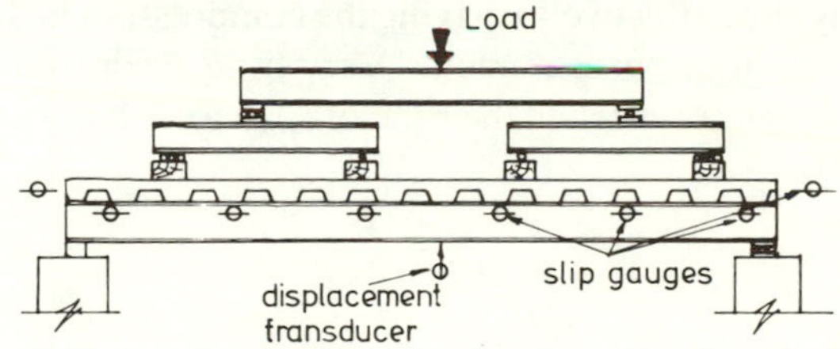


FIG. 2.20. Testing arrangement for composite beam.

TABLE 2.2
RESULTS OF COMPOSITE BEAM TESTS

| *Concrete* | *Deck* | *Composite action* | *Direction of corrugation* | *Span (m)* |
|---|---|---|---|---|
| Lightweight | CF46 | 87% | T | 4·5 |
| Lightweight | Holorib | 90% | T | 4·5 |
| Normal weight | Holorib | 89% | T | 4·5 |
| Lightweight | Metecno | 80% | T | 5·5 |
| Lightweight | QL60 | 86% | T | 6·0 |
| Normal weight | CF46 | 85% | T | 6·0 |
| Normal weight | Metecno | 100% | P | 6·0 |
| Lightweight | Metecno | 100% | P | 6·0 |

strengths used were those obtained from the push-out tests carried out by the authors and quoted by Nelson (1983).

The initial tests showed that the amount of interaction actually achieved was much greater than that anticipated in design. For example, the slab that had been designed to achieve an interaction of 50%, the minimum permissible under the proposed guide, in fact developed an interaction of 93%. These interaction values have been based upon the ultimate plastic moment calculated in accordance with the CONSTRADO guide. The beam was designed with approximately 50% of the studs necessary to achieve the full plastic moment and yet, when tested, sustained a moment of 93% of this moment. The degree of stiffness was not of the same order as this interaction and the stiffness was less than that predicted by linear interaction theory.

Several other conclusions were derived from the test series. The inclusion of different decks did not appear to affect the strength and stiffness to a large degree, nor did the type of concrete (normal or light-weight) affect the deformation (Fig. 2.21). The estimation of the breadth of slab that might be effectively carrying the compression in the system was also checked by strain measurement. Very little strain reduction due to shear lag was found, even when the ratio of span to slab breadth was as low as 3 (Fig. 2.22).

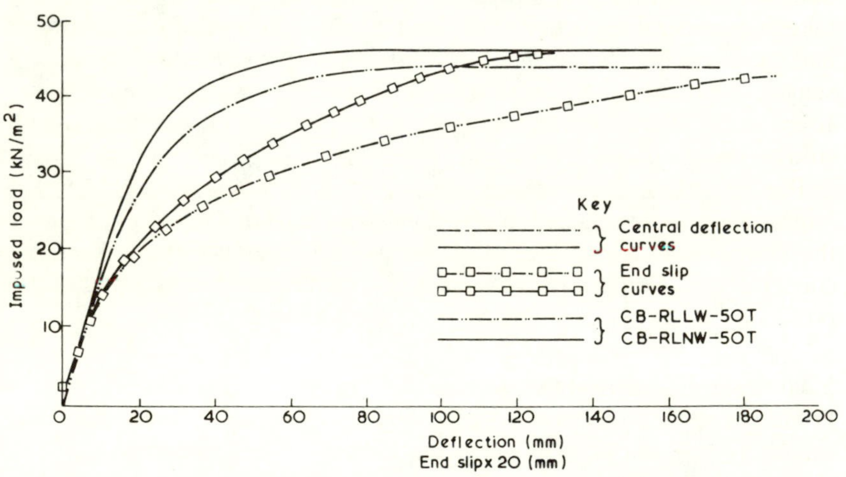


FIG. 2.21. Comparison of normal-weight and lightweight concrete.

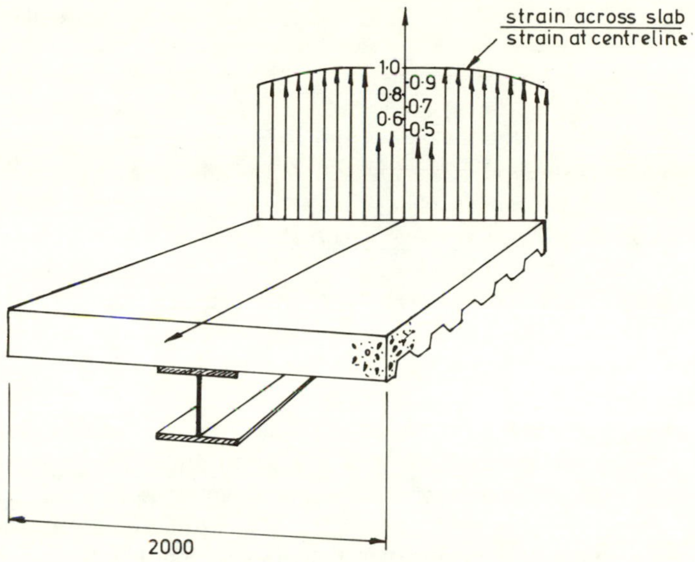


FIG. 2.22. Distribution of strain across slab.

From measurement of slip along the beam it has also been found that this does not vary linearly; from Fig. 2.23 it can be deduced that the last stud takes a considerable portion of the load. Figure 2.24 shows this stud after it had been broken out of the failed beam; the considerable curvature suggests that the stud is very flexible. This differs from the findings of Johnson (1970) for studs in solid slabs, and points to a definite variation in stud behaviour between solid and voided slabs.

The results of these tests have shown that, although linear partial interaction theory appears to give conservative results for ultimate load, the stiffness of slabs designed in this manner does not show the same conservatism. Consequently, a study has commenced to investigate the possibility of developing more accurate methods of analysis.

### 2.4.3 Recent Analytical Developments

Folded plate methods have been used by modelling the discrete studs as a smeared connector plate of equivalent shear modulus. So far this has been shown to underestimate the stiffness to a considerable extent, and there is considerable difficulty in establishing a suitable connector modulus, given

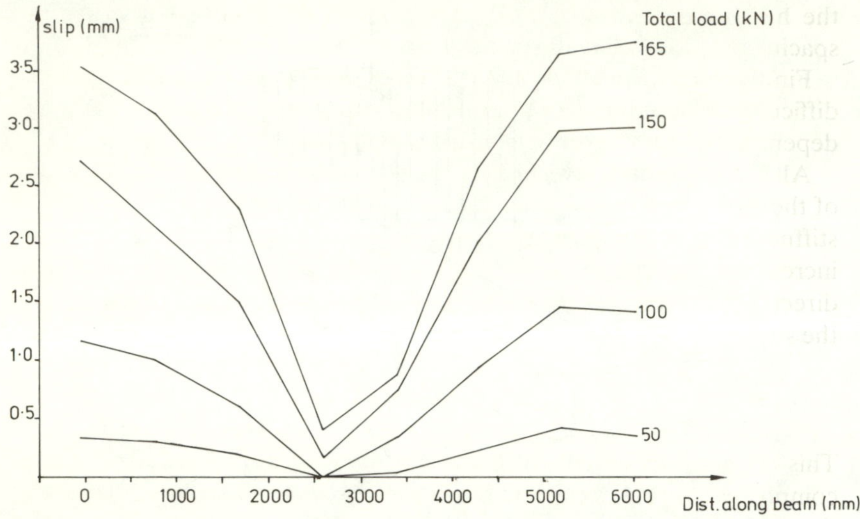


FIG. 2.23. Slip variation along the composite beam.

FIG. 2.24. Deformation of end stud.

the highly non-linear variation in slip along the beam. A non-uniform spacing of studs would also create difficulty from this point of view.

Finite element methods are also being developed but suffer from the difficulty in determining the exact stud stiffness as this appears to vary, depending on the degree of bending compression.

All of the methods investigated so far do not accurately model the action of the shear stud in the beam, and it is therefore concluded that the stud stiffness is of vital importance in the analysis of beam behaviour. This is of increasing importance as the stiffness of the connection may be related directly to the long-term deformation due to creep of the concrete around the stud.

## 2.5 CONCLUSION

This chapter has described the current state of the art in the design of composite flooring decks and has concentrated upon the three phases of the structural action that must be considered by the designer. At present, much of the design is empirical, and some designs are based on performance testing; current design procedures tend to be conservative in some instances and unconservative in others.

The rapidly increasing adoption of such flooring systems in practice has resulted in an intensification of the supporting research effort. Currently, the authors are involved in studying the effects of continuity upon the flooring decks and in assessing their vibration characteristics.

## REFERENCES


AMERICAN IRON AND STEEL INSTITUTE (1968) *Cold-formed steel design manual.*

BRITISH STANDARDS INSTITUTION (1965) *Composite construction in structural steel and concrete. Simply supported beams in building.* CP 117: Part 1.

BRITISH STANDARDS INSTITUTION (1982) *Structural use of steel in building. Code of Practice for design of floors with profiled steel sheeting.* BS 5950: Part 4.

CONSTRADO (1983a) *Steel framed multi-storey buildings, design recommendations for composite floors and beams using steel decks. Section 1, Structural.* October.

CONSTRADO (1983b) *European Recommendations for Steel Construction. The design of profiled steel sheeting.* ECCS–TC7.

EUROPEAN CONVENTION FOR CONSTRUCTIONAL STEELWORK (1977) *European recommendations for the testing of profiled metal sheets.* ECCS-XVII-77-2E, April, Document 20.

GOLDBERG, J. E. and LEVE, H. L. (1957) *Theory of prismatic folded plate structures*, IABSE, **17**, 59–86.

GRANT, J. A., FISHER, J. W. and SLUTTER, R. G. (1977) Composite beams with formed steel deck, *AISI Eng. J.*, First Quarter.

HARDING, P. W. (1986) *The development of composite flooring systems*, Research Report, University College, Cardiff.

JOHNSON, R. P. (1970) Research on steel–concrete composite beams, *Proc. ASCE*, **96**(ST3), 445–59, March.

JOHNSON, R. P. (1975) *Composite structures of steel and concrete. Vol. 1: Beams, columns, frames and applications in building*, CONSTRADO.

JOHNSON, R. P. and MAY, I. M. (1975) Partial-interaction design of composite beams, *Struct. Eng.*, **53**, August.

KARMAN, T. VON, SECHLER, E. E. and DONNELL, L. H. (1932) The strength of thin plates in compression, *Trans. ASCE*, **54**, 53.

NELSON, T. R. W. (1983) *Weld-through deck application and UK design data manual*.

NEWMARK, N. M., SIESS, C. P. and VIEST, I. M. (1951) Tests and analysis of composite beams with incomplete interaction, *Proc. Soc. Exper. Stress. Anal.*, **9**(1), 75–92.

PORTER, M. L. and EKBERG, C. E. (1976). Design recommendations for steel deck floor slabs, *J. Struct. Div., ASCE*, **102**(ST11), 2121–36.

PORTER, M. L., EKBERG, C. E., GREIMANN, L. F. and ELLEBY, H. A. (1976) Shear-bond analysis of steel deck-reinforced slabs, *J. Struct. Div., ASCE*, **102**(ST12), 2255–68.

PRASANNAN, S. and LUTTRELL, L. D. (1984) *Flexural strength formulations for steel-deck composite slabs*, Research Report, West Virginia University, Morgantown.

SCHUSTER, R. M. and EKBERG, C. E. (1970) *Commentary on the tentative recommendations for the design of cold-formed steel decking as reinforcement for concrete floor slabs*, Research Report, Iowa State University, Ames.

TIMOSHENKO, S. P. and GERE, J. (1982) *Theory of elastic stability*, McGraw-Hill.

TIMOSHENKO, S. P. and GOODIER (1947) *Strength of materials*, McGraw-Hill.

WINTER, G. (1947) Strength of thin steel compression flanges, *Trans. ASCE*, **112**.

WRIGHT, H. D., EVANS, H. R. and HARDING, P. W. (1984) The use of push-out tests to simulate shear stud connectors in composite beam construction, *Proc. Conf. on the Design of Concrete Structures, the Use of Model Analysis*, Building Research Establishment.

WRIGHT, H. D., EVANS, H. R. and HARDING, P. W. (1987) The use of profiled steel sheeting in floor construction, *J. Construct. Steel Res.*, **7**(4), 279–95.

*Chapter 3*

# BEHAVIOR AND DESIGN OF COMPOSITE BEAMS WITH WEB OPENINGS

DAVID DARWIN

*Department of Civil Engineering, University of Kansas, Lawrence, Kansas, USA*

*SUMMARY*

*Current research on the behavior and strength of composite beams with web openings is reviewed. The studies include beams with both solid and ribbed slabs. Research has demonstrated that the concrete slab contributes significantly to both flexural and shear strength at an opening. A practical design procedure is presented and compared with test results. Guidelines for detailing are presented.*

## NOTATION

| | |
|---|---|
| $a_o$ | Length of opening |
| $\bar{a}$ | Depth of concrete compression zone |
| $b_e$ | Effective width of concrete slab |
| $b_f$ | Flange width of steel section |
| $d$ | Depth of steel section |
| $d_h, d_l$ | Distance from top of steel section to centroid of concrete force at high or low moment end of opening |
| $e$ | Eccentricity of opening, positive up |
| $f'_c$ | Compressive (cylinder) strength of concrete |
| $h_o$ | Depth of opening |
| $h_r$ | Nominal height of deck rib |
| $s_t, s_b$ | Depth of top and bottom steel tees |

| | |
|---|---|
| $t_e$ | Effective thickness of concrete slab |
| $t_f$, $t_w$ | Flange or web thickness of steel section |
| $t_s$ | Total concrete slab thickness |
| $t'_s$ | Thickness of concrete slab above the rib |
| $v_c$ | Maximum concrete shear stress |
| $w_r$ | Average width of concrete rib |
| $x$ | Neutral axis location below top of flange in steel section |
| $z$ | Distance between points about which secondary moments are calculated |
| $A_f$ | Area of flange |
| $A_s, A_{sn}, A_{st}$ | Gross or net area of steel section and top tee steel area |
| $A_{sc}$ | Cross-sectional area of shear stud |
| $A_{vc}$ | Effective concrete shear area |
| $E_c$ | Modulus of elasticity of concrete |
| $F_u$ | Tensile strength of stud |
| $F_y, F_{yr}$ | Yield strength or reduced yield strength of steel |
| $H_s$ | Length of stud connector after welding |
| $M$ | Moment at opening center line |
| $M_{bh}, M_{bl}, M_b$ | Secondary bending moment in bottom tee at high or low moment end, or at both ends when $P = 0$ |
| $M_m, M_n$ | Maximum or nominal moment capacity at opening center line |
| $M_{th}$, $M_{tl}$ | Secondary bending moment in top tee at high or low moment end |
| $M_u$ | Factored bending moment at opening center line |
| $N, N_o, N_r$ | Number of shear connectors between high moment end of opening and support, or over the opening, or in one rib |
| $P, P_b, P_t$ | Axial force in a tee, or bottom tee, or top tee |
| $P_c, P_{ch}, P_{cl}$ | Axial force in concrete under pure bending, or under pure shear at high or low moment end of opening |
| $Q_n$ | Shear connector capacity |
| $R$ | Strength reduction factor for shear connector |
| $T'$ | Tensile force in net steel section |
| $V$ | Shear at opening |
| $V_b, V_t$ | Shear in bottom or top tee, maximum values $V_b$ (max), $V_t$ (max) |
| $V_m, V_n$ | Maximum or nominal shear capacity at opening |
| $V_{pt}$ | Plastic shear capacity of top tee steel web |
| $V_t$ (sh) | Pure shear capacity of top tee |
| $V_u$ | Factored shear at opening |

| | |
|---|---|
| $\alpha_b, \beta_b, \gamma_b$ | Variables used to calculate $V_b$ (max) |
| $\alpha_t, \beta_t, \gamma_t$ | Variables used to calculate $V_t$ (max) |
| $\tau$ | Shear stress in steel |

## 3.1 INTRODUCTION

The use of web openings in composite beams to allow the passage of utilities can improve the overall economy of a building by reducing the story height. However, web openings can substantially weaken a beam. In many cases, this reduced strength is countered by the use of reinforcement around the opening, a procedure which may more than double the cost of a beam.

The strength at an opening must be calculated accurately to determine whether reinforcement is necessary. An important aspect of the calculation is to account for the contribution of the concrete slab to member capacity. This contribution can be significant.

While rational design procedures for web openings in steel sections have been well established for many years, major research on web openings in composite beams dates only from 1977 (Darwin, 1984). A great deal has been accomplished in a brief time period. Studies include composite beams with both solid and ribbed slabs, and work is still under way.

This chapter summarizes the current state-of-the-art in the behavior and design of web openings in composite beams. Efficient techniques now exist that allow web openings to be designed in a rational manner. The chapter includes the presentation of a practical strength design technique. Detailing guidelines are also presented.

## 3.2 BEAM BEHAVIOR

This section describes the response of composite beams with web openings as they are loaded to failure. Knowledge of member response is useful in understanding the design techniques discussed in the chapter.

### 3.2.1 Forces Acting at the Opening

The forces acting at an opening are illustrated in Fig. 3.1 for a beam in positive bending. As illustrated in the figure, the bending moment increases from the left to the right. At the center line of the opening, the beam is subjected to a bending moment, $M$, and a shear, $V$. The portion of

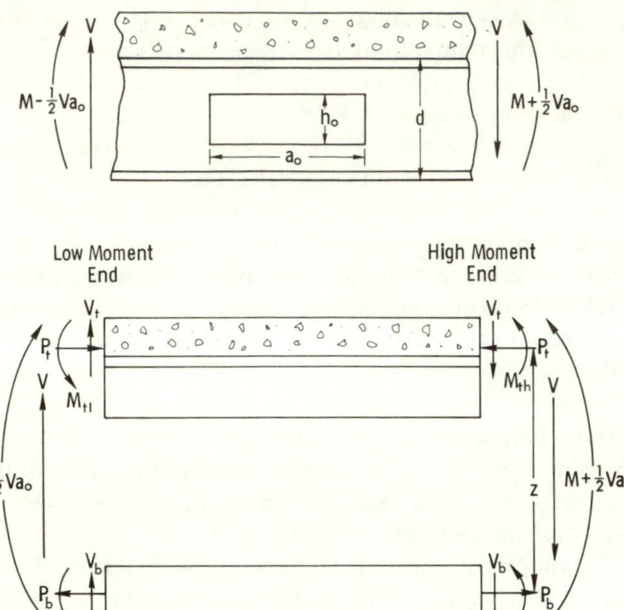


FIG. 3.1. Forces acting at web opening (after Donahey and Darwin, 1986).

the beam above the opening, or top tee, is subjected to an axial compressive force $P_t$, a shear $V_t$, and secondary bending moments at the low and high moment ends $M_{tl}$ and $M_{th}$, respectively. The portion of the beam below the opening, or bottom tee, is subjected to an axial tensile force $P_b$, a shear $V_b$, and secondary bending moments at the low and high moment ends $M_{bl}$ and $M_{bh}$, respectively. Based on equilibrium, these forces are related as follows:

$$P_b = P_t = P \tag{3.1}$$

$$V = V_b + V_t \tag{3.2}$$

$$V_b a_o = M_{bl} + M_{bh} \tag{3.3}$$

$$V_t a_o = M_{tl} + M_{th} \tag{3.4}$$

$$M = Pz + M_{th} + M_{bh} - \frac{Va_o}{2} \tag{3.5}$$

in which $a_o$ = the length of the opening, and $z$ = the distance between the points about which the secondary bending moments are calculated.

### 3.2.2 Response to Loading

The response of beams with solid slabs will be discussed first, followed by a description of beams with ribbed slabs.

The deformation and failure modes of composite beams with web openings depend upon the ratio of moment to shear, $M/V$, at the opening (Granade, 1968; Clawson and Darwin, 1980, 1982a; Cho, 1982).

Typical deformation and failure modes are illustrated in Fig. 3.2 for beams with solid slabs. For beams with high $M/V$ ratios (Fig. 3.2(a)), the

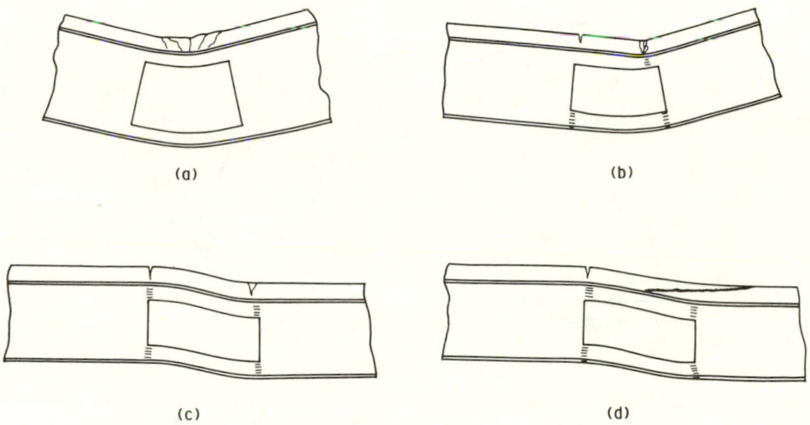


FIG. 3.2. Failure modes at web openings in beams with solid slabs: (a) pure bending, $V = 0$; (b) high $M/V$; (c) lower $M/V$, concrete crushes; (d) low $M/V$, diagonal tension failure in concrete (Darwin, 1984).

mode of deformation is primarily one of flexure. The steel yields in tension while the concrete crushes in compression. The secondary bending moments and shear do not play an important role.

As the moment/shear ratio decreases, shear and secondary bending become progressively more important. This results in an increasing differential, or Vierendeel, deformation through the opening (Fig. 3.2(b), (c), (d)). The top and bottom tees exhibit a well-defined change in curvature, as shown in Fig. 3.2(c) and (d). Transverse cracks form in the slab above the low moment end of the opening, due to the secondary bending moment. Failure is governed by crushing of the concrete or, more likely, a diagonal tension failure in the concrete at the high moment end. The beams exhibit a high amount of slip between the concrete and the steel, even when the section is designed for full composite action (Clawson and Darwin, 1980, 1982a). The beams also exhibit longitudinal cracking of the slab over the steel section. The added capacity provided by the concrete can greatly increase the capacity of a section.

An important observation obtained from tests is that the first yield in the steel does not give an accurate measurement of the section capacity. Clawson and Darwin (1980, 1982a) found that the load at first yield varied

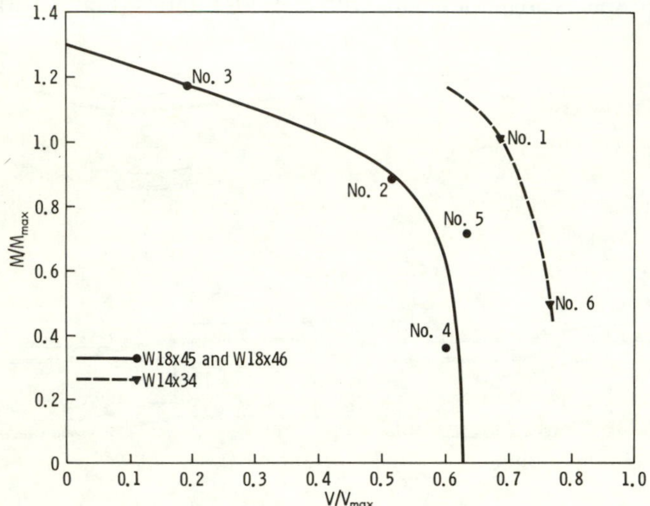


FIG. 3.3. Moment–shear interaction for test specimens with solid slab (Clawson and Darwin, 1982a).

from 17% to 46% of the failure load. As a general rule, the failure of the beams was ductile; that is, they obtained large deformations and sizeable concrete cracking (both longitudinal and transverse) prior to failure.

The tests of beams with solid slabs demonstrate that the interaction between moment capacity and shear capacity is rather weak. As shown in Fig. 3.3, the moment capacity at web openings is relatively unaffected by shear until the shear reaches the maximum shear capacity.

More recent research using ribbed slabs (Wong and Redwood, 1980; Redwood and Wong, 1982; Redwood and Poumbouras, 1983; Darwin, 1984; Donahey and Darwin, 1986) shows that composite beams with ribbed slabs are similar in many ways to beams with solid slabs. There are, however, a number of significant differences. The ribbed slabs have a lower compression area and a lower shear connector capacity. In virtually all cases, failure is preceded by rib cracking or rib separation, as illustrated in Figs 3.4 and 3.5. The rib separation may occur at the high moment end of the opening or in a region of the beam near the high moment end but away from the opening. Rib separation seems to occur in beams subjected to both high and low $M/V$ ratios (Figs 3.4 and 3.5, respectively). For beams with the ribs transverse to the steel section, failure appears to be governed by a failure in the concrete at or around the shear studs. These 'stud' failures vary from a pull-out failure at the high moment end of the opening to a shoving or shear failure at the low moment end of the opening.

As with the solid slabs, beams with ribbed slabs exhibit a large amount of slip between the concrete and the steel, as well as transverse cracking over the low moment end of the opening and longitudinal cracking over the steel section prior to failure. Beams with ribbed slabs and openings with moderate to low $M/V$ ratios also exhibit a tendency of the slab to 'bridge' over the opening. The thicker the slab, the more apparent is the bridging (Fig. 3.6). Failure of beams in these cases is clearly preceded by a pull-out failure of the studs, with or without transverse rib cracking at the high moment end or in the region of the beam away from the high moment end. As with the solid slabs, the beams with the transverse ribs exhibit only a weak interaction between shear and moment capacity at the opening.

Ribbed slabs with longitudinal ribs directly above the steel section behave in a manner that is very close to that of solid slabs. In the limited number of beams of this type that have been tested, failure has been manifested by a longitudinal shear failure between the rib and the surrounding deck (Fig. 3.7).

Although ribbed slabs contain less concrete than solid slabs, the slab continues to play an important role in member capacity at an opening and clearly participates in carrying the load. Also, as in the case of the beams with solid slabs, the first yield of the steel is not a good measure of strength. Recent tests at the University of Kansas indicate that yield occurred at values of load that range from 19% to 52% of the ultimate capacity, virtually the same range as obtained for the solid slabs (Donahey and Darwin, 1986).

A number of parameters control opening capacity (Clawson and Darwin, 1980). For a given geometry, a change in steel strength seems to cause the biggest increase in the strength at an opening. A change in concrete strength has less effect. The deeper and the longer the opening,

FIG. 3.4. Failure at web opening in beam with ribbed slab, high $M/V$ (Donahey and Darwin, 1986).

FIG. 3.5. Failure at web opening in beam with ribbed slab, low $M/V$ (Donahey and Darwin, 1986).

FIG. 3.6. Bridging of slab at opening (Donahey and Darwin, 1986).

FIG. 3.7. Longitudinal rib failure (Donahey and Darwin, 1986).

the lower the capacity. A downward or negative eccentricity (opening centered below the midheight of the steel section) will generally increase the maximum shear capacity, but will decrease the maximum moment capacity at an opening. An upward or positive eccentricity will have the opposite effect. Shear connectors play an important role in both solid and ribbed slabs. As the number of connectors is increased over the opening and between the opening and the point of zero moment, the capacity of the section will be increased.

The choice of slab type can have a marked effect on strength at an opening because of the combined effect of shear connector capacity and effective slab thickness. The shear connector capacity is generally more critical than the effective slab thickness. Solid slabs provide more strength than ribbed slabs, and ribbed slabs with the ribs parallel to the steel sections (longitudinal ribs) provide more strength than slabs with ribs transverse to the steel section.

Solid slabs and slabs with longitudinal ribs can accommodate more shear connectors than slabs with transverse ribs. In addition, the strength of shear connectors decreases progressively from solid slabs to slabs with longitudinal ribs to slabs with transverse ribs (Grant, Fisher and Slutter, 1977). As a result, for solid slabs, shear connector capacity is generally great enough to force a crushing or diagonal tension failure in the slab (Fig. 3.2), while for ribbed slabs, strength usually is limited by shear stud capacity (Figs 3.5–3.7). This strongly suggests that for ribbed slabs, steps in design or construction that increase the shear connector capacity will also increase the strength at web openings.

### 3.2.3 Serviceability—Deflection and Cracking

The effect of a single opening on the total deflection of a beam under service load is small. However, the local deflection at an opening can be relatively high, and may lead to transverse cracking near the low moment end of an opening at loads as low as 21% of the ultimate (Donahey and Darwin, 1986). Local deformation and cracking are most severe in areas of low $M/V$ ratio.

Shored construction will result in earlier cracking than unshored construction, because, with unshored construction, deformation due to the slab dead load will occur before the concrete has set. The slab will participate only in carrying loads in addition to the beam dead load. This will raise the cracking load as a percentage of the ultimate load.

Donahey and Darwin (1986) found that cutting the opening after the slab is placed is likely to cause a transverse crack over the opening. That crack, however, appears to have little effect on member response when load is applied.

## 3.3 DESIGN OF OPENINGS

### 3.3.1 Strength at Openings

Models developed to represent the strength of composite beams at web openings (Todd and Cooper, 1980; Clawson and Darwin, 1980; Redwood and Wong, 1982; Donoghue, 1982; Poumbouras, 1983; Redwood and Poumbouras, 1984; Donahey and Darwin, 1986) do so through the construction of moment–shear interaction diagrams, as illustrated in Fig. 3.8. The models have a number of similarities, although the procedures for constructing the diagrams differ. As a minimum, the analysis usually starts with the determination of the pure moment capacity, $M_m$, and the pure shear capacity, $V_m$, at the opening. Once these values have been obtained, the simpler models, such as might be used in design, connect the two points with a curve or a series of straight line segments to represent the interaction diagram (Clawson and Darwin, 1980; Redwood and Wong, 1982; Donoghue, 1982; Redwood and Poumbouras, 1984; Donahey and Darwin, 1986).

The more detailed models generate the full diagram, point by point, based on an analytical representation (Todd and Cooper, 1980; Clawson and Darwin, 1982b; Poumbouras, 1983; Donahey and Darwin, 1986). For these models, the interaction diagrams are developed by calculating the moment, $M_n$, that can be carried as the shear, $V_n$, is increased from zero to $V_m$. For each value of shear, portions are assigned to the top and bottom tees. The axial force and secondary bending moments in the bottom tee are then obtained. An axial force of equal magnitude but opposite sign (eqn (3.1)) is assigned to the top tee. The shear in the top tee is distributed between the concrete and steel at the high and low moment ends, based on the assumptions inherent in the model.

For models applicable to unreinforced (Todd and Cooper, 1980; Clawson and Darwin, 1980, 1982b; Redwood and Wong, 1982; Redwood and Poumbouras, 1984; Donahey and Darwin, 1986) and reinforced (Donoghue, 1982; Donahey and Darwin, 1986) openings there is general agreement on the strength of the sections under pure bending. The models differ in how they handle the high shear cases and how they represent the interaction diagram between pure bending and pure shear.

For their strength model (Fig. 3.8), Clawson and Darwin (1980, 1982b) assume full composite behavior for beams with solid slabs and generate the interaction diagram by first determining the distribution of shear between the top and bottom tees for the case of zero axial force within the tees ($P_b = P_t = 0$). This ratio of $V_t$ to $V_b$ is then held constant as the value of total shear, $V_n$, is increased from zero to the maximum value, $V_m$. The corresponding bending moment, $M_n$, is maximized for each value of $V_n$, since the procedure is an equilibrium method which produces a lower bound solution.

For their simplified model, Clawson and Darwin (1980) generate only the pure bending and pure shear strengths and then connect the two with an ellipse (Fig. 3.8).

Redwood and Wong (1982) and Redwood and Poumbouras (1984) calculate the maximum bending and shear strengths. Using a procedure developed by Redwood (1968), Redwood and Wong (1982) calculate the maximum moment that can be sustained at the maximum shear. Redwood and Poumbouras (1984) obtain the moment corresponding to the maximum shear, using eqn (3.5). In each case, this generates a vertical line on the right side of the interaction curve (Fig. 3.9). The curve is closed with an ellipse. United States Steel (1984) has prepared a design aid based on the Redwood and Wong (1982) model.

Donoghue (1982) developed a simplified design procedure for com-

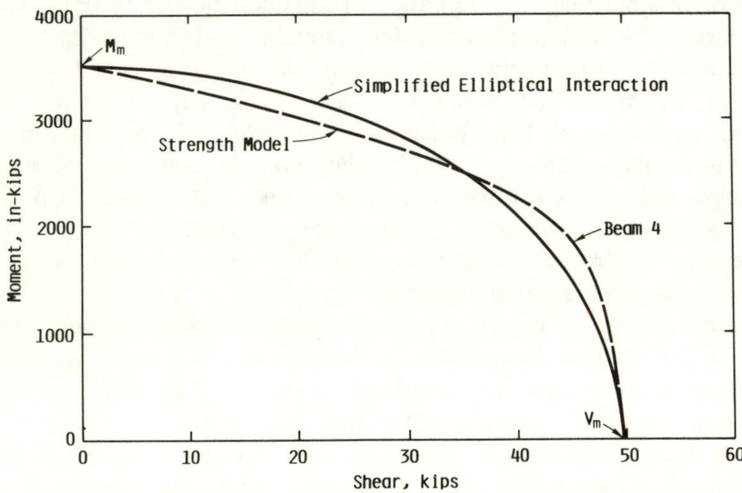


FIG. 3.8. Full strength model and simplified moment–shear interaction diagrams for Clawson and Darwin (1980, 1982b).

posite beams with reinforced web openings. The procedure neglects the contribution of the concrete slab to the shear strength of the section.

Donahey and Darwin (1986) have developed both a detailed strength model and several simplified design representations. The strength model is an extension of the Clawson and Darwin (1980, 1982b) model. The new model is more accurate than the original and applies to composite beams with ribbed slabs as well as beams with solid slabs. The new model can also account for reinforcement of the opening. For their simplified design procedures, which apply to unreinforced openings, Donahey and Darwin

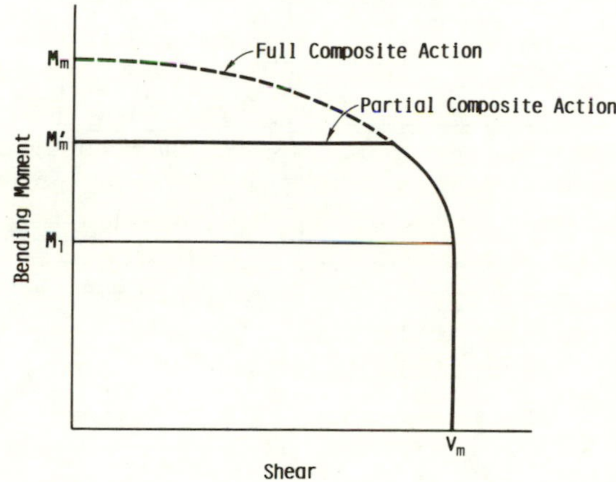


FIG. 3.9. Moment–shear interaction diagram for Redwood and Poumbouras (1984).

(1986) calculate the maximum bending and shear strengths and connect the points with a cubic curve. Their three design procedures differ in how the pure shear capacity, $V_m$, is calculated. The following section presents the most accurate of their three simplified design procedures.

### 3.3.2 Design Procedure

As illustrated in Figs 3.3, 3.8 and 3.9, the interaction between moment and shear capacity at an opening is relatively weak. The interaction curve can be represented by an appropriate curve if the pure moment and pure shear strengths, $M_m$ and $V_m$, can be established accurately.

Fig. 3.10 illustrates openings in composite beams with a solid slab, a ribbed slab with the ribs transverse to the beam, and a ribbed slab with the ribs parallel to the beam. The openings are of length $a_o$ and depth $h_o$, and may have an eccentricity $e$ (positive upward) with respect to the centerline

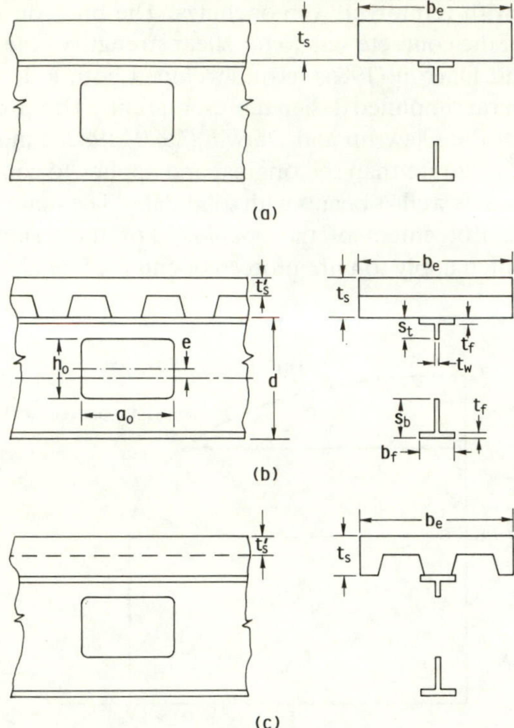


FIG. 3.10. Beam and opening configurations: (a) solid slab; (b) ribbed slab with transverse ribs; (c) ribbed slab with longitudinal ribs (Donahey and Darwin, 1986).

of the steel section. The slab thicknesses, $t_s$ and $t_s'$, effective slab width*, $b_e$, and steel section dimensions, $d$, $b_f$, $t_f$, $t_w$, $s_t$ and $s_b$, are as shown.

The procedure presented here (Donahey and Darwin, 1986) has been calibrated to a wide range of tests, including the three cases illustrated in Fig. 3.10.

The interaction equation relates the nominal moment and shear strengths, $M_n$ and $V_n$, with the maximum moment and shear capacities, $M_m$ and $V_m$ (Fig. 3.11).

$$\left(\frac{M_n}{M_m}\right)^3 + \left(\frac{V_n}{V_m}\right)^3 = 1 \qquad (3.6)$$

*Expressions in Appendix I.

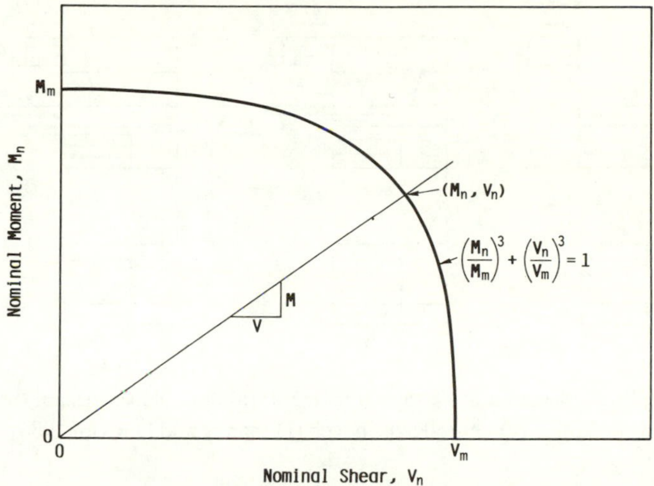


FIG. 3.11. Design interaction diagram (Donahey and Darwin, 1986).

*3.3.2.1 Maximum Moment Capacity, $M_m$*

The maximum or pure moment capacity of a composite beam at a web opening, $M_m$, is obtained using flexural strength procedures, which are now summarized. Fig. 3.12 illustrates stress diagrams for sections in pure bending.

For a given beam and opening configuration, the force in the concrete, $P_c$, is limited to the lowest of the concrete compressive strength, the shear connector capacity, or the yield strength of the net steel section.

$$P_c \leq 0{\cdot}85 f_c' b_e t_e \tag{3.7a}$$

$$\leq NRQ_n \tag{3.7b}$$

$$\leq T' = F_y A_{sn} \tag{3.7c}$$

in which $t_e = t_s$ for solid slabs, $= t_s'$ for ribbed slabs with transverse ribs, and $= (t_s + t_s')/2$ for ribbed slabs with longitudinal ribs; $A_{sn}$ = net steel area $= A_s - t_w h_o$; $N$ = number of shear connectors from end of opening to support; $R$ = strength reduction factor for shear connectors in ribbed slabs* (= 1 for solid slabs); and $Q_n$ = shear connector capacity*.

*Expressions in Appendix I.

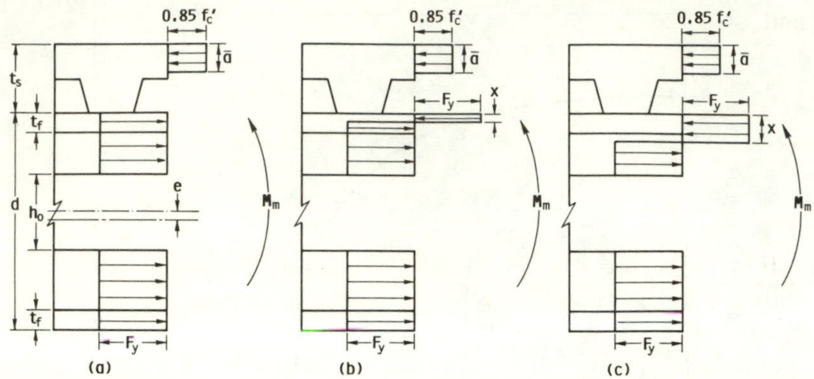


FIG. 3.12. Stress diagrams at maximum moment: (a) neutral axis in slab; (b) neutral axis in flange; (c) neutral axis in web (Donahey and Darwin, 1986).

The maximum moment capacity, $M_m$, depends on which of the inequalities in eqn (3.7) governs.

If $P_c = T'$ (eqn (3.7c) and Fig. 3.12(a)),

$$M_m = T'\left(\frac{d}{2} + \frac{t_w h_o e}{A_{sn}} + t_s - \frac{\bar{a}}{2}\right) \tag{3.8}$$

in which $\bar{a} = P_c/0{\cdot}85 f'_c b_e$ for solid slabs and ribbed slabs for which $\bar{a} \leq t'_s$.

If $\bar{a} > t'_s$, as it can be for ribbed slabs with longitudinal ribs, the term $(t_s - \bar{a}/2)$ in eqn (3.8) must be replaced with the appropriate expression for the distance between the top of the steel flange and the centroid of the concrete force.

If $P_c < T'$ (eqn (3.7a) or (3.7b)), a portion of the steel section will be in compression. The neutral axis may be in either the flange or the web of the top tee. The location of the neutral axis can be determined on the basis of the inequality

$$P_c + F_y A_f \lesseqgtr (A_{sn} - A_f) F_y \tag{3.9}$$

in which $A_f$ = the flange area = $b_f t_f$.

If the left side of eqn (3.9) exceeds the right side, the neutral axis will be in the flange (Fig. 3.12(b)), at a distance $x$ from the top of the flange.

$$x = \frac{1}{2 b_f F_y}(A_{sn} F_y - P_c) \tag{3.10}$$

and

$$M_{\mathrm{m}} = F_{\mathrm{y}}\left(\frac{A_{\mathrm{sn}}\,d}{2} + t_{\mathrm{w}}h_{\mathrm{o}}e - b_{\mathrm{f}}x^2\right) + P_{\mathrm{c}}\left(t_{\mathrm{s}} - \frac{\bar{a}}{2}\right) \tag{3.11a}$$

$$= T'\left(\frac{d}{2} + \frac{t_{\mathrm{w}}h_{\mathrm{o}}e - b_{\mathrm{f}}x^2}{A_{\mathrm{sn}}}\right) + P_{\mathrm{c}}\left(t_{\mathrm{s}} - \frac{\bar{a}}{2}\right) \tag{3.11b}$$

If the right side of eqn (3.9) is greater than the left side, the neutral axis will be in the web (Fig. 3.12(c)), at a distance $x$ from the top of the flange.

$$x = \frac{A_{\mathrm{sn}} - 2A_{\mathrm{f}}}{2t_{\mathrm{w}}} - \frac{P_{\mathrm{c}}}{2F_{\mathrm{y}}t_{\mathrm{w}}} + t_{\mathrm{f}} \tag{3.12}$$

and

$$M_{\mathrm{m}} = F_{\mathrm{y}}\left[\frac{A_{\mathrm{sn}}\,d}{2} + t_{\mathrm{w}}h_{\mathrm{o}}e - (b_{\mathrm{f}} - t_{\mathrm{w}})\,t_{\mathrm{f}}^2 - t_{\mathrm{w}}x^2\right] + P_{\mathrm{c}}\left(t_{\mathrm{s}} - \frac{\bar{a}}{2}\right) \tag{3.13a}$$

$$= T'\left[\frac{d}{2} + \frac{t_{\mathrm{w}}h_{\mathrm{o}}e - (b_{\mathrm{f}} - t_{\mathrm{w}})t_{\mathrm{f}}^2 - t_{\mathrm{w}}x^2}{A_{\mathrm{sn}}}\right] + P_{\mathrm{c}}\left(t_{\mathrm{s}} - \frac{\bar{a}}{2}\right) \tag{3.13b}$$

#### *3.3.2.2 Maximum Shear Capacity, $V_{\mathrm{m}}$*

The procedure used to calculate the maximum shear capacity distinguishes the various design procedures from each other. The procedure described here is the most accurate current design model (Donahey and Darwin, 1986).

The maximum shear capacity, $V_{\mathrm{m}}$, is obtained by considering the load state in which the axial loads in the top and bottom tees, $P_{\mathrm{t}}$ and $P_{\mathrm{b}}$, equal zero (Fig. 3.13). This provides a very close approximation of the true pure shear capacity, but is not precisely pure shear, since $M_{\mathrm{tl}} \neq M_{\mathrm{th}}$. Thus the moment at the opening centerline has a small but finite value.

The procedure consists of summing the individual shear capacities of the top and bottom tees:

$$V_{\mathrm{m}} = V_{\mathrm{b}}(\max) + V_{\mathrm{t}}(\max) \tag{3.14}$$

$V_{\mathrm{b}}$ (max) and $V_{\mathrm{t}}$ (max) are obtained under the combined effects of shear and secondary bending. As a result, the interaction between shear and axial stresses must be considered. Since the shear is largely carried by the web of

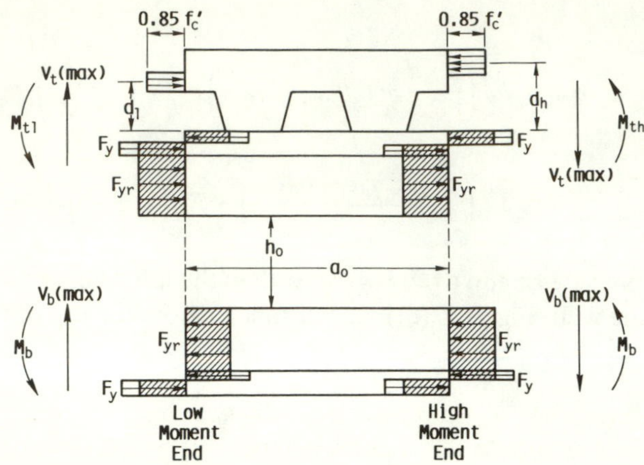


FIG. 3.13. Stress diagram at maximum shear (Donahey and Darwin, 1986).

the top and bottom tees, the effect of shear on the axial stresses within the webs is important.

The interaction between shear stress, $\tau$, and axial stress, $F_{yr}$, can be represented using the von Mises yield criterion:

$$3\tau^2 + F_{yr}^2 = F_y^2 \tag{3.15a}$$

or

$$F_{yr} = (F_y^2 - 3\tau^2)^{1/2} \tag{3.15b}$$

The solution of the interaction problem can be simplified using a linear approximation to eqn (3.15b).

$$F_{yr} = 1{\cdot}2F_y - \sqrt{3}\tau \tag{3.16}$$

Moment equilibrium for the bottom tee is expressed in eqn (3.3):

$$V_b a_o = M_{bl} + M_{bh} = 2M_{b(P=0)} \tag{3.3}$$

Representing the secondary moments in eqn (3.3) in terms of the stresses, $F_y$ (in the flange) and $F_{yr}$ (in the web), and using eqn (3.16) for $F_{yr}$, with $\tau = V_b/t_w s_b$, a closed-form expression may be obtained for the shear capacity of the bottom tee:

$$V_b(\max) = F_y\left(\frac{\beta_b - \sqrt{\beta_b^2 - 4\alpha_b\gamma_b}}{2\alpha_b}\right) \tag{3.17}$$

in which $\alpha_b = 3 + 2\sqrt{3}(a_o/s_b)$

$$\beta_b = 2\sqrt{3}(b_f - t_w)\left(s_b - t_f + \frac{t_f^2}{s_b}\right) + 2{\cdot}4\sqrt{3}t_w s_b + 2a_o(b_f + 0{\cdot}2t_w)$$

$$\gamma_b = (b_f - t_w)^2 t_f^2 + 1{\cdot}44 t_w^2 s_b^2 + 2{\cdot}4 t_w (b_f - t_w)(s_b^2 - s_b t_f + t_f^2)$$

For the top tee, moment equilibrium is expressed in eqn (3.4):

$$V_t a_o = M_{tl} + M_{th} \tag{3.4}$$

An expression for $V_t$ (max) can be obtained by a procedure similar to that used for the bottom tee. For the top tee, the forces in the concrete at the high and low moment ends, $P_{ch}$ and $P_{cl}$, must be considered in addition to the steel stresses.

$P_{ch}$ is limited by concrete capacity, shear connector capacity, and top tee tensile strength:

$$P_{ch} \leq 0{\cdot}85 f_c' b_e t_e \tag{3.18a}$$

$$\leq NRQ_n \tag{3.18b}$$

$$\leq F_y A_{st} \tag{3.18c}$$

The concrete force at the low moment end depends on $P_{ch}$ and the shear connector capacity over the opening:

$$P_{cl} = P_{ch} - N_o RQ_n \geq 0 \tag{3.19}$$

The solution of eqn (3.4) for $V_t$ (max) yields an expression of the same form as eqn (3.17):

$$V_t(\max) = F_y \left(\frac{\beta_t - \sqrt{\beta_t^2 - 4\alpha_t \gamma_t}}{2\alpha_t}\right) \tag{3.20}$$

in which $\alpha_t = 3 + 2\sqrt{3}\,(a_o/s_t)$

$$\beta_t = 2\sqrt{3}(b_f - t_w)\left(s_t - t_f + \frac{t_f^2}{s_t}\right) + 2{\cdot}4\sqrt{3}t_w s_t + 2a_o(b_f + 0{\cdot}2t_w)$$

$$+ \frac{2\sqrt{3}}{s_t F_y}(P_{ch} d_h - P_{cl} d_l) + \frac{\sqrt{3}}{F_y}(P_{ch} - P_{cl})$$

$$\gamma_t = (b_f - t_w)^2 t_f^2 + 1{\cdot}44 t_w^2 s_t^2 + 2{\cdot}4 t_w (b_f - t_w)(s_t^2 - s_t t_f + t_f^2)$$

$$+ \frac{2(b_f + 0{\cdot}2 t_w)}{F_y}(P_{ch} d_h - P_{cl} d_l) - \frac{(P_{ch}^2 + P_{cl}^2)}{2F_y^2}$$

$$+ \frac{[(b_f - t_w)t_f + 1{\cdot}2 t_w s_t]}{F_y}(P_{ch} - P_{cl})$$

$$d_h = t_s - \frac{0{\cdot}5 P_{ch}}{0{\cdot}85 f_c' b_e} \tag{3.21}$$

$$d_l = \frac{0{\cdot}5 P_{cl}}{0{\cdot}85 f_c' b_e} \text{ for solid slabs} \tag{3.22a}$$

$$d_l = t_s - t_s' + \frac{0{\cdot}5 P_{cl}}{0{\cdot}85 f_c' b_e} \text{ for ribbed slabs with transverse ribs} \tag{3.22b}$$

For ribbed slabs with longitudinal ribs, $d_l$ is based on the centroid of the compressive force in the concrete, considering all ribs that lie within the effective width, $b_e$ (Fig. 3.10(c)). Conservatively, it can be obtained using eqn (3.22a), replacing $b_e$ by the sum of the minimum rib widths for the ribs that lie within $b_e$.

Equation (3.20), like eqn (3.17), is based on the assumption that all of the shear carried by a tee is carried by the steel web. This assumption yields consistent results for the bottom tee. However, it may be overconservative for the top tee, in which the concrete slab may also carry shear. This is accounted for by comparing $V_t$ (max) with the maximum plastic shear capacity of the top tee web:

$$V_{pt} = \frac{F_y}{\sqrt{3}} t_w s_t \tag{3.23}$$

If $V_t$ (max) from eqn (3.20) exceeds $V_{pt}$, then the top tee web is assumed to be fully yielding in shear ($F_{yr} = 0$) and $P_{ch}$ is further limited:

$$P_{ch} \leq F_y (b_f - t_w) t_f \tag{3.24}$$

If eqn (3.24) controls $P_{ch}$ instead of eqn (3.18), $P_{cl}$, $d_h$ and $d_l$ are recalculated using eqns (3.19), (3.21) and (3.22). Resolving eqn (3.4) yields

$$V_t(\max) = \frac{1}{a_o}\left[(P_{ch}d_h - P_{cl}d_l) + \frac{t_f}{2}(P_{ch} - P_{cl}) + \frac{F_y}{2}(b_f - t_w)t_f^2 - \frac{(P_{ch}^2 + P_{cl}^2)}{4F_y(b_f - t_w)}\right] \tag{3.25}$$

Finally, $V_t$ (max) is limited by the pure shear capacity of the top tee, $V_t$ (sh):

$$V_{pt} \le V_t(\max) \le V_t(\mathrm{sh}) \tag{3.26}$$

$$V_t(\mathrm{sh}) = V_{pt} + \nu_c A_{vc} \tag{3.27}$$

in which $\nu_c$ = maximum concrete shear stress*; $A_{vc}$ = effective concrete shear area, $= 3t_s^2$ for solid slabs, $= 3t_s t_s'$ for slabs with transverse ribs, and $= 3t_s(t_s + t_s')/2$ for slabs with longitudinal ribs.

*3.3.2.3 Interaction Procedure*

Once $M_m$ and $V_m$ are obtained, the nominal capacities, $M_n$ and $V_n$, can be obtained for a given ratio of factored moment to factored shear, $M_u/V_u$, using eqn (3.6) (Fig. 3.11) as follows:

$$\left(\frac{M_n}{M_m}\right)^3 + \left(\frac{V_n}{V_m}\right)^3 = 1 \tag{3.6}$$

$$\left(\frac{M_n}{M_m}\right)^3 \left(\frac{V_m}{V_n}\right)^3 + 1 = \left(\frac{V_m}{V_n}\right)^3 \tag{3.28}$$

Solving for $V_n$ and letting $M_n/V_n = M_u/V_u = M/V$ yields

$$V_n = V_m\left[\frac{\left(\frac{M}{V}\right)^3}{\left(\frac{M_m}{V_m}\right)^3} + 1\right]^{-1/3} \tag{3.29}$$

*Expression in Appendix I.

$$M_n = V_n\left(\frac{M}{V}\right) \tag{3.30}$$

$$M_n = M_m\left[\frac{\left(\frac{M_m}{V_m}\right)^3}{\left(\frac{M}{V}\right)^3} + 1\right]^{-1/3} \tag{3.31}$$

*3.3.2.4 Comparison with Test Results*

Fig. 3.14 compares the results obtained by the design procedure with the results of 35 tests. The tests include solid slabs and ribbed slabs with both transverse and longitudinal ribs. For all beams, the average value of $V_n$ (test)/$V_n$ is 1·067, with a standard deviation of 0·082. The procedure is more conservative for the 13 beams with solid slabs ($V_n$ (test)/$V_n$ = 1·106, standard deviation = 0·087) than for the 22 beams with ribbed slabs ($V_n$ (test)/$V_n$ = 1·043, standard deviation = 0·071).

The procedure compares favorably with the detailed strength model developed by Donahey and Darwin (1986). The procedure is applicable to a wider range of cases than the Redwood and Poumbouras (1984) model, which is overconservative for beams with solid slabs. The conservatism of Redwood and Poumbouras (1984) results from a limitation of the maximum top tee shear capacity, $V_t$ (max), to the plastic shear capacity of

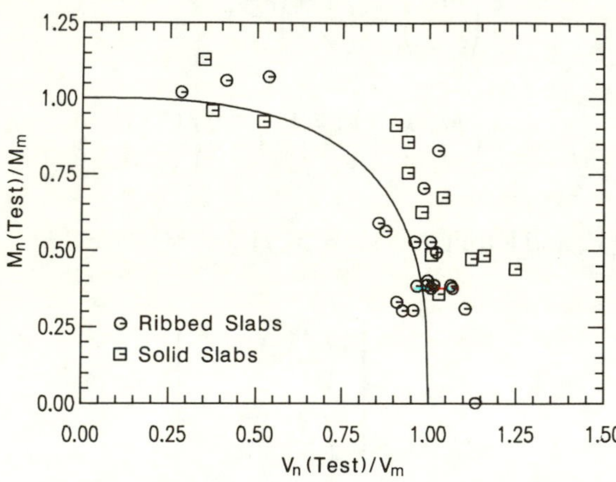


FIG. 3.14. Comparison of design procedure with test results (Donahey and Darwin, 1986).

the top tee web, $V_{pt}$. Test results uniformly demonstrate that the slab will carry a portion of the top tee shear.

### 3.3.3 Detailing

The strength and performance of a composite beam with a web opening can be enhanced through detailing practice. Some aspects of the detailing are reflected in the design procedure presented in the previous sections. Although additional research is still needed, a number of observations can be made, based on the tests completed to date.

As discussed in Section 3.2.2, and reflected in the design procedure, the strength at an opening is highly dependent upon the shear connector capacity. Thus, increasing the number of shear studs and using the maximum possible length of shear stud in ribbed slabs will increase the strength at an opening. Although not reflected in the design expressions, an increased density of shear connectors should be used toward the direction of increasing as well as decreasing moment from the high moment end of the opening. The additional shear connectors are warranted due to the bridging that tends to occur in the slabs (Fig. 3.6). It would be good practice to use at least two studs per foot (0·3 m) for a distance equal to the depth of the section, $d$, or the length of the opening, $a_o$, whichever is greater, from the high moment end of the opening toward the direction of increasing moment.

The tendency of the slabs to crack both transversely and longitudinally suggests the need to increase the reinforcing steel in the slab over the opening. The increased reinforcing steel will not prevent the cracks from forming, but will limit the crack widths. Transverse and longitudinal reinforcement ratios of 0·0025 in the vicinity of the opening (i.e., within a distance $d \geq a_o$) are suggested.

Beams with longitudinal ribs may fail due to a shear failure between the rib and the surrounding deck (Fig. 3.7). The nature of the failure suggests that additional transverse reinforcing steel that crosses the crack surface will improve the post-crack performance. A reinforcement ratio of 0·0025, as recommended above, will limit slip. To be effective, this reinforcement must be below the heads of the shear studs.

## 3.4 CONCLUDING REMARKS

A review of the current research on composite beams with web openings has been presented. A great deal of progress has been made in a short time toward developing a rational and accurate design procedure. A procedure that has been calibrated against all of the existing test results has been presented. The contribution of the concrete slab to shear strength, as well as flexural strength, at an opening is considered. Taking advantage of this contribution to shear strength should significantly lower the cost and simplify the construction of web openings in composite beams.

## REFERENCES

AISC (1980) *Manual of Steel Construction*, 3th Edn, American Institute of Steel Construction, Chicago, Illinois.

AISC (1986) *Load and Resistance Factor Design Manual of Steel Construction*, 1st Edn, American Institute of Steel Construction, Chicago, Illinois.

CHO SOON HO (1982) An investigation on the strength of composite beams with web openings, MS Arch. Eng. thesis, Hanyong University, Seoul, Korea (Dec.).

CLAWSON, W. C. and DARWIN, D. (1980) Composite beams with web openings, *SM Report No. 4*, University of Kansas Center for Research, Lawrence, Kansas, USA (Oct.).

CLAWSON, W. C. and DARWIN, D. (1982a) Tests of composite beams with web openings, *J. Struct. Div., ASCE*, **108** (ST1), 145–62.

CLAWSON, W. C. and DARWIN, D. (1982b) Strength of composite beams with web openings, *J. Struct. Div., ASCE*, **108**(ST3), 623–41.

DARWIN, D. (1984) Composite beams with web openings, *Proc. National Engineering Conf.*, AISC, Tampa, Florida (Mar.).

DONAHEY, R. C. and DARWIN, D. (1986) Performance and design of composite beams with web openings, *SM Report No. 18*, University of Kansas Center for Research, Lawrence, Kansas, USA (April).

DONOGHUE, C. M. (1982) Composite beams with web openings: design, *J. Struct. Div., ASCE*, **108**(ST12), 2652–67.

GRANADE, C. J. (1968) An investigation of composite beams having large rectangular openings in their webs, MS thesis, University of Alabama, USA.

GRANT, J. A., FISHER, J. W. and SLUTTER, R. O. (1977) Composite beams with formed steel deck, *AISC Engg J.*, **14**(1), 24–43.

POUMBOURAS, G. (1983) Modification of a theory predicting the shear strength of composite beams with large web openings, *Project Report No. U83-20*, Department of Civil Engineering and Applied Mechanics, McGill University, Montreal, Quebec, Canada (Apr.).

REDWOOD, R. G. (1968) Plastic behaviour and design of beams with web openings, *Proc. 1st Canad. Struct. Engg Conf.*, Canadian Steel Industries Construction Council, Toronto, Canada (Feb.), 127–38.

REDWOOD, R. G. and WONG, P. P. K. (1982) Web holes in composite beams with steel deck, *Proc. 8th Canad. Struct. Engg Conf.*, Canadian Steel Construction Council, Vancouver, Canada (Feb.).

REDWOOD, R. G. and POUMBOURAS, G. (1983) Tests of composite beams with web holes, *Canad. J. Civ. Engg*, **10**(4), 713–21.

REDWOOD, R. G. and POUMBOURAS, G. (1984) Analysis of composite beams with web openings, *J. Struct. Engg, ASCE*, **110**(9), 1949–58.
TODD, D. M. and COOPER, P. B. (1980) Strength of composite beams with web openings, *J. Struct. Div., ASCE*, **106**(ST2), 431–44.
UNITED STATES STEEL (1984) *Rectangular, Concentric and Eccentric Unreinforced Web Penetrations in Composite Steel Beams: A Design Aid*, ADUSS 27–8532–01, Pittsburgh, Pennsylvania, USA (Oct.).
WONG, P. P. K. and REDWOOD, R. G. (1980) Pilot test of a hollow composite beam containing a web opening, *Structural Engineering Report No. 80–6*, McGill University, Montreal, Quebec, Canada (Dec.).

## APPENDIX I

The design procedure presented here was developed using the following expressions for $b_e$, $R$, $Q_n$, and $v_c$. See Notation for undefined symbols.

Effective width of slab (AISC, 1986):

$$b_e \leq 1/4 \text{ of the beam span} \tag{3.32a}$$

$$\leq \text{center to center spacing of beams} \tag{3.32b}$$

Strength reduction factors for shear connectors in ribbed slabs (AISC, 1980, 1986):

$$R = \left(\frac{0{\cdot}85}{\sqrt{N_r}}\right)\left(\frac{w_r}{h_r}\right)\left(\frac{H_s}{h_r} - 1{\cdot}0\right) \leq 1{\cdot}0 \text{ for beams with transverse ribs, and} \tag{3.33}$$

$$= 0{\cdot}6\left(\frac{w_r}{h_r}\right)\left(\frac{H_s}{h_r} - 1{\cdot}0\right) \leq 1{\cdot}0 \text{ for beams with longitudinal ribs} \tag{3.34}$$

in which $h_r$ = nominal rib height; $H_s$ = length of stud connector after welding, $\leq (h_r + 3\,\text{inches [76 mm]})$ in computations, although the actual length may be greater; $N_r$ = number of stud connectors on a beam in one rib, not to exceed 3 in computations, although more than three studs may be installed; and $w_r$ = average width of concrete rib.

Shear connector capacity (AISC, 1980, 1986):

$$Q_n = 0{\cdot}5A_{sc}\sqrt{f'_c E_c}\ \text{(kip)}\ [= 0{\cdot}0005A_{sc}\sqrt{f'_c E_c}\ \text{(kN)}] \tag{3.35}$$

in which $A_{sc}$ = cross-sectional area of stud, in$^2$ [mm$^2$]; $f'_c$ = concrete compressive strength, kip/in$^2$ [MPa]; $E_c$ = concrete modulus of elasticity, kip/in$^2$ [MPa].

Limitation on shear connector capacity (AISC, 1986; Donahey and Darwin, 1986):

$$RQ_n \leq A_{sc} F_u \tag{3.36}$$

in which $F_u$ = tensile strength of stud.

Maximum concrete shear stress (Clawson and Darwin, 1980, 1982b):

$$\begin{aligned} v_c &= 0{\cdot}11\sqrt{f'_c}\ (\text{kip/in}^2) \\ &= 0{\cdot}29\sqrt{f'_c}\ (\text{MPa}) \end{aligned} \tag{3.37}$$

*Chapter 4*

# COMPOSITE GIRDERS WITH DEFORMABLE CONNECTION BETWEEN STEEL AND CONCRETE

V. KŘÍSTEK and J. STUDNIČKA

*Faculty of Civil Engineering, Czech Technical University, Prague, Czechoslovakia*

*SUMMARY*

*The influence of a deformable connection on the stress distribution and deflections of composite girders and vertical tubular structures is studied and various methods of analysis are presented. Simplified methods of analysis are proposed and are intended for use as design tools. The effect of the deformability of the connection between the steel and concrete parts is demonstrated on typical composite structures. The difference in structural performance between composite girders with deformable connection and composite tubular structures is highlighted. Recommendations and design formulae derived from parametric studies are presented.*

## NOTATION

| | |
|---|---|
| $a$ | Vertical distance |
| $a_1, a_2$ | See eqn (4.52) |
| $b_1, b_2$ | See eqn (4.52) |
| $b_n$ | Effective width |
| $e$ | Distance between centroids of concrete deck and steel girder |
| $k_1$ | Coefficient of deformability |
| $m$ | Interaction moment |
| $n$ | Term number of Fourier series |
| $p$ | Loading |

| | |
|---|---|
| $q$ | Longitudinal shear flow |
| $r$ | Radius |
| $s$ | Circumferential coordinate |
| $t$ | Thickness |
| $u, v, w$ | Displacements |
| $x$ | Longitudinal coordinate |
| $z$ | Vertical distance measured from centroid of steel section |
| $A$ | Cross-sectional area of steel girder |
| $A_c$ | Concrete area |
| $A_s$ | Shear area of steel section |
| $E$ | Young's modulus of elasticity for steel |
| $E_c^*$ | Effective modulus for concrete |
| $F$ | See eqn (4.47) |
| $G$ | Shear modulus for steel |
| $H$ | Height of a structure |
| $I$ | Second moment of area |
| $J$ | Compliance function |
| $K$ | Deformability characteristics of the connection |
| $K_1$ | See eqn (4.31) |
| $L$ | Span |
| $M$ | Bending moment |
| $M_n$ | See eqn (4.12) |
| $N$ | Axial force |
| $N_c$ | Axial force in concrete deck |
| $P_n$ | See eqn (4.9) |
| $Q$ | Shear force |
| $R$ | Balancing force |
| $T_n$ | See eqn (4.9) |
| $U_{a,n}$ | See eqn (4.19) |
| $U_{b,n}$ | See eqn (4.16) |
| $U_{c,n}$ | See eqn (4.20) |
| $V_n$ | See eqn (4.11) |
| $\left.\begin{matrix}\beta_{1,n}\\ \beta_{2,n}\end{matrix}\right\}$ | See eqn (4.54) |
| $\epsilon_a$ | Axial strain |
| $\eta$ | Loading ratio |
| $\kappa$ | Coefficient of distribution of shear stress |
| $\sigma$ | Normal stress in steel girder |
| $\sigma_c$ | Normal stress in concrete deck |

| | |
|---|---|
| $\tau$ | Shear stress |
| $\phi$ | Creep coefficient |
| $\psi$ | Ratio, see eqn (4.66) |
| $\Delta u$ | Longitudinal slip |
| $\Phi$ | Rotation angle of the element, see eqn (4.14) |
| $\bar{\Phi}$ | Rotation, see eqn (4.2c) |

## 4.1 INTRODUCTION

In the design of composite structures the connection between a steel girder and a concrete slab used to be made with rather stiff connectors, i.e. channels, angles, stirrups, bars and hoops, etc. (Fig. 4.1a). In order that the steel beam and the slab act as a composite structure with full interaction, the connectors must have adequate strength and stiffness. Such rigid connectors provide that there are no horizontal or vertical separations at the interface, i.e. no slip or uplift. However, all connectors, particularly the studs which are commonly used as connectors at the present time, are flexible to some extent and therefore only partial interaction exists (Fig. 4.1b). This implies that slip occurs at the interface between the concrete flange and the steel beam, and so causes a discontinuity of strain that has to be taken into account in the analysis. For most connectors used in practice, failure by vertical separation is unlikely and any uplift would

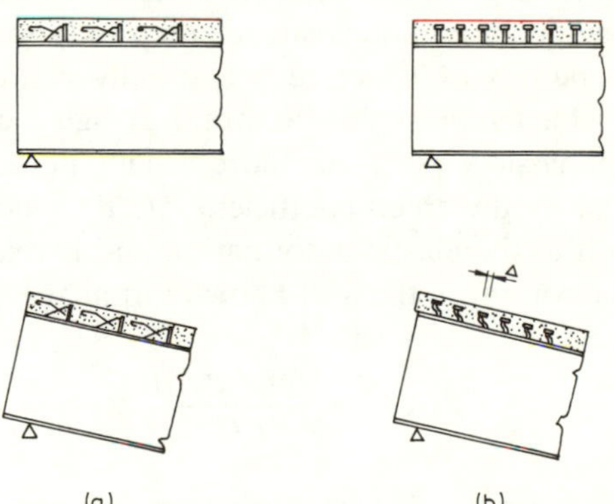


FIG. 4.1. (a) Stiff connectors with perfect connection showing no slip between steel girder and concrete slab after loading; (b) deformable connectors showing slip between steel girder and concrete slab after loading.

have only negligible effect on the behaviour of the composite structure. It is therefore usually sufficient to consider only slip in the study of the effects of partial interaction.

The problem of the deformable connection is more severe when fewer connectors are used than the number required for full interaction between the steel and concrete parts of a composite girder. The use of a reduced number of studs is advantageous when the stiffness and strength of the composite girder with full interaction are excessive, when precast concrete slabs are used, or when the concrete deck is cast on metal decking with corrugations that run across the flange of the steel beam, etc. (see Johnson and May, 1975). It is evident that the deformability of connection represents a considerable problem in the analysis and design of composite structures.

The main object of this chapter is to discuss the progress in the methods of structural analysis of composite structures with deformable connection and to emphasise some recent developments that are likely to be of practical use.

## 4.2 METHODS OF ANALYSIS

The composite structure consists of two components: steel and concrete. In order to obtain adequate accuracy in the analysis of composite structures, realistic constitutive relations for materials should be used in calculations. This applies mainly to concrete, the behaviour of which is considerably affected by long-term deformations due to creep and shrinkage.

The creep properties of concrete are usually characterised by the function $J(t,t')$ which represents the strain at age $t$ caused by a unit uniaxial stress acting since age $t'$ or, more usually in the Standard Code recommendations, by the creep coefficient $\phi(t,t')$ which is the ratio of creep deformation to the elastic deformation and is related to the compliance function according to the well-known formula

$$J(t,t') = \frac{1+\phi(t,t')}{E(t')} \tag{4.1}$$

Several practical models for predicting creep exist at present. The compliance function according to the BP Model (Bažant and Panula, 1978) is better physically justified and has a smaller statistical error than other existing models.

It should also be noted that there is no general agreement among engineers as to the definition of the conventional elastic modulus of concrete. The modulus obtained from the Standard Code recommendations corresponds approximately to two hours of load duration and represents approximately one-half of the true instantaneous modulus (Bažant and Wittmann, 1983). A considerable error can be caused by combining incompatible values of elastic modulus $E(t')$ and creep coefficient $\phi(t,t')$. Therefore the creep coefficient must be determined from the same compliance function, $J(t,t')$, as $E(t')$.

Here, it should be emphasised again that inaccuracies in material characterisation usually cause by far the most serious error in the results of structural analysis. Therefore the effort spent on determination of material properties should be commensurate to that devoted to the structural analysis itself.

For practical purposes the most popular approach to the analysis of composite structures is the quasi-elastic effective modulus method. For this model, creep of concrete is accounted for by reducing the modulus of concrete by a factor $[1+\phi(t,t')]$.

The effective modulus method gives good results only when the concrete stress does not vary significantly during the period under investigation and when the ageing of the concrete is negligible, as in old concrete. To proceed more accurately the approach according to the Trost–Bažant method can be recommended.

To date extensive work has been carried out to analyse composite structures and several concepts have been proposed.

The effect of the deformability of connection between a symmetrical I shaped simply supported steel girder and concrete slab was investigated by Yam (1981). The method is based on differential equations and gives for the prescribed flexibility of the connection, the position of neutral axis and also the stress distribution in the cross section. Similarly Johnson (1975) presented a theoretical study on the flexible connection, assuming an ideal cross section composed of a steel and concrete layers.

Despite the fact that the plastic methods have been permitted in most of the Standards, the problem of inelastic behaviour of composite structures exhibiting a slip between the concrete slab and the steel beam is not yet fully understood and formulated in rigorous form. Yam (1981) has pointed out that for an inelastic behaviour of simply supported beams, the loss of interaction due to slip has negligible effect on the ultimate moment of resistance, and that for continuous beams, the loss of interaction has little effect on the stress distribution and negligible effect on ultimate moments.

Recently Aribert and Abdel Aziz (1985) presented a new calculation model for composite steel and concrete beams, which is valid up to collapse, and takes into account both the effects of slip and uplift at interface between the slab and steel girder. This model uses relatively sophisticated laws of behaviour: non-monotonous variation law of concrete in compression, weak resistance of concrete in tension, and the phenomenon of strain-hardening for steel. As for the connectors, they are characterised by variable stiffness in shear and in tension.

A numerical study of the elastic longitudinal bending behaviour of composite box girder bridges, in which the use of flexible shear connectors results in incomplete interaction between the slab and steel beam, has been presented by Moffat and Dowling (1978). The results of the study provided background for the formulation of design rules included in BS 5400: Part 5.

Hence, to predict the structural performance of composite structures (with deformable connection between steel and concrete components), the quasi-elastic analysis—which accounts for the intricate character of the actual behaviour of the deformable connection at different stress levels—appears to be an acceptable and adequate approach. Based on rigorously formulated assumptions, the quasi-elastic analysis represents a clear approach and enables simple formulae to be formulated and obtained for use as a design tool.

Among the many different models for the analysis of composite structures are elasticity solutions based on finite element or finite difference methods, exact methods based on folded plate theory, and approximate methods based on simplified structural behaviour.

### 4.2.1 General Methods

#### *4.2.1.1 The Finite Element Method*

The finite element method has reached considerable perfection and has become a practically universal method for the solution of the problems of mechanics in recent years. The continuum is replaced by an assembly of finite elements interconnected at nodal points. Stiffness matrices are developed for the finite elements on the basis of assumed displacement patterns, and after that an analysis based on the direct stiffness method may be performed to determine nodal point displacements and thence the internal stresses in the elements. As the method is well documented in the literature, no attempt will be made to review it in detail here.

Although a finite element solution is capable of giving a comprehensive and adequate picture of the stress distribution, this solution requires the use of large computers and is too costly, particularly if the repeated analysis is required at the preliminary design stage.

*4.2.1.2 Folded Plate Analysis*

The folded plate theory was originally developed for the analysis of isotropic thin-walled prismatic structures (e.g. roofs and bridges) (see Goldberg and Leve, 1957; DeFries-Skene and Scordelis, 1964; Křístek, 1979a). This method also provides a powerful tool for the analysis of composite beams and can be employed to determine the effects of deformable connection between the concrete deck and the steel girder. Steel–concrete composite girders are usually of constant cross-section and hence the folded plate theory is an ideally suited technique for predicting the stress distribution. A prismatic folded plate structure is a shell consisting of rectangular plates mutually supporting along their longitudinal edges and simply supported at the two ends by transverse diaphragms. The complexity of the cross-section as well as the ratio of the cross-sectional dimension to the span are irrelevant because the solution is not based on the elementary theory of bending but on the elasticity plane stress theory and the plate bending theory. Bearing in mind the assumptions of the theory of elasticity, the folded plate theory is in fact an exact method because it considers the structure in its actual form as an assembly of plate elements forming together a real spatial system. The method has the advantage that it can treat any general set of joint loads or displacements.

Each folded plate element is subjected to element forces and displacements at its longitudinal edges. There are two edges, hence the plate has eight degrees of freedom, eight edge element forces and edge displacements.

Due to the diaphragms at both ends, which are rigid in their own plane and perfectly flexible in the direction normal to it, the distribution of all applied forces and displacements along a ridge or joint is taken as a harmonic of order $n$. This makes it possible to treat the ridge as a nodal point and to operate with single forces and displacements instead of functions. If the conditions of static equilibrium and geometric compatibility are satisfied at the nodal point they will be automatically maintained along the entire ridge; this results in a considerable saving in computer time.

At the first step of a solution, an element stiffness matrix $(8 \times 8)$ is written, relating the element forces in the relative system to the corresponding displacements. Then a displacement transformation matrix

$(8 \times 8)$ is defined, relating the element displacements in the relative coordinate system and in the fixed system. Applying this transformation matrix, the element stiffness matrix in the fixed coordinate system may be found. The stiffness matrix for the whole structure can now be assembled from the element stiffness matrices.

The objective of the general stiffness method is to find all unknown joint displacements. Once these are known, it is possible to determine all internal forces and stresses acting in the structure.

Statically more complicated folded plate structures (e.g. continuous and clamped beams, girders with flexible support frame bents, etc.) can be analysed by means of the force method (Scordelis, 1971; Křístek, 1979a). This approach consists in removing all redundant constraining effects, i.e. supports, diaphragms, etc., so that a simply supported folded plate structure with end diaphragms is obtained. This stucture represents the primary system for the solution by the force method. The removed constraining effects are replaced by the redundants, the magnitudes of which are determined later from the deformation conditions.

A method developed by Křístek (1983) deals with the analysis of folded plates with arbitrary boundary conditions corresponding to the effect of deformability of end diaphragms. The method may be applied to folded plates without any diaphragms as well as to structures with deformable end diaphragms, which is the case with many bridge girders.

Since the structure is considered in its actual form, the folded plate theory represents an exact method suitable for analysing the behaviour of composite beams and simultaneously other important effects such as a shear lag in the concrete deck or the state of stresses near the concentrated forces. This method takes full advantage of the use of digital computers. The calculation consists mostly of routine matrix operations where the number of solution steps is greatly reduced owing to the small size of the stiffness matrices, in banded form, and therefore requires only a small amount of computer time.

For the folded plate analysis, the composite girder is idealised as an assemblage of rectangular elements connected along the longitudinal joints. Individual parts of the composite girder (i.e. the concrete deck, flanges and web of the steel girder, Fig. 4.2a) are idealised by regular folded plate elements, as shown in Fig. 4.2b. The connection of the concrete deck to the top flange must model as truthfully as possible the actual structural performance; hence the connection should be rigid against vertical strain as well as against transverse flexure although, to provide zero longitudinal normal stresses, it should be without any stiffness

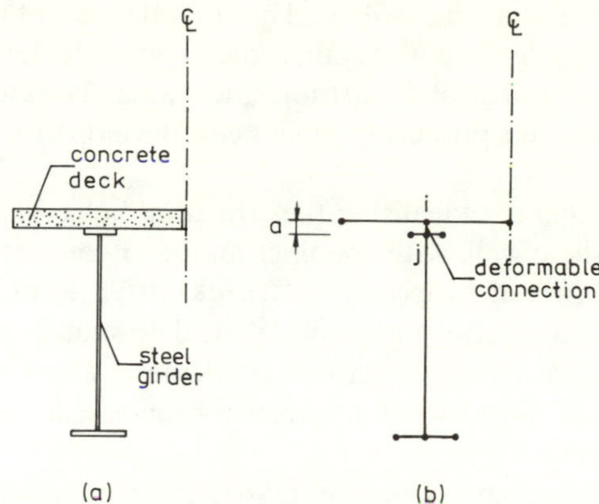


FIG. 4.2. (a) Cross-section of composite girder; (b) folded plate idealization.

in the longitudinal direction. However, the shear deformability of such an element should represent the flexibility of the connectors and their spacing.

To satisfy the required conditions it can be written (Fig. 4.2b)

$$v_j = v_i - \bar{\Phi} a \tag{4.2a}$$

$$w_j = w_i = w \tag{4.2b}$$

$$\bar{\Phi}_j = \bar{\Phi}_i = \bar{\Phi} \tag{4.2c}$$

$$u_j = u_i - \frac{\mathrm{d}w}{\mathrm{d}x} a + qK \tag{4.2d}$$

in which $v_i$, $w_i$ and $u_i$ and $v_j$, $w_j$ and $u_j$ are the horizontal, vertical and longitudinal displacements respectively of joint $i$ and $j$ (see Fig. 4.2b); $\bar{\Phi}_i$ and $\bar{\Phi}_j$ are the joint rotations in the plane of the cross-section; $q$ is the longitudinal shear flow acting between the concrete deck and the steel girder; and $a$ is the vertical distance between joints $i$ and $j$ (usually the half-thickness of the concrete deck). These relations should be taken into account when assembling the stiffness matrix for the whole structure.

Another way to model the connection of the concrete deck to the top flange of the steel girder consists of the use of a 'special folded plate element' having the required properties. Various approaches may be used:

1. A well-known orthotropic element used in the finite strip method, the stiffness matrix of which is available elsewhere, renders the connec-

tion properties satisfactorily and has four degrees of freedom at each longitudinal edge as have regular folded plate elements.

2. The stiffness matrix of the orthotropic folded plate element, allowing for the required properties, has been developed recently (Hájek, 1985).
3. A special element idealising the structural behaviour of framework panels or shear walls with openings for the folded plate analysis of tall buildings has been proposed by Křístek (1979b). This element can be regarded as a suitable one to model the deformable connection of the concrete deck to the steel girder, particularly when the connection is made with studs which exhibit a rather pronounced flexural pattern of deformation.
4. DeFries-Skene and Scordelis (1964) have presented the ordinary theory for folded plates based on the following two assumptions for the individual plate elements: (a) slab action is defined by the behaviour of transverse one-way slab strips spanning between longitudinal joints; (b) membrane stresses may be calculated by the elementary beam theory applied to individual plate elements. The membrane stiffness matrix is found by applying four typical deformation patterns to a plate element. A similar procedure, with generalised cross-sectional characteristics, may also be used to determine the membrane stiffness of the special element.

To incorporate the connection of the concrete deck to the top flange, some minor changes are necessary in the folded plate programs generally used; to avoid these, a current isotropic folded plate element, with properties governed by the shear deformability derived from the connection conditions, may be used approximately instead of the orthotropic element. To fulfil the requirement of rigidity against transverse flexure, a thicker element may be used and the Young's modulus reduced. It has been proved that for most practical cases the increase in the cross-sectional stiffness due to the addition of the connecting element is negligible and could be easily eliminated by a slight reduction of the area of the top steel flange.

### 4.2.2 Simplified Methods

#### *4.2.2.1 Analysis of Girder-type Composite Structures with Deformable Connectors*

A hand calculation of composite girders with deformable connectors was proposed by Křístek and Studnička (1982). This simple method employs harmonic analysis, and the results are presented in the form of simple formulae; the suggested procedure is capable of estimating the influence of varying girder parameters, particularly the rigidity of the connection between the concrete deck and the steel girder, and is intended for use as a design tool.

A composite beam consisting of a steel girder and an effective breadth $b$ of the concrete deck, loaded by an external vertical loading $p$, is considered (Fig. 4.3).

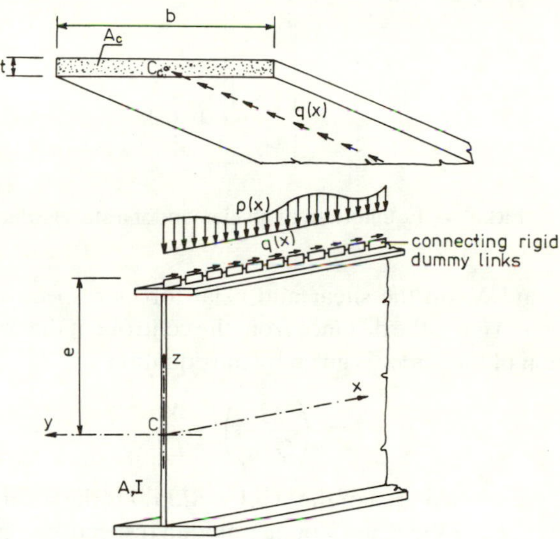


FIG. 4.3. Steel girder interconnected through rigid dummy links with the concrete deck.

The top flange of the steel girder is interconnected through the rigid dummy links with the centroidal axis of the concrete deck, see Fig. 4.3. A longitudinal shear flow $q$ acts between the girder and the deck. The vertical loading $p$ can be assumed to act directly on the top of the steel girder.

The conditions governing the equilibrium of an element of the steel girder shown in Fig. 4.4 are

$$Q' + p = 0 \tag{4.3}$$

$$Q - M' + qe = 0 \tag{4.4}$$

$$N' + q = 0 \tag{4.5}$$

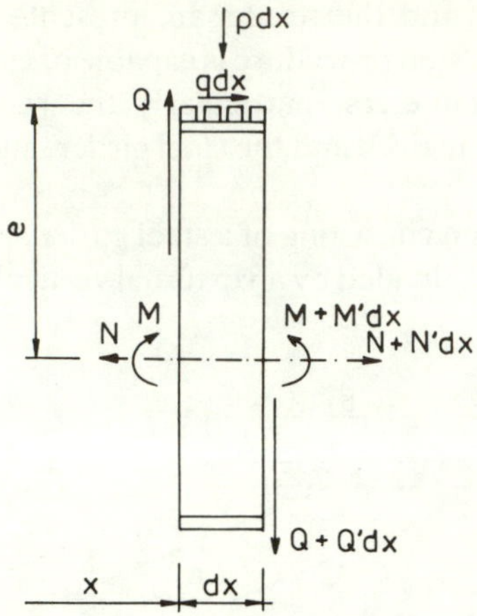


FIG. 4.4. Equilibrium of an element of a steel girder.

in which $Q$ and $N$ are the shear and axial forces respectively, $M$ is the bending moment and $e$ the distance from the centroid of the concrete deck. The deflection of the girder is given by the equation

$$v'' = \left(\frac{Q}{GA_s}\right)' - \frac{M}{EI} \tag{4.6}$$

in which $A_s$ is the shear area of the steel section, $I$ is the second moment of area and $G$ and $E$ are shear and Young's moduli respectively. It follows from eqns (4.3), (4.4) and (4.6) that the deflection

$$v^{iv} = \frac{p - q'e}{EI} - \frac{p''}{GA_s} \tag{4.7}$$

Let us assume a simply supported girder of constant cross-section and span $L$ (i.e. the boundary conditions are $v(0) = v(L) = M(0) = M(L) = 0$). It is convenient to express the loading and the desired shear flow by Fourier series

$$p(x) = \sum_{n=1}^{\infty} P_n \sin\frac{n\pi x}{L}\,;\; q(x) = \sum_{n=1}^{\infty} T_n \cos\frac{n\pi x}{L} \tag{4.8}$$

in which the coefficients are

$$P_n = \frac{2}{L}\int_0^L p(x)\sin\frac{n\pi x}{L}\,dx;\; T_n = \frac{2}{L}\int_0^L q(x)\cos\frac{n\pi x}{L}\,dx \tag{4.9}$$

For a uniformly distributed loading,

$$P_n = -\frac{2p}{\pi n}[(-1)^n - 1]$$

The solution of eqn (4.7) is conveniently found in the form

$$v = \sum_{n=1}^{\infty} V_n \sin\frac{n\pi x}{L} \tag{4.10}$$

Substituting eqns (4.8) and (4.10) into eqn (4.7):

$$V_n = \frac{P_n(1+\alpha_n) + T_n\dfrac{n\pi}{L}e}{\dfrac{n^4\pi^4}{L^4}EI} \tag{4.11}$$

where the coefficient $\alpha_n = (n^2\pi^2/L^2)(EI/GA_s)$ represents the influence of shear strains.

The bending moments in the steel girder follow from eqns (4.3) and (4.6) as:

$$M = \sum_{n=1}^{\infty} M_n \sin\frac{n\pi x}{L} = -EI\left(v'' + \frac{p}{GA_s}\right) \tag{4.12}$$

The coefficient $M_n$ is given by

$$M_n = \frac{P_n + T_n\dfrac{n\pi}{L}e}{\dfrac{n^2\pi^2}{L^2}} \tag{4.13}$$

Since the rotation angle of the element is

$$\Phi = v' - \frac{Q}{GA_s} \tag{4.14}$$

the longitudinal displacement of the upper ends of the links due to bending of the steel girder is given by

$$u_b = \sum_{n=1}^{\infty} U_{b,n} \cos\frac{n\pi x}{L} = \Phi e \tag{4.15}$$

The coefficients $U_{b,n}$ are given by

$$U_{b,n} = \frac{P_n + T_n \frac{n\pi}{L} e}{\frac{n^3 \pi^3}{L^3} EI} e \tag{4.16}$$

The axial strain of the steel girder is given by the relation

$$\epsilon_a = u_a' = \frac{N}{EA} \tag{4.17}$$

where $A$ is the cross-sectional area of the steel girder.

Using eqns (4.5) and (4.8), eqn (4.17) may be rewritten in the form

$$\epsilon_a = -\frac{L}{EA} \sum_{n=1}^{\infty} \frac{T_n}{n\pi} \sin \frac{n\pi x}{L} \tag{4.18}$$

where the longitudinal displacement $u_a = \Sigma_{n=1}^{\infty} U_{a,n} \cos(n\pi x/L)$ due to axial strain of the steel girder is characterised by the coefficients

$$U_{a,n} = T_n \frac{L^2}{n^2 \pi^2 EA} \tag{4.19}$$

The concrete deck is axially loaded by shear flow $q$ (Fig. 4.3); its longitudinal displacement, $u_c = \Sigma_{n=1}^{\infty} U_{c,n} \cos(n\pi x/L)$, is described in a manner analogous to eqn (4.19) by the coefficients

$$U_{c,n} = -T_n \frac{L^2}{n^2 \pi^2 E_c^* A_{c,n}} \tag{4.20}$$

in which $E_c^*$ is the effective modulus for concrete and $A_{c,n} = b_n t$ is the concrete area (given as the product of the effective breadth of the concrete deck corresponding to the $n$th term of the Fourier series and the deck thickness $t$).

Since the connection between the steel girder and the concrete deck cannot be considered as perfect, longitudinal slip $\Delta u$ which occurs at the joint is proportional to the shear flow acting there. In the linear stage it may be written

$$\Delta u = qK = K \sum_{n=1}^{\infty} T_n \cos \frac{n\pi x}{L} \tag{4.21}$$

where $K$ is the deformability characteristic of the connection. This may be defined for the design purpose as

$$K = \frac{p}{k} = \frac{\text{spacing of connectors}}{\text{connector modulus}}$$

The following relation for longitudinal displacements along the joint of the steel girder and the concrete deck is obtained:

$$u_{\mathrm{b}} + u_{\mathrm{a}} = u_{\mathrm{c}} - \Delta u \tag{4.22}$$

From eqns (4.15) and (4.19–4.21), a formula for the coefficients expressing the desired shear flow $q$ can be written as

$$T_n = \frac{-P_n eL}{n\pi\left[e^2 + \frac{I}{A}\left(1 + \frac{EA}{E_{\mathrm{c}}^* A_{\mathrm{c},n}}\right) + \frac{Kn^2\pi^2 EI}{L^2}\right]} \tag{4.23}$$

It follows from this formula that the shear deformations of the girder influence only the deflections (see eqn (4.11)) but the magnitude of the shear flow acting between the steel girder and the concrete deck remains unchanged.

Using formula (4.23), it is possible to express the coefficients for deflection (eqn (4.11)) and the bending moment transferred by the steel girder (eqn (4.13)). Axial forces acting in the steel girder (force $N$) as well as in the concrete deck (force $N_{\mathrm{c}}$) due to shear flow $q$ can be determined using eqns (4.5) and (4.8); it holds true that

$$N_{\mathrm{c}} = -N = \frac{L}{\pi}\sum_{n=1}^{\infty}\frac{T_n}{n}\sin\frac{n\pi x}{L} \tag{4.24}$$

again in terms of the coefficients $T_n$.

Longitudinal normal stresses in the steel section are given by the well-known formula

$$\sigma = \frac{N}{A} - \frac{M}{I}z \tag{4.25}$$

in which $z$ is the vertical distance measured from the centroidal axis of the steel section (positive upwards, see Fig. 4.3).

The normal stress at the middle plane of the concrete deck is expressed by the formula

$$\sigma_c = \frac{N_c}{A_c} \tag{4.26}$$

Expressing the loading and all functions describing the behaviour of the composite girder in the form of Fourier series, the solution for each term of the series is made separately with great advantage and the results are simply summed. The accuracy of the results, i.e. the difference between the solution obtained and the results that would be derived for $n \to \infty$, depends on the type of loading and on the character of the internal force or deformation which is compared. The deflections are very well approximated by a very small number of terms (see Table 4.1); the expressions for bending moments, axial forces and longitudinal normal stresses also need only a few terms.

These deductions follow from eqns (4.11), (4.13) and (4.23), from which it is also evident that the more deformable the connection between the concrete deck and the steel girder, the better the convergence of deflections and stresses in the concrete.

The simplified method presented is intended for use as a design tool since the calculations involved make it suitable for the repeated analysis required to determine optimum proportions, e.g. by varying the number and type of connectors, the depth of steel girder, dimensions of the deck, etc. In this way, a considerable saving in cost can be made. The method enables the results to be predicted from hand calculation but may also, for added convenience, be easily programmed for a pocket calculator.

The approximate analysis of a composite continuous girder for possible support conditions can be carried out in two successive steps. In this

TABLE 4.1
AN EXAMPLE OF CONVERGENCE OF SOLUTION IN A TYPICAL CASE

| *Function analysed (at midspan)* | *Number of terms used* | | | | |
|---|---|---|---|---|---|
| | *1* | *3* | *5* | *7* | *9* |
| Deflections % | 100·3 | 100 | 100 | 100 | 100 |
| Bending moments %<br>Axial forces %<br>Longitudinal stresses % | 103·2 | 99·4 | 100·2 | 100 | 100 |

approach, the distribution of bending moment and shear force are first determined by conventional continuous beam analysis which accounts for the expected reduction in stiffness due to the occurrence of tensile stresses in the concrete deck in the support regions. This approximate analysis yields the locations of the inflection points at which the bending moment is zero. The portions of the girder with positive bending moments may then be treated as simply supported beams and the analysis presented can then be carried out independently for each individual portion of the girder; this would result in an acceptable solution for this problem.

### *4.2.2.2 Analysis of Tubular Composite Structures*

The connection between the steel and concrete components of the cross-section of girder-type structures has to be sufficient to provide the desired composite action (see Section 4.2.2.1). This follows from the fact that the centroids of the steel and concrete components are rather distant and therefore the possible weakness of connectors would manifest itself in a relative slip, resulting in an independent flexural performance of both components and in considerable decrease of the flexural rigidity as well as of the load-carrying capacity of the system.

A quite different situation arises at structural arrangements where the centroids of steel and concrete components are coincident. A structure composed of two coaxial hollow cylinders (Fig. 4.5a), which can form, for example, the cross-section of a TV tower (Fig. 4.5b), can be demonstrated as a typical example (Křístek and Studnička, 1985).

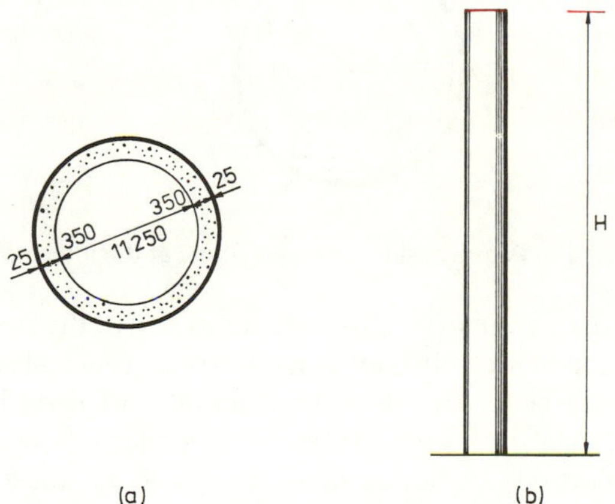


FIG. 4.5. Composite tower-type concentric structure.

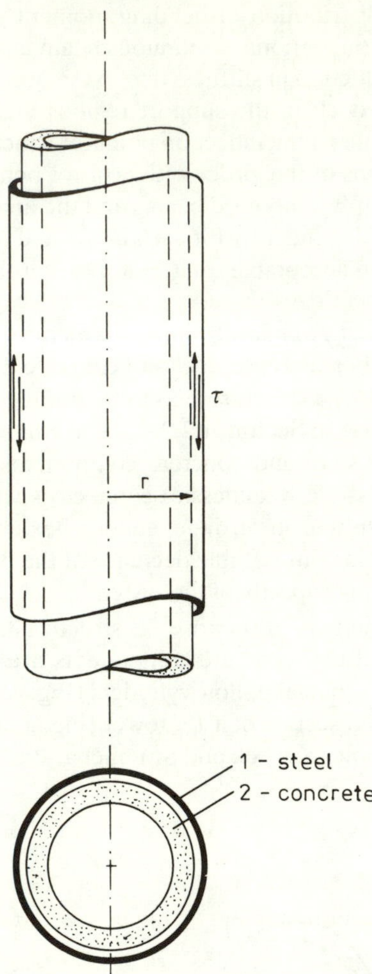


FIG. 4.6. Vertical shear stresses acting at the interface.

Let us consider a composite structure composed of the steel tube placed as the outer component (subscript 1) with the concrete (subscript 2) placed inside. Due to the action of external horizontal loading $p$ (the wind loads, earthquake effects etc.), the structure is subjected to bending. The horizontal deflections of both cylinders, $v_1$ and $v_2$, must be equal ($v_1 = v_2 = v$) and the composite action is provided by the presence of interaction forces originating at the interface between both parts.

The interaction forces have two components: the radial one and the shear stress $\tau$ acting in the vertical direction (Fig. 4.6). The resultant effect of the radial component of the interaction forces is represented by the portion $p_2$ of the external horizontal loading $p$ which is to be taken by the concrete (internal) part of the structure. The steel part thus takes the portion

$$p_1 = p - p_2 \tag{4.27}$$

of the external loading.

The vertical shear stresses $\tau$ acting at the interface between the structural components are caused by restraining the tendency of the cross-sections of both cylinders to rotate independently. It is assumed that the interaction vertical shear stresses $\tau(s)$ are proportional to the relative slip at the interface according to the formula

$$\Delta u = \tau K_1 \tag{4.28}$$

in which $K_1$ ($m^3N^{-1}$) is the coefficient of deformability of the connection.

It is postulated that

$$\Delta u = (\Phi_1 - \Phi_2)z(s) \tag{4.29}$$

where $\Phi_1$ and $\Phi_2$ are rotations of cross-sections, and $z(s)$ is the coordinate of a point on the interface, measured in the plane of bending (Fig. 4.7). Hence,

$$z(s) = r \sin\frac{s}{r} \tag{4.30}$$

By applying eqns (4.28–4.30) it is possible to express the continuously distributed interaction moment acting between the cylinders as

$$m = \int_0^{2\pi r} \tau(s)z(s)\,\mathrm{d}s = \frac{\Phi_1 - \Phi_2}{K_1}\int_0^{2\pi r} z^2(s)\,\mathrm{d}s$$

$$= (\Phi_1 - \Phi_2)\frac{r^2}{K_1}\int_0^{2\pi r} \sin^2\frac{s}{r}\mathrm{d}s = \frac{\Phi_1 - \Phi_2}{K_1}\pi r^3 \tag{4.31}$$

or briefly

$$\Phi_1 - \Phi_2 = mk_1 \tag{4.32}$$

in which $k_1 = K_1/\pi r^3$ ($N^{-1}$) is the coefficient of deformability in the relative rotation of cross-sections.

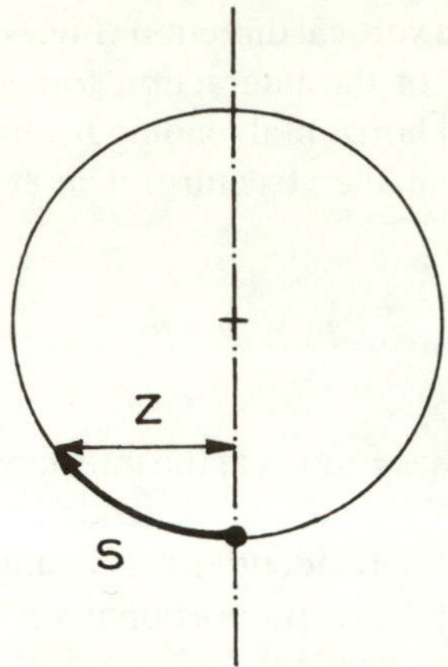


FIG. 4.7. Coordinates of a point at the interface.

As the horizontal displacements of both cylinders are equal, the deflection lines are given by eqns

$$v^{\text{iv}} = \frac{p_1 - m'}{E_1 I_1} - \frac{p_1''}{G_1 A_{s1}} \tag{4.33}$$

$$v^{\text{iv}} = \frac{p_2 + m'}{E_2 I_2} - \frac{p_2''}{G_2 A_{s2}} \tag{4.34}$$

where $E_1$ and $E_2$ are the moduli of elasticity (or the effective moduli), $I_1$ and $I_2$ the second moments of area, $G_1$ and $G_2$ the shear moduli, and $A_{s1}$ and $A_{s2}$ the shear cross-sectional areas.

Equation (4.33) governs deflections of the steel part which is loaded by the portion of external loading $p_1$ and influenced by the effect of the interaction moment. Similarly, eqn (4.34) is valid for deflections of the concrete part.

Rotations of cross-sections (considering the shear deformations) can be expressed as

$$\Phi_1 = v' - \frac{Q_1}{G_1 A_{s1}} \tag{4.35}$$

$$\Phi_2 = v' - \frac{Q_2}{G_2 A_{s2}} \tag{4.36}$$

in which $Q_1$ and $Q_2$ are shear forces.

It follows from the conditions of equilibrium that

$$Q_1 = -p_1 \tag{4.37}$$

$$Q_2 = -p_2 \tag{4.38}$$

Equations (4.27) and (4.32)–(4.38) form a system of eight fundamental equations for unknown functions $v, m, p_1, p_2, \Phi_1, \Phi_2, Q_1, Q_2$, which can be simplified by eliminating the functions which are not desired. The boundary conditions, which are known at the ends, are employed to determine the constants in the general solution of the system of basic equations. Structures of variable cross-section may be modelled by considering that the length of the structure is divided into a number of intervals of constant cross-section, and the conditions which are valid at the connection of intervals are applied.

In order to frame the range of influence of the deformability of connection, two limiting cases can be discussed:

(1) the perfect connection (full interaction, $k_1 = 0$);
(2) no connection (with a free slip at the interface, $k_1 \rightarrow \infty$).

(1) For the perfect connection (see eqn (4.32)), i.e. for $k_1 = 0$, $\Phi_1$ equals $\Phi_2$ and from eqns (4.35) and (4.36) we obtain

$$\frac{Q_1}{G_1 A_{s1}} = \frac{Q_2}{G_2 A_{s2}} \tag{4.39}$$

By substituting eqns (4.37) and (4.38) for eqn (4.39), it is found that in the case of perfect connection, the external loading is divided between the steel and concrete parts in proportion to their shear rigidities, i.e.

$$\frac{p_1}{p_2} = \frac{G_1 A_{s1}}{G_2 A_{s2}} \tag{4.40}$$

By substituting this relationship for eqns (4.33) and (4.34) we obtain, after rearranging the equation governing deflection,

$$v^{\text{iv}} = \frac{p}{\bar{E}I} - \frac{p''}{\bar{G}A_s} \tag{4.41}$$

in which $\bar{E}I = E_1 I_1 + E_2 I_2$ is the overall flexural stiffness and $\bar{G}A_s = G_1 A_{s1} + G_2 A_{s2}$ is the overall shear stiffness of the composite structure. Relationship (4.41) is identical to the well-known equation governing deflections of a non-homogeneous girder.

(2) If there is no connection restraining the vertical slip at the interface (i.e. for $k_1 \to \infty$), the interaction moment $m$ equals zero and the problem is described by the following equations:

$$p_1 + p_2 = p \tag{4.$\overline{27}$}$$

$$v^{\text{iv}} = \frac{p_1}{E_1 I_1} - \frac{p_1''}{G_1 A_{s1}} \tag{4.$\overline{33}$}$$

$$v^{\text{iv}} = \frac{p_2}{E_2 I_2} - \frac{p_2''}{G_2 A_{s2}} \tag{4.$\overline{34}$}$$

Thus the division of the external loading $p$ into portions $p_1$ and $p_2$ (i.e. loads on the individual structural parts) depends on the flexural as well as the shear rigidities in the case without any connection.

In order to get a simple rule for a rough estimation of division of the external loading between two structural parts, in a tall slender structure the shear deformation terms in eqns (4.$\overline{33}$) and (4.$\overline{34}$) may be neglected and an approximate relation, given below, results:

$$\frac{p_1}{p_2} \simeq \frac{E_1 I_1}{E_2 I_2} \tag{4.42}$$

Comparing formulae (4.40) and (4.42), it is obvious that the rate of connection does not play a decisive role in the performance of the structure. For thin-walled cylinders, the following approximate relationships are valid.

$$\begin{aligned} I_1 &\simeq \pi \left( r + \frac{t_1}{2} \right)^3 t_1 & \qquad I_2 &\simeq \pi \left( r - \frac{t_2}{2} \right)^3 t_2 \\ A_{s1} &\simeq 2\pi \left( r + \frac{t_1}{2} \right) t_1 \kappa & \qquad A_{s2} &\simeq 2\pi \left( r - \frac{t_2}{2} \right) t_2 \kappa \end{aligned} \tag{4.43}$$

in which $t_1$ and $t_2$ are thicknesses of the walls and $\kappa$ is the coefficient expressing influence of the distribution of shear stresses on deformation of structure. For the thin-walled cylinders it is reasonable to use $\kappa = 0{\cdot}5$.

Formula (4.40) thus gives

$$\frac{p_1}{p_2} = \frac{G_1}{G_2} \frac{\left(r + \frac{t_1}{2}\right) t_1}{\left(r - \frac{t_2}{2}\right) t_2} = \eta_1 \tag{4.44}$$

and formula (4.42) gives

$$\frac{p_1}{p_2} = \frac{E_1}{E_2} \frac{\left(r + \frac{t_1}{2}\right)^3 t_1}{\left(r - \frac{t_2}{2}\right)^3 t_2} = \eta_2 \tag{4.45}$$

It can be written that

$$\frac{\eta_1}{\eta_2} = \frac{G_1 E_2}{E_1 G_2} \left( \frac{r - \frac{t_2}{2}}{r + \frac{t_1}{2}} \right)^2 = \frac{1 + \nu_2}{1 + \nu_1} \left( \frac{r - \frac{t_2}{2}}{r + \frac{t_1}{2}} \right)^2$$

which for the steel–concrete structure reduces to

$$\frac{\eta_1}{\eta_2} = 0{\cdot}88 \left( \frac{r - \frac{t_2}{2}}{r + \frac{t_1}{2}} \right)^2 \tag{4.46}$$

It is evident that $\eta_2 > \eta_1$ and this inequality grows with increasing wall thicknesses. This means that the steel component of the structure takes a little higher portion of the external loading when there is no connection between the cylinders than is the case with perfect connection. The difference, however, is rather small and it can be concluded that the composite action, contrary to the case for girder-type structures, is not essential for coaxial structures.

The closed-form solution of the above differential equations is usually not convenient for practical use. On the other hand, a series-type solution represents a very rapid and versatile design tool. Harmonic analysis can be applied directly to the case of simply supported girders, and tower-type structures may be analysed by first establishing a simply supported beam (Křístek, 1979b) obtained by creating a mirror image of the structure about its foundation.

Let a force $R$ be applied at the midspan cross-section of the substitute structure which balances all loads acting on the substitute structure (Fig. 4.8). The substitute structural system as a whole is in a state of equilibrium. Assume now that the clamping of the bottom of the actual structure is released and that the end cross-sections of the substitute structure, $x = 0$ and $x = L = 2H$, are complemented by dummy simple supports (Fig. 4.8). No reactions appear in the dummy simple supports. The states of stress of the substitute and actual structures are identical, and the displacements of both structures differ only in that there is deflection of the substitute structure at its midspan ($x = 0{\cdot}5L = H$) whereas the actual structure is clamped. Harmonic analysis can be applied to such a simply supported substitute structure of span $L$. Special boundary conditions corresponding to the simply supported substitute structure allow all the desired functions in the form of Fourier series with unknown amplitudes:

$$
\begin{aligned}
v(x) &= \sum_{n=1}^{\infty} V_n s_n(x) & m(x) &= \sum_{n=1}^{\infty} A_n c_n(x) \\
p_1(x) &= \sum_{n=1}^{\infty} P_{1,n} s_n(x) & p_2(x) &= \sum_{n=1}^{\infty} P_{2,n} s_n(x) \\
\Phi_1(x) &= \sum_{n=1}^{\infty} F_{1,n} c_n(x) & \Phi_2(x) &= \sum_{n=1}^{\infty} F_{2,n} c_n(x) \\
Q_1(x) &= \sum_{n=1}^{\infty} T_{1,n} c_n(x) & Q_2(x) &= \sum_{n=1}^{\infty} T_{2,n} c_n(x)
\end{aligned}
\tag{4.47}
$$

in which

$$
s_n(x) = \sin\frac{n\pi x}{L} = \sin\alpha_n x \qquad c_n(x) = \cos\frac{n\pi x}{L} = \cos\alpha_n x \tag{4.48}
$$

Since the external loading $p(x)$ and the balancing force $R = 2\int_0^H p(x)\,\mathrm{d}x$ can also be expressed by a Fourier series

$$
\sum_{n=1}^{\infty} P_n \sin\frac{n\pi x}{L} \tag{4.49}
$$

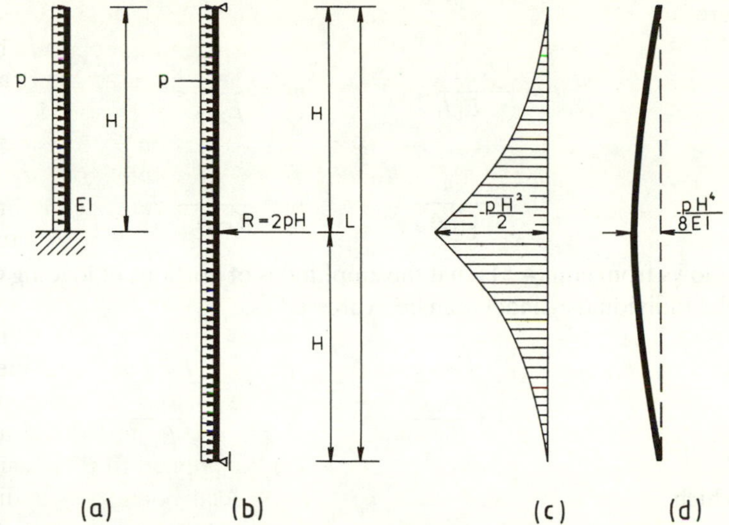


FIG. 4.8. Representation of a vertical structure by a substitute simple beam.

in which

$$P_n = \frac{2}{L}\int_0^L p(x)\sin\frac{n\pi x}{L}\,\mathrm{d}x + \frac{2R}{L}\sin\frac{n\pi}{2}$$
$$= \frac{2}{H}\int_0^H p(x)\sin\frac{n\pi x}{L}\,\mathrm{d}x + \frac{2R}{L}\sin\frac{n\pi}{2} \tag{4.50}$$

the entire analysis can be conducted for each term of the series independently and the results simply added together.

By substituting the eqns (4.47) for eqns (4.27) and (4.32)–(4.38), the following are obtained:

$$\begin{aligned}
P_{1,n} + P_{2,n} &= P_n \\
V_n\alpha_n^4 &= a_1(P_{1,n} + A_n\alpha_n) + b_1P_{1,n}\alpha_n^2 \\
V_n\alpha_n^4 &= a_2(P_{2,n} - A_n\alpha_n) + b_2P_{2,n}\alpha_n^2 \\
F_{1,n} &= V_n\alpha_n - b_1T_{1,n} \\
F_{2,n} &= V_n\alpha_n - b_2T_{2,n} \\
T_{1,n}\alpha_n &= P_{1,n} \\
T_{2,n}\alpha_n &= P_{2,n} \\
F_{1,n} - F_{2,n} &= A_nk_1
\end{aligned} \tag{4.51}$$

where

$$a_1 = \frac{1}{E_1 I_1} \qquad a_2 = \frac{1}{E_2 I_2}$$
$$b_1 = \frac{1}{G_1 A_{s1}} \qquad b_2 = \frac{1}{G_2 A_{s2}} \tag{4.52}$$

It follows from eqn (4.51) that the amplitudes of portions of loading taken by the individual cylinders can be expressed as

$$P_{1,n} = \frac{P_n}{1 + \dfrac{\beta_{1,n}}{\beta_{2,n}}} \qquad P_{2,n} = \frac{P_n}{1 + \dfrac{\beta_{2,n}}{\beta_{1,n}}} \tag{4.53}$$

in which

$$\beta_{1,n} = a_1 \left(1 - \frac{b_1}{k_1}\right) + b_1 \left(\alpha_n^2 - \frac{a_2}{k_1}\right)$$
$$\beta_{2,n} = a_2 \left(1 - \frac{b_2}{k_1}\right) + b_2 \left(\alpha_n^2 - \frac{a_1}{k_1}\right) \tag{4.54}$$

The amplitudes of Fourier series expressing the interaction moment are

$$A_n = \frac{1}{\alpha_n k_1}(b_2 P_{2,n} - b_1 P_{1,n}) \tag{4.55}$$

It is possible to express the internal forces in the structure; the bending moments are given by the following formulae:

For the external (steel) cylinder:

$$M_1(x) = \sum_{n=1}^{\infty} \left(\frac{P_{1,n}}{\alpha_n^2} + \frac{A_n}{\alpha_n}\right) s_n(x) \tag{4.56}$$

For the internal (concrete) cylinder:

$$M_2(x) = \sum_{n=1}^{\infty} \left(\frac{P_{2,n}}{\alpha_n^2} - \frac{A_n}{\alpha_n}\right) s_n(x) \tag{4.57}$$

Similarly, the shear forces are obtained as given below:
For the external cylinder:

$$Q_1(x) = \sum_{n=1}^{\infty} \frac{P_{1,n}}{\alpha_n} c_n(x) \tag{4.58}$$

For the internal cylinder:

$$Q_2(x) = \sum_{n=1}^{\infty} \frac{P_{2,n}}{\alpha_n} c_n(x) \tag{4.59}$$

In the limiting case of complete composite action ($k_1 = 0$) it follows from eqn (4.54) that

$$\frac{\beta_{1,n}}{\beta_{2,n}} = \frac{b_1}{b_2}$$

Hence, according to eqn (4.53), it can be shown that

$$\frac{P_{1,n}}{P_{2,n}} = \frac{b_2}{b_1} = \frac{G_1 A_{s1}}{G_2 A_{s2}} \tag{4.60}$$

which is in agreement with eqn (4.40).

If there is no composite action ($k_1 \rightarrow \infty$), the following formula is obtained from eqn (4.54):

$$\frac{\beta_{1,n}}{\beta_{2,n}} = \frac{a_1 + b_1 \alpha_n^2}{a_2 + b_2 \alpha_n^2} \tag{4.61}$$

For tall structures the terms $\alpha_n^2 = (n\pi/L)^2$ reach small values, particularly for the low harmonics. Therefore it is possible to write approximately in this case that

$$\frac{\beta_{1,n}}{\beta_{2,n}} = \frac{a_1}{a_2} \tag{4.62}$$

i.e., the loading is distributed between both parts in proportion to their flexural stiffnesses:

$$\frac{P_{1,n}}{P_{2,n}} = \frac{a_2}{a_1} = \frac{E_1 I_1}{E_2 I_2} \tag{4.63}$$

which is obviously in agreement with formula (4.42).

As an example, the composite tower in Fig. 4.5 can be analysed. The material and geometrical characteristics of the structure are $A_1 = 940\,544\ \text{mm}^2$, $A_{s1} = 470\,272\ \text{mm}^2$, $I_1 = 1{\cdot}6858 \times 10^{13}\ \text{mm}^4$, $E_1 = 210\,000\ \text{MPa}$, $G_1 = 81\,000\ \text{MPa}$, $A_2 = 12{\cdot}755 \times 10^6\ \text{mm}^2$, $A_{s2} = 6{\cdot}3755 \times 10^6\ \text{mm}^2$, $I_2 = 2{\cdot}1453 \times 10^{14}\ \text{mm}^4$, $E_2 = 30\,500\ \text{MPa}$, $G_2 = 13\,270\ \text{MPa}$, $a_1 = 2{\cdot}8247 \times 10^{-19}\ \text{N}^{-1}\ \text{mm}^{-2}$, $a_2 = 1{\cdot}5283 \times 10^{-19}\ \text{N}^{-1}\ \text{mm}^{-2}$, $b_1 = 2{\cdot}6252 \times 10^{-11}\ \text{N}^{-1}$, $b_2 = 1{\cdot}1816 \times 10^{-11}\ \text{N}^{-1}$, $H = 170\ \text{m}$.

(i) For the case of complete composite action, according to eqn (4.60) we obtain

$$\frac{P_{1,n}}{P_{2,n}} = 0{\cdot}45$$

Thus, the external cylinder takes the portion of loading

$$P_{1,n} = 0{\cdot}31P_n$$

and the internal cylinder carries the portion:

$$P_{2,n} = 0{\cdot}69P_n$$

(ii) For the case with no composite action, using eqn (4.63), it is possible to write

$$\frac{P_{1,n}}{P_{2,n}} = 0{\cdot}54$$

and the portions taken by individual structural parts are

$$P_{1,n} = 0{\cdot}35P_n$$

$$P_{2,n} = 0{\cdot}65P_n$$

(iii) For the composite action provided by studs placed in locations of $400 \times 400$ mm it is possible to consider the coefficient of deformability of the connection by the value of $K_1 = 1\ \text{mm}^3\ \text{N}^{-1}$. Thus

$$k_1 = 1/3{\cdot}14 \times 6000^3 = 1{\cdot}47 \times 10^{-12}\ \text{N}^{-1}$$

and by substituting into eqn (4.54) we obtain (for $n = 1$)

$$\beta_{1,1} = -7{\cdot}49 \times 10^{-18}\ \text{N}^{-1}\ \text{mm}^{-2}; \quad P_{1,1} = 0{\cdot}31P_1$$

$$\beta_{2,1} = -3{\cdot}34 \times 10^{-18}\ \text{N}^{-1}\ \text{mm}^{-2}; \quad P_{2,1} = 0{\cdot}69P_1$$

and also, for $n = 15$, we obtain

$$P_{1,15} = 0{\cdot}32P_{15}$$

$$P_{2,15} = 0{\cdot}68P_{15}$$

This result confirms that such a weak connection is sufficient to provide the structural performance which is identical to that of the structure with complete connection.

## 4.3 SUMMARY OF STUDIES USING THE ABOVE CONCEPT

In order to assess the effect of deformability of the connection of the concrete deck to the steel girder it is possible to use the results of a study investigating typical bridge composite girders (Křístek and Studnička, 1982). The stresses in the concrete deck and in the flanges of a steel girder as well as the midspan deflections are plotted against the connector deformability in Fig. 4.9. The shaded area indicates the range of deformability corresponding to the usual stud connectors ($K = 0{\cdot}5 \times 10^{-3}$ to $3 \times 10^{-3}$ mm$^2$ N$^{-1}$).

The relationship between an increase of stress in the steel girder and a decrease of stress in the concrete deck and the growing flexibility of connectors is shown in Fig. 4.9a,b. It is found that the values of stresses in the bottom flange (which usually determine the girder dimensions) are much less sensitive to a change in connector deformability than the stresses in the top flange. It is apparent that the stresses in the bottom flange indicate a tendency to remain approximately the same as those corresponding to the perfect connection. A difference of less than 2% is observed in practically all the cases. However, the more pronounced increase in stresses in the top flange (which is substantially smaller than for those in the bottom flange) does not influence the design of the steel girder dimensions. A reduction of stresses in the concrete deck (approximately 10%) will be welcomed in the majority of practical cases because of the presence of additional stresses caused by temperature and shrinkage. These results are in agreement with the findings of Yam (1981).

The deflections increase slightly with growing connection flexibility (approximately 7%, see Fig. 4.9c), but for composite bridge beams, which are always of considerable flexural stiffness, this increase is not very relevant.

In order to achieve some analytical expressions for the influence of the

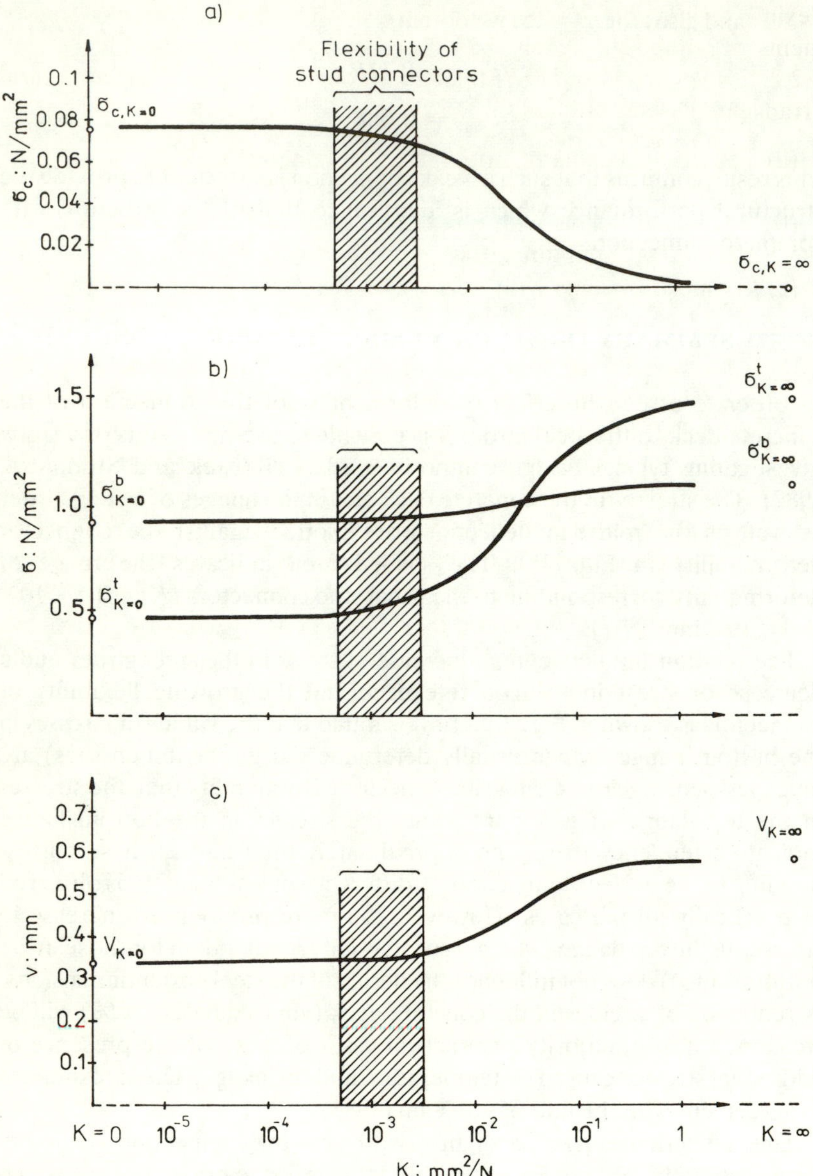


FIG. 4.9. Stress and deflection at midspan for a simply supported composite beam with connection of changing deformability: (a) stress in composite deck; (b) stresses in steel girder ($t$ = top flange, $b$ = bottom flange); (c) deflection.

flexible connection, a wide parametric study covering various arrangements of composite beams, based on the simplified method in Section 4.2.2.1, was carried out (Křístek and Studnička, 1982). Three structural arrangements were studied:

(a) composite beams of bridge type with concrete decks of constant thickness;
(b) composite beams with universal rolled steel girders and concrete decks of constant thickness;
(c) composite beams with universal rolled steel girders and concrete decks with transverse ribs.

*Case (a)*. In the first stage of the study, the proportions of the composite beams were chosen to correspond to a wide variety of practical arrangements: span lengths $L$ were 10–50 m, non-symmetrical steel cross-sections were of depth $(1/25–1/10)L$. The results, based on the analysis of a variety of arrangements with two typical values of connector deformability characteristic, $K = 0$ (the perfect connection) and $K = 0{\cdot}003\ \mathrm{mm^2\ N^{-1}}$ (connection by studs) were generalised and applied in the Italian Standard (1985). According to this standard the deflection of a composite girder with deformable connections may be safely approximated by the following relationship:

$$v = v_0 \omega \tag{4.64}$$

in which $v_0$ is the deflection of composite girder with the perfect connection, and $\omega$ is a coefficient expressing the influence of the connector deformability, which may be expressed as

$$\omega = 1 + 237\psi^2 K \tag{4.65}$$

where

$$\psi = \frac{A_c}{A_c + (E/E_c^*)A} \tag{4.66}$$

It has been found that the range of increases in deflections due to connection flexibility, for composite beams of the type discussed having dimensions commonly used in civil engineering structures, is limited to approximately 10%. This is in agreement with the published conclusions of other investigators.

*Case (b).* In accordance with the results of a parametric study (Křístek and Studnička, 1982), the Italian Standard (1985) recommends that deflections of composite beams with rolled steel girders be predicted by eqn (4.64) with the coefficient

$$\omega = 1 + (27 + 417\psi^2)K \tag{4.67}$$

*Case (c).* Similarly, the deflections of composite beams having universal rolled steel beams and concrete decking with corrugations that run across the flange of the steel beam, may again be approximated by eqn (4.64) applying for $\psi < 0{\cdot}25$,

$$\omega = 1 + 83K \tag{4.68a}$$

and for $\psi \geq 0{\cdot}25$

$$\omega = 1 + [83 + 1110(\psi - 0{\cdot}25)^2]K \tag{4.68b}$$

As an example, the deflections predicted according to eqns (4.65)–(4.68) for a typical value of connector flexibility $K = 0{\cdot}003\ \text{mm}^2\ \text{N}^{-1}$ are plotted in terms of the ratio $\psi$ in Fig. 4.10, together with curves showing the stress variation in the bottom flange of the steel girder.

Comparing the results approximated by eqns (4.65)–(4.68), it is seen that the effects of deformable connections are particularly important in composite beams with a greater concrete area. The effects are more pronounced in beams with shallow universal steel girders than for composite beams of the bridge type, and more in composite beams with transverse ribs than in girders having a deck of constant thickness.

The effect of incomplete interaction on the stiffness of continuous beams was investigated by Yam and Chapman (1972). They have concluded that the difference between complete and incomplete interactions in maximum deflection is usually only 10%. Similarly, by comparing the stress distributions for the section under the point load and that over the central support, Yam (1981) has concluded that, for the section under sagging moment, the effect of incomplete interaction on the extreme fibre stresses is not decisive. However, for the section under hogging moment, the loss of interaction results in a noticeable reduction in the tensile strain at the top of the slab. It can be noted here that the loss of interaction is in fact beneficial in that it alleviates the cracking of concrete.

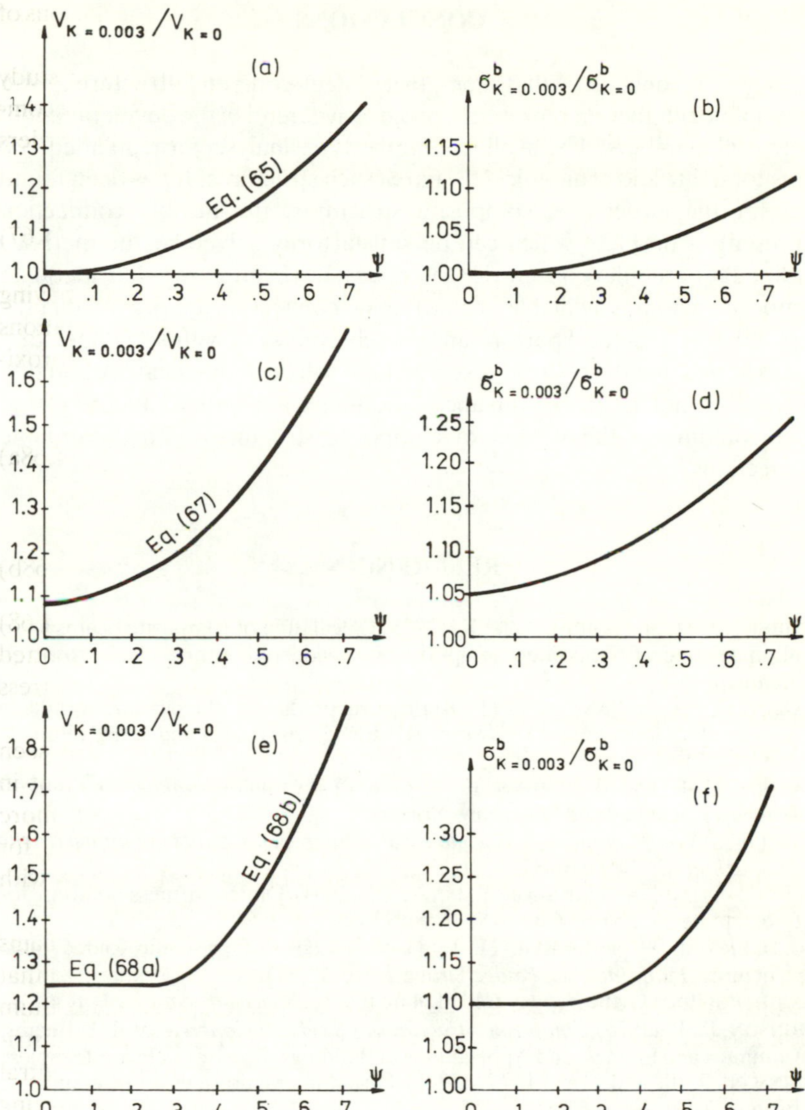


FIG. 4.10. Effect of extent of concrete area for composite beams with deformable connection ($K = 0{\cdot}003\ \text{mm}^2\ \text{N}^{-1}$): (a) deflections and (b) stresses in bottom flange for bridge-type beams; (c) deflections and (d) stresses in bottom flange for universal beams and a concrete deck of constant thickness; (e) deflections and (f) stresses in bottom flange for universal beams and a concrete deck with transverse ribs.

## 4.4 CONCLUSIONS

It may be concluded that composite steel–concrete structures with incomplete interaction represent a progressive trend in the development of structural members. The available methods of analysis form an adequate basis for a safe and economical design of such structures. It has been found that for the girder-type composite structures, the flexible connection represents a problem which can be satisfactorily solved by the methods outlined. For tubular concentric structures it has been shown that adequate composite action is available even when no connectors are provided and a very efficient structural performance is achieved even without connectors.

It is hoped that the results presented here will be of practical use and that they will stimulate further advances and thus contribute to a more global understanding of the subject of composite structures with deformable connections.

## REFERENCES


ARIBERT, J. M. and ABDEL AZIZ, K. (1985) Calculation of composite beams up to ultimate state with the effect of uplift at steel–concrete interface, *Construction Métallique* (4).

BAŽANT, Z. P. and PANULA, L. (1978) Practical prediction of creep and shrinkage of concrete, *Materials and Structures*, RILEM, Paris, Parts I and II (69), Parts V and VI (72).

BAŽANT, Z. P. and WITTMANN, F. H. (1983) *Creep and Shrinkage in Concrete Structures*, J. Wiley and Sons, New York.

CNR (1985) *Travi Composte di Acciaio e Calcestruzzo. Istruzioni per l'impiego nelle Costruzioni*, CNR 10016–85.

DEFRIES-SKENE, A. and SCORDELIS, A. C. (1964) Direct stiffness solution for folded plates, *J. Struct. Div., ASCE*, **90**(ST4).

GOLDBERG, J. E. and LEVE, H. L. (1957) Theory of prismatic folded plate structures, *Publ. Int. Ass. Bridge Struct. Engg*, **17**(87).

HÁJEK, P. (1985) Orthotropy in folded plate theory, *Space Structures*, in press.

JOHNSON, R. P. (1975) *Composite Structures of Steel and Concrete*, Vol. 1, Beams, Columns, and Frames, and Applications in Buildings, Crosby Lockwood Staples.

JOHNSON, R. P. and MAY, I. M. (1975) Partial-interaction design of composite beams, *Struct. Engr*, **53**(8).

JOHNSON, R. P. and SMITH, D. G. E. (1975) Design rules for the control of deflections in composite beams, *Struct. Engr*, **53**(9).

KŘÍSTEK, V. (1979a) *Theory of Box Girders*, J. Wiley and Sons, New York.

KŘÍSTEK, V. (1979b) Folded plate approach to analysis of shear wall systems and frame structures, *Proc. Instn Civ. Engrs*, Part 2(67).

KŘÍSTEK, V. (1983) Folded plates with deformable end cross-sections, *Proc. Instn Civ. Engrs*, Part 2(75).

KŘÍSTEK, V. and STUDNIČKA, J. (1982) Analysis of composite girders with deformable connectors, *Proc. Instn Civ. Engrs*, Part 2(73).

KŘÍSTEK, V. and STUDNIČKA, J. (1985) Působení ocelobetonové věže uzavřeného průřezu při příčném zatížení, *Inženýrské stavby* (1).

MOFFAT, K. R. and DOWLING, P. J. (1978) The longitudinal bending behaviour of composite box girder bridges having incomplete interaction, *Struct. Engr*, **56B**(3).

SCORDELIS, A. C. (1971) Analytical solutions for box girder bridges, *Proc. Conf. Modern Developments in Bridge Design and Construction*, Cardiff, UK.

YAM, L. C. P. (1981) *Design of Composite Steel Concrete Structures*, Surrey University Press.

YAM, L. C. P. and CHAPMAN, J. C. (1972) The inelastic behaviour of continuous composite beams of steel and concrete, *Proc. Instn Civ. Engrs* (53).

# *Chapter 5*

# STEEL FRAMES WITH CONCRETE INFILLS

T. C. LIAUW

*Department of Civil and Structural Engineering, University of Hong Kong, Hong Kong*

## *SUMMARY*

*A literature review of the state-of-the-art in infilled frames is presented.*

*On the basis of analytical and experimental studies, the structural behaviour of steel frames loaded laterally and containing concrete infills with or without shear connectors at the interfaces is described. The stress distributions in the structures, with particular attention given to the infill/frame interface stresses, are studied by means of nonlinear finite element analysis.*

*In the light of these analytical results and the experimental observations, a plastic theory is developed for the predictions of strength and failure mode of an infilled frame with or without shear connectors. The collapse load predictions are compared with results obtained from experiments and the nonlinear finite element analysis. Simplified design rules and examples with the aid of design charts are given.*

## NOTATION

| | |
|---|---|
| $F_b$ | Axial force of the beam at the compressive corner |
| $F_c$ | Axial force of the column at the compressive corner |
| $F_n$, $F_s$ | Normal and shear forces at the interface |
| $H$ | Applied lateral load |
| $H_i$, $H_{i+1}$ | Applied lateral load at the $i$th and $(i+1)$th storey respectively |
| $H_u$ | Collapse shear |

| | |
|---|---|
| $h$ | Storey height |
| $h_i, h_{i+1}$ | Storey height of $i$th and $(i+1)$th storey respectively |
| $k_n, k_s$ | Normal and shear stiffness of interface |
| $l$ | Span of the infilled frame |
| $M_p$ | Plastic moment of the frame |
| $M_{pb}$ | Plastic moment of the beam |
| $M_{pc}$ | Plastic moment of the column |
| $M_{pj}$ | Plastic moment of the joint, i.e. the smaller value of $M_{pb}$ and $M_{pc}$ |
| $m$ | Relative strength parameter, $\sqrt{4M_p/\sigma_c th^2}$ |
| $m_b$ | Relative strength parameter for beams, $\sqrt{4M_{pb}/\sigma_c th^2}$ |
| $m_{bc}$ | Relative strength parameter for beams and columns, $\sqrt{2(M_{pb}+M_{pc})/\sigma_c th^2}$ |
| $m_c$ | Relative strength parameter for columns, $\sqrt{4M_{pc}/\sigma_c th^2}$ |
| $n$ | Number of storeys |
| $S_i, S_{i+1}$ | Storey shear of the $i$th and $(i+1)$th storey respectively |
| $s$ | Shear strength of interface connection |
| $t$ | Thickness of the infilled panel |
| $\alpha_b l$ | Length from the loaded corner to the plastic hinge on the beam |
| $\alpha_c h$ | Length from the loaded corner to the plastic hinge on the column |
| $\beta h$ | Length of contact between the column and the panel in the diagonal crushing mode |
| $\Delta_f$ | Gap width at the interface due to initial lack of fit |
| $\Delta_n, \Delta_s$ | Relative normal displacement and relative shear displacement (i.e. slip) of interface |
| $\Delta_r, \Delta_r'$ | Previous and new residual slips at interface |
| $\theta$ | Angle between the diagonal of the infilled panel and the horizontal |
| $\lambda$ | Load factor |
| $\mu$ | Coefficient of friction at interface |
| $\sigma_c$ | Crushing stress of the panel material |
| $\phi$ | Deflected angle of the collapsing storey |
| $(\ldots)_i$ | That of the $i$th storey |

## 5.1 INTRODUCTION

Rigidly jointed steel frames have been used extensively in building construction. However, a frame system becomes structurally less efficient when it is subjected to large lateral load, such as strong wind, earthquake or explosion. The situation becomes worse when the structural height is increased, resulting in parabolically increasing bending force which has to be resisted by the frames.

In the past few decades, efforts to develop more efficient structural systems with better resistance against strong lateral loads have yielded good results. Among these systems, the one which is least understood but has great potential is the structure known as infilled frame, a structure combining the frame with the infill within the frame, which has been shown to exhibit excellent behaviour.

### 5.1.1 Infilled Frame: Composite Structure

Although the contribution of the infills to the lateral stiffness and stability of frame structures has long been recognised, the infills are usually considered non-structural and it is a common practice to neglect their effects in structural analysis and design. The reluctance of practising engineers to take the contribution of the infills into consideration has been, firstly, due to lack of knowledge concerning the composite behaviour of infilled frames, secondly, due to lack of practical methods for predicting the stiffness and strength, and thirdly, due to the possibility of future removal of the infill for access, doors or windows.

An infilled frame with no connectors or no bonding at any of the interfaces between the frame and the infill is called a non-integral infilled frame. An infilled frame with shear connectors or strong bonding at all the interfaces is called an integral infilled frame. In this chapter, the type of infilled frame to be discussed is confined to single bay steel frames with concrete infills. One of the experimental integral infilled frames is shown in Fig. 5.1, and the shear connectors used in the experimental models together with their shear-slip characteristic are shown in Fig. 5.2.

### 5.1.2 Literature Review

The following review of infilled frames is not confined to steel frames with concrete infills, since different materials have been used in the past by different investigators. The review is divided under separate headings for clarity.

#### *5.1.2.1 Early Research*

An investigation into the behaviour of brick masonry infilled frames subjected to lateral loads was conducted by Polyakov (1950, 1957) in the USSR. Pin-jointed single-storey and multistorey steel frames with masonry infills were used in the tests, and separations between the frame and the

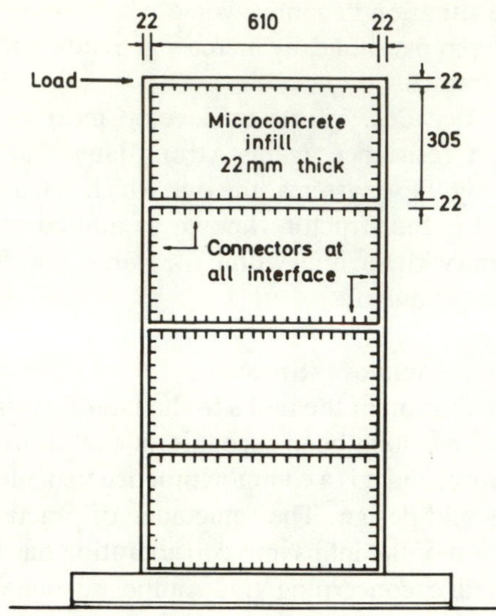


FIG. 5.1. Details of model C2 (dimensions in mm).

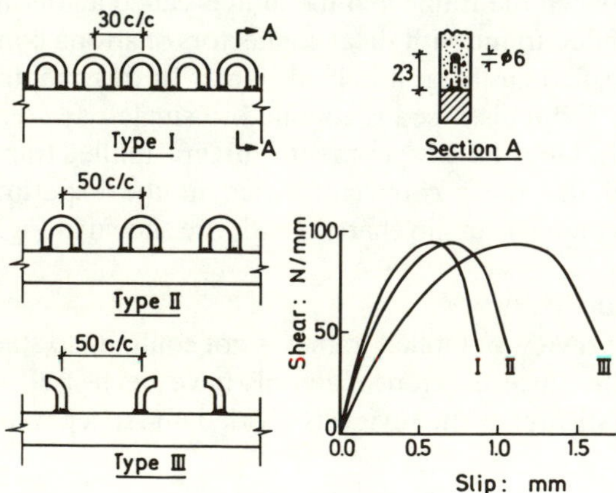


FIG. 5.2. Details of the connectors and shear-slip curves of the connections.

infill were observed except at the two compressive corners. This led to the conceptual proposal of replacing the infill with a diagonal bracing strut (Fig. 5.3a), and empirical formulae were given to predict the stiffness and strength of the infilled frames.

A project was initiated in the USA after the Second World War to study the atomic blast resistance of various structures, one of which was the infilled frame. Whitney, Anderson and Cohen (1955) proposed that an infilled frame be treated as a simple cantilever beam in which the columns and the infill were represented respectively by the flanges and the web of an

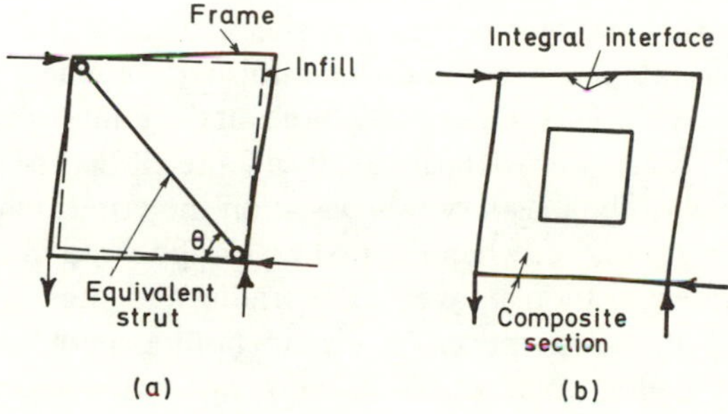


FIG. 5.3. Representations of infill frames: (a) equivalent strut analogy; (b) equivalent frame analogy.

I-beam. Benjamin and Williams (1957) participated in the project using reinforced concrete and steel frames with various types of infill including concrete, reinforced concrete, concrete blocks and brickwork. Analytical methods proposed were either based on the simple theory of strength of materials or on empirical formulae.

In the UK, Thomas (1953) illustrated the stiffening effect of the brick infills, using the test results of encased rectangular steel frames with various types of weak and strong infill. He found that the stiffness of the frame was increased considerably, even with a weak infill, and he proposed to allow for the racking strength of the walls to avoid the necessity of providing special connections or bracing in structural frameworks subjected to lateral forces. Similar tests were performed by Wood (1958) who noticed that due to the presence of the infills, the mode of failure was changed to a new composite mode with a compression band in the brickwork causing the formation of plastic hinges in the frame.

*5.1.2.2 Non-integral Infilled Frames and Diagonal Strut Analogy*
More research interest on non-integral infilled frames was aroused in the 1960s. Researchers realised that the economic and structural advantages could be significant if the infills were incorporated into the frame as a composite structure. The focus of attention during this period was to establish a simple solution to predict the ultimate strength of non-integral infilled frames.

Sachanski (1960) investigated the interaction between a reinforced concrete frame and a masonry infill. Both materials were assumed to be of two elastic homogeneous materials and the problem was treated as one of plane-stress. By introducing a finite number of contact joints at the interface boundary, the unknown normal and shear forces at these joints were solved from simultaneous equations established by the conditions of displacement compatibility. Once these forces at the joints were known, the stresses in the masonry were determined by a finite difference method. The carrying capacity of the masonry was based on the criterion of maximum principal tensile stress, i.e. when the first crack appeared in the masonry. Apparently the separation between the frame and the infill had been neglected, and the assumption of elastic homogeneous material for masonry and the first crack criterion were not realistic and large discrepancies with test results were noted.

In the study of single frames with brickwork and concrete infills, Holmes (1961) proposed the infill be replaced by an equivalent strut (Fig. 5.3a) having a width equal to one-third of the diagonal length. This proposal was too simplistic since the possible variation of structural parameters other than the diagonal length was ignored. The ultimate deflection and the ultimate load were calculated from the shortening of the equivalent strut based on the maximum concrete strain at failure. Again, the variation of concrete strains along the diagonal was neglected. Holmes (1963) extended the method to analyse two-storey infilled frames without improvement.

Stafford Smith (1966) studied single square-shape steel frames with mortar infills and found that the equivalent strut width increased with the increase in the relative stiffness of the frame. A non-dimensional parameter ($\lambda h$) was used as an indication of the relative stiffness of the frame against the infill as follows:

$$\lambda h = \left(\frac{E_c t h^3 \sin 2\theta}{4 E_s I}\right)^{1/4}$$

where $E_c$ and $E_s$ are the Young's modulus of the infill and the frame respectively, and $I$ is the second moment of area of the frame members. The length of contact, $\alpha$, between the frame and the panel was related to $\lambda h$ in a single transcendental equation:

$$\frac{\alpha}{h} = \frac{\pi}{2\lambda h}$$

With the length of contact obtained, a finite difference analysis gave the equivalent strut width in terms of the relative stiffness parameter $\lambda h$. For strength prediction in the case of infill failure, the collapse load was calculated assuming it to be in equilibrium with the contact stresses.

Stafford Smith (1967) extended the method to multistorey rectangular infilled frames, in which the length of contact between the beam and the infill was experimentally found to be almost constant and equal to half the length of the beam. In a later study, Stafford Smith and Carter (1969) proposed the variation of the modulus of elasticity of the infill material with the stress level, with the consequence of the equivalent strut width being decreased as the loading was increased.

Mainstone (1971) tested steel frames with infills of full-size brickwork, model brickwork and micro-concrete. The test results showed wide variation which could not be accounted for by any theoretical analysis existing at the time, and no conclusive proposal was given.

Kadir and Hendry (1975) tested steel frames with model brickwork infills under racking load. The tests confirmed the separation of the interface between the frames and the infills which acted as compressive diagonal struts. However, when a high bond strength mortar was used, the behaviour of the panels was observed to be similar to that of reinforced concrete infills built integrally with the frames, producing much higher strength and stiffness as a result of no separation and more uniform stress distribution at the infill/frame interfaces.

Liauw and Lee (1977) approached the diagonal strut analogy with the use of the strain energy method to establish the sectional area of the equivalent diagonal strut. The method enabled the predictions of stiffness and strength of the structure with or without opening in the infill. However, the strength predictions were too low compared with the experimental results.

During the period, investigators were confronted with the complex interaction problem of infilled frames. They either found unsurmountable difficulties in using classical methods of stress analysis, or resorted to simplistic methods. Although the basic function and the main advantage of the infill were generally recognised, the employment of elastic theory or empirical solution for an essentially nonlinear problem was not satisfactory.

### *5.1.2.3 Integral Infilled Frames and Equivalent Frame Method*

Liauw (1970, 1973) studied the elastic behaviour of stress distribution in single integral infilled frames, on the basis of stress functions in the form of Fourier series, and verified the results by tests on photoelastic models using the stress-frozen technique. The study showed that a major portion of the shearing load was carried by the panel, and the normal stress distribution at the interface was not uniform.

Mallick and Garg (1971) pointed out that the structural behaviour of infilled frames could be very much improved by providing connectors along the structural interface, and expressed their preference for integral infilled frames for three reasons: (1) integral infilled frames gave sufficient warning before failure in the appearance of diagonal cracks at a load less than the ultimate load; (2) integral infilled frames were stiffer than non-integral infilled frames; (3) the provision of connectors reduced the risk of lack of fit. However, contrary to the test data, they failed to see that the integral infilled frames had higher strengths and concluded that there was no marked difference in the ultimate load of the two types of infilled frames.

Liauw (1972) proposed an analysis using the equivalent frame method (Fig. 5.3b) for stiffness prediction of integral infilled frames with or without openings. The frame members together with the infills were treated as composite members and the whole structure was transformed into an equivalent frame. Liauw and Lee (1977) pointed out that, when shear connectors were provided, the composite action of the frame and the infill invalidated the concept of diagonal strut analogy, and the use of an equivalent frame analogy was proposed.

### *5.1.2.4 Linear Finite Element Analysis*

As the finite element method became popular in the 1960s, its versatility has led to widespread application in structural mechanics. An attempt to analyse non-integral infilled frames by the finite element method, assuming elasticity of materials, was given by Mallick and Severn (1967). In the analysis, separation and slip at the structural interface (Fig. 5.4a) were taken into consideration. However, it appeared that the finite size and the axial deformation of the frame members had been neglected.

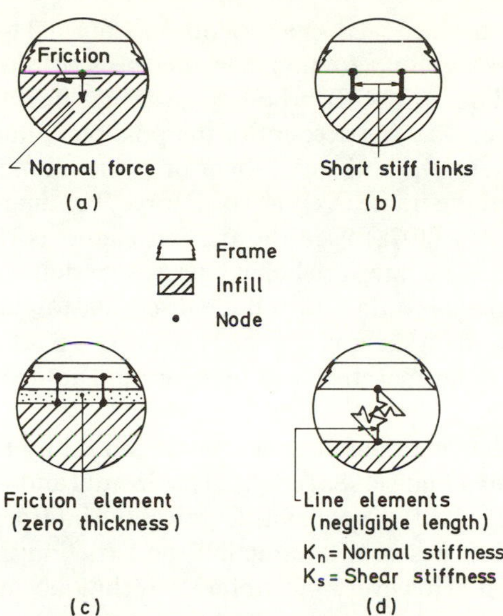


FIG. 5.4. Representations of interface in finite element analysis: (a) friction and normal force used by Mallick and Severn (1967); (b) short stiff link used by Riddington and Stafford Smith (1977); (c) friction element used by King and Pandey (1978); (d) line element used by Dawe and Young (1985).

Barua and Mallick (1977) and Mallick and Garg (1971) refined the method by taking into account the axial deformation of the frame, the effect of which was shown to be negligible. In the latter paper, when connectors were provided at the interface, the interaction forces between the frame and the infill were assumed to consist only of normal forces, i.e. the shear forces at the interface were neglected. This assumption was not justifiable, because the main contribution of the connectors at the interface is in the transmission of the lateral shear from the frame to the infill. Again, the elasticity of materials was assumed in both papers.

In the study of elastic stress distribution, Riddington and Stafford Smith (1977) introduced short stiff linking members (Fig. 5.4b) as interface elements whereby the linking members were removed when the interface had separated. Along the length of contact, two possibilities were allowed: (1) no slip—the short stiff linking member forced both nodes at the interface to have identical displacement so that friction was developed with no slip; and (2) no friction—the linking matrix was of a form which forced the two nodes to have equal displacements only normal to the interface so that slip occurred with no friction. On the basis of this study, Stafford Smith and Riddington (1978) produced a design method for masonry-infilled steel frames, taking account of the possible failure of the infill in either diagonal tension, horizontal shear or corner compression, and the possible failure of the frame in either axial force, bending or shear.

King and Pandey (1978) used the friction elements (Fig. 5.4c) at the interface, whereby the interface behaviour was modelled by adjusting the element properties according to the idealised elasto-plastic friction–slip characteristic based on shear box tests. However, the relief of friction due to the reduction of normal stress or the separation at the interface had not been considered.

Essentially following the interface elements used by Liauw and Kwan (1982), Dawe and Young (1985) adopted horizontal and vertical two-node line elements (Fig. 5.4d) of negligible length at the interface to model bond forces and tension ties in one direction, and friction and shear ties in another direction. However, the iterative technique used in the finite element analysis was confined to the state prior to initial major cracking of the infill.

During this period, the attempts to apply the finite element method to infilled frames were mostly limited to linear analysis. This was inconsistent with the main objective of finding the ultimate strength of the structures, because linear analysis cannot reflect the actual behaviour of the structure due to the cracking and the crushing of the infill. In order to take into consideration the material nonlinearity and the whole range of behaviour of an infilled frame, nonlinear finite element analysis must be employed.

#### *5.1.2.5 Nonlinear Finite Element Analysis*

Early in the 1960s, the nonlinear phase preceding the collapse of infilled frames had been recognised, and the nonlinearity of load–deflection curves and the significance of stress redistribution near collapse were pointed out by Mainstone (1962) and Wood (1968).

However, during the period prior to 1980, the analytical attempts were essentially confined to elastic analysis until Liauw and Kwan (1982) proposed the use of nonlinear finite element analysis (see Section 5.3) to deal with the whole range of problems in infilled frames. The finite element formulation and the iterative procedure which were presented took account of: (1) the material nonlinearity in compression, up to crushing and the cracking of the infill; (2) the nonlinear shear–slip characteristic at the interface where connectors were provided; (3) the friction–slip–separation characteristic at the interface where connectors were not provided; (4) the initial lack of fit at the interface; and (5) the yielding of the frame with idealised elasto-plastic characteristic. The nonlinear behaviour of concrete infills in steel frames without shear connectors (Liauw and Kwan, 1984a) and with shear connectors (Kwan and Liauw, 1984) was theoretically and experimentally studied, whereby the entire range of load–deflection, the stiffness and the strength of the structure and the stress redistribution toward collapse, including crushing and cracking patterns, were presented.

Dhanasekar, Page and Kleeman (1985) also presented a nonlinear finite element formulation in respect of brick masonry in steel frames. The infill was modelled as a continuum using macroscopic stress–strain relations and a failure surface criterion in the shape of three intersecting elliptic cones, obtained from biaxial tests on brick masonry panels. The iterative procedure in the finite element analysis was used in which material nonlinearity, progressive cracking and sliding of the masonry joints and at the interface had been taken into account. The mortar joint at the interface was modelled using a six-node isoparametric line element, and the frame was modelled using a two-node beam element assumed to remain elastic. According to most experiments conducted by other investigators, however, the assumption of an elastic frame at all times appears to be unjustified.

#### *5.1.2.6 Plastic Theory*

Since the equivalent strut analogy was proposed in the 1950s in the research on infilled frames, little attention was given to the bending strength of the frame members of the structure, until Wood (1978) proposed a plastic theory in which the stress redistribution near collapse and the importance of the bending strength of the frame were recognised. Four different failure modes were identified from experimental observations. The failure modes and the corresponding collapse shears were evaluated in terms of the bending strength of the frame and the crushing stress of the panel. However, different interface conditions of the infilled frames were not differentiated. Such ambiguity led to unreasonable consequences: in the case of non-integral infilled frames, separation at the structural interface was ignored and the shear stress (friction) at the interface became unrealistically large, whereas in the case of integral infilled frames the shear strength of the interface was neglected. A penalty factor, empirically found from tests, was proposed to reduce the effective crushing strength of the infill in order to account for the imperfect plasticity of concrete infill.

Making use of the knowledge gained from the nonlinear finite element analysis and the evidence gained from laboratory experiments, Liauw and Kwan (1983a, 1983b) developed a plastic theory to deal with non-integral and integral infilled frames, taking account of the stress redistribution due to the development of cracks and crushing of the infill toward collapse, and of the shear strength at the infill/frame interface provided by the shear connectors. Essentially, the interface stresses obtained previously from nonlinear finite element analysis were simplified and applied to the frames which were analysed in accordance with the plastic theory. A simple design procedure was later developed with the aid of design charts (Liauw and Kwan, 1984b) and the various plastic analyses with three different types of interface condition were amalgamated into a unified plastic analysis (Liauw and Kwan, 1985b).

The study of single-bay infilled frames was extended to multibay infilled frames by Lo and Liauw (1984) and by Liauw and Lo (1985) in a series of experimental investigations using models having up to two bays and four storeys. The tests showed that the strengths of multibay infilled frames were increased approximately by 50% in the two-bay models.

Dawe and McBride (1985) reported an experiment on six full-scale masonry-infilled steel frames loaded gradually by a horizontal shear force. The comparisons of the test results with the analyses proposed by previous investigators showed that the best agreement was given by the collapse load prediction based on Liauw and Kwan (1983a, 1983b) when the crushing strength of concrete infill was reduced by Wood's penalty factor to account for the quasi-plasticity of the masonry infill.

#### *5.1.2.7 Dynamic Behaviour*

Mallick and Severn (1968) investigated the dynamic characteristics of infilled frames both with and without shear connectors, whereby the energy dissipation capacity and the damping ratio were obtained from half-cycle load tests. The natural frequencies and mode shapes were also studied both experimentally and analytically. The infilled frames were analysed as a shear-type structure in which the axial forces were ignored, and as a bending-type structure in which the bending forces were ignored. The test results for natural frequencies and mode shapes showed that the bending model was superior to the shear model for multistorey structures having a height/span ratio of 3 or greater.

Klinger and Bertero (1978) developed a macroscopic mathematical model to predict the essential features of the cyclic load–deflection characteristics of reinforced concrete frames infilled with reinforced concrete, clay-block and concrete-block masonry. The macroscopic model basically replaced each infill panel with a pair of equivalent diagonal struts so that an infilled frame was idealised as a cross-braced frame. The mechanical behaviour of the equivalent diagonal struts was predefined in accordance with the experimentally observed infill behaviour, e.g. initial stiffness and strength, decreased strength with increased deformation, and decreased stiffness on reloading. The macroscopic model offered efficiency in nonlinear dynamic analysis, but its successful use depended on the representation of the equivalent diagonal struts in place of the physical behaviour of the infill.

Kahn and Hanson (1979) studied experimentally the cyclic behaviour of reinforced concrete frames with three different reinforced concrete infills: cast-in-place wall, single precast wall and interlinked multiple precast wall, all being connected to the frame. The test results indicated that the cast-in-place wall gave the best results in terms of strength, whereas the linked multiple precast wall was the least efficient in all aspects.

Liauw (1979) reported tests on non-integral and integral concrete-infilled steel frames subjected to incremental dynamic loading with increasing frequency. The tests showed the significant improvement in stiffness and strength due to the provision of shear connectors in the structures with both solid infill and infill having an opening. The beneficial effects of the shear connectors were further confirmed by Liauw and Kwan (1985a) in the experimental study of hysteretic characteristic, energy dissipation capacity and degradation rate of infilled frames with different interface conditions.

Bertero & Brokken (1983) studied the effects of retrofitting masonry and lightweight concrete infills within a 1/3-scale reinforced concrete frame subjected to monotonic and cyclic loading. Four different types of infill were used: (1) hollow clay masonry; (2) hollow concrete masonry; (3) solid brick masonry with exterior welded wire fabric covered in cement mortar and anchored to the frame; and (4) 150 mm thick reinforced lightweight concrete. For reasons of economy, ease of construction and favourable mechanical characteristics, the third type of infill was regarded as the most promising configuration, although the tests appeared to show the last type of infill as having the best performance.

Palsson, Goodno, Craig and Will (1984) studied the influence of heavy precast concrete claddings on the dynamic response of a tall steel frame building. The interaction of the frame with the cladding was investigated for moderate wind and earthquake loadings, using a linear elastic finite element model for the frame combined with a linear or nonlinear model for the cladding connection subsystem. The study confirmed that the dynamic response was substantially changed from a bare frame representation to the one with a heavy exterior façade. It was expressed that neglecting the influence of claddings might not be conservative because the dynamic characteristics of the overall structure could be altered to such a degree by the added stiffness of the claddings that the sensitivity of the structure to certain earthquake loading might be increased substantially.

Because the response sensitivity to certain earthquake loading of an infilled frame can be substantially increased as a consequence of the added stiffness of the infill, Liauw, Tian and Cheung (1986) proposed a solution in which a sliding base device would be incorporated between the base of the infilled frame and the foundation. The function of the sliding base device was like that of a safety valve, whereby the transmission of vibration force from a strong earthquake to the superstructure would be limited by the frictional coupling of the device, and like that of a damper whereby the repeated sliding would dissipate the vibration energy and reduce the vibration level. The substructure method was proposed for the analysis of the structure–sliding base system, in which the infilled frame as the upper substructure and the sliding base as the lower substructure were considered separately in the first instance. After the correct mathematical models for the dynamic characteristics of both substructures had been established separately through a procedure of analysis and/or experiment, the two substructures were coupled for response analysis. The analysis and experiment on two infilled frames demonstrated clearly that the response of the structure with a fixed base increased linearly with increased magnitude of the base excitation, whereas the response of the structure with a sliding base was deflected to approximately a constant magnitude after sliding occurred, even when the base excitation was increased.

### 5.1.3 Scope

During the past three decades, the general trend of research in infilled frames has developed from elastic theory to plastic theory, and from linear analysis to nonlinear analysis. Such a trend has been made possible by a series of developments in the philosophy of engineering concept of structures, in the better understanding of basic characteristics of engineering materials and in the rapidly improving computational capability and numerical techniques.

The scope of presentation in this chapter is confined to single-bay steel frames with concrete infills, lightly reinforced to control shrinkage, with and without shear connectors at the infill/frame interface. The general characteristics of the infilled frames are described for an initial understanding of such structures, and the approach and the important findings of nonlinear finite element analysis are presented for a deeper understanding. On this basis, the plastic theory for integral and non-integral infilled frames is developed, and the theoretical predictions on the strengths and failure modes of the infilled frames are compared with results obtained from both nonlinear finite element analysis and experiments. Finally, it is possible to move one step further from analysis to design, for which a simple design procedure and design charts are given.

## 5.2 GENERAL BEHAVIOUR

The general behaviour of infilled frames subjected to lateral load at the top level is briefly described as follows.

### 5.2.1 Non-integral Infilled Frames (Models A)

The analytical and experimental load–deflection curves of the models for deflections up to 2% of the height are shown in Fig. 5.5. There was initial lack of fit between the infilled panels and the frame because of shrinkage of the infill material.

During testing, separation at the tensile corners occurred almost immediately after the models were loaded, so that the panels were in contact with the frame only in the vicinity of the compressive corners. However, with increased load the interface configuration became stable after the frame had gained firm contact with the panels.

At greater load, stiffness gradually decreased when the compressive corners of the panels yielded. The models reached their peak strength when the corners were crushed. Crushing of the infill appeared to occur progressively outwards from the corners. During crushing, obvious signs of yielding at the steel columns were also observed. After peak load, the models continued to sustain substantial loading (more than 85% of peak load) for a large range of deflection.

Both models A-2 and A-3 failed with the same collapse mechanism. The development of the collapse mechanism at various stages as observed in the test is shown in Fig. 5.6. Practically no cracks were produced in the infills of either model before failure.

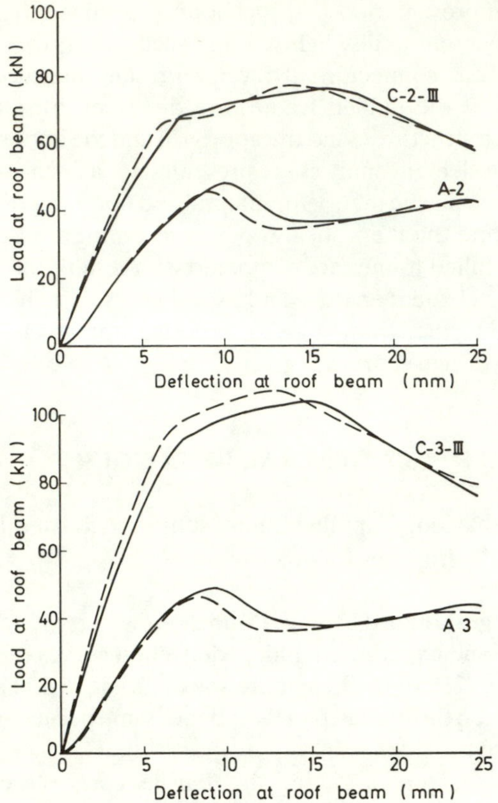


FIG. 5.5. Load–deflection curves.

### 5.2.2 Fully-integral Infilled Frames (Models C)

The C-models generally had higher stiffness and strength, as shown in Fig. 5.5, and maintained their strength up to large deflections leading to large energy absorption before failure. Numerous cracks in the infills at 45° to the beams developed continuously from about one-quarter peak load onwards. The stiffness dropped gradually as the compressive corners of the infilled panels and the infill/beam connections yielded. The models failed in shear at the infill/beam connections and by crushing of the panels at the compressive corners. Obvious signs of yielding of the columns were also observed. All the C-models failed with the same collapse mechanism, and the development of the collapse mechanism at various stages is shown in Fig. 5.6.

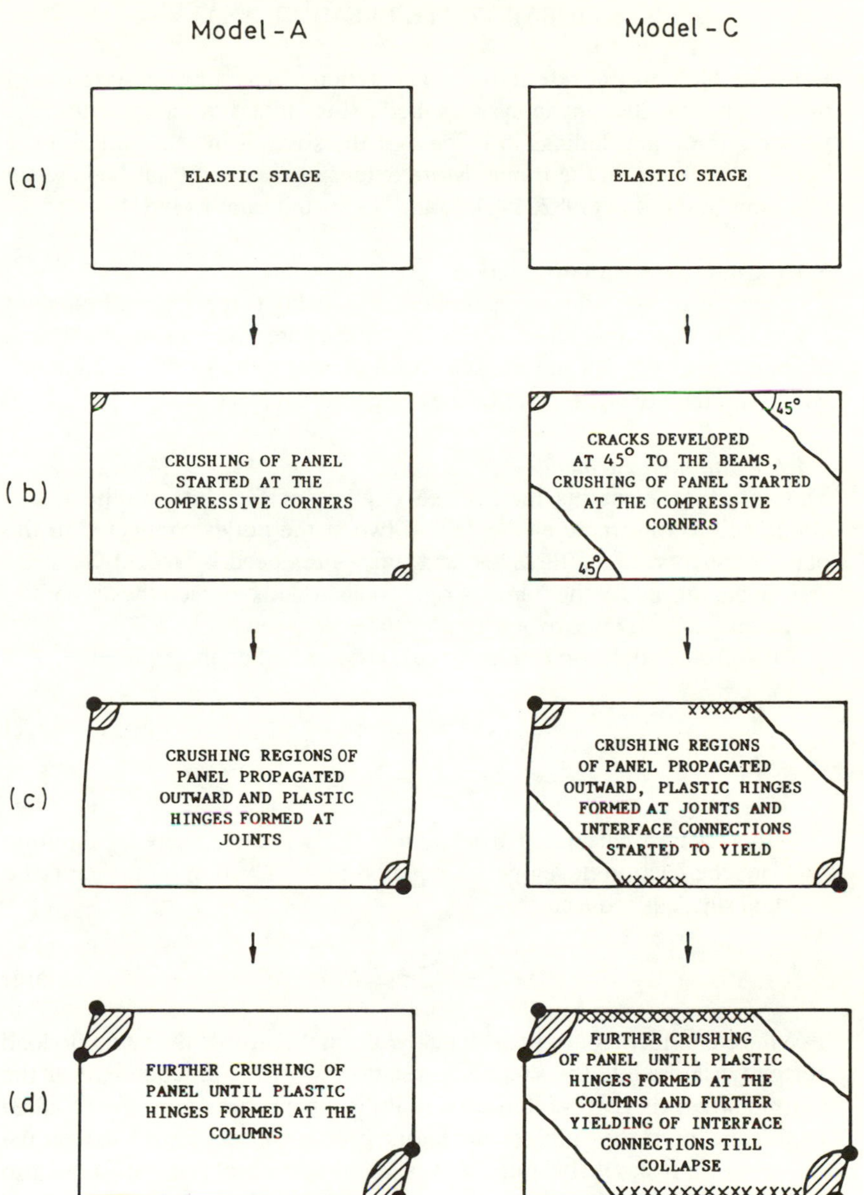


FIG. 5.6. Development of collapse mechanism at various stages.

## 5.3 NONLINEAR FINITE ELEMENT ANALYSIS

For completeness of presentation, this section gives a brief description of nonlinear finite element analysis applied to the infilled frame structure and presents the main findings in respect of the stresses in the infill and the bending moments in the frame. More extensive treatment has been given by Liauw and Kwan (1982, 1984a), and Kwan and Liauw (1984).

### 5.3.1 Finite Element Formulation

The infilled frame structure is modelled by using three types of element (Fig. 5.7): the panel elements which constitute the infills, the frame elements which constitute the skeletal frame, and the interface elements which simulate the behaviour of the structural interface.

#### *5.3.1.1 Interface Element*

The interface elements have three degrees of freedom at the nodes connected to the frame elements and two at the nodes connected to the panel elements. Since the skeletal frame is replaced by frame elements residing at the axis of the frame, a rigid rod is added between the axis of the frame and the interface to account for the eccentricity (Fig. 5.7).

The interface behaviour, Fig. 5.8(a), is described by the equations

$$F_n = k_n(\Delta_n + \Delta_f) \tag{5.1}$$

$$F_s = k_s(\Delta_s - \Delta_r) \tag{5.2}$$

If the friction calculated from eqn (5.2) is greater than the limiting friction, the friction developed is equated to the limiting friction and the residual slip $\Delta_r$ is reduced to

$$\Delta_r' = \Delta_s - (F_s/k_s) \tag{5.3}$$

Where connectors are provided, the normal stiffness is assumed to be perfectly rigid while the shear–slip relation, which is independent of the normal stress, is represented by a multilinear interpolation, Fig. 5.8(b). Unlike the interface without connectors, both initial lack of fit and residual slip are zero. It should be noted that $k_s$ is now the secant shear stiffness, the definition of which is shown in Fig. 5.8(b). It is also assumed that after shear failure, which occurs when the magnitude of slip exceeds a certain limit, the interface behaves as if there were no connectors.

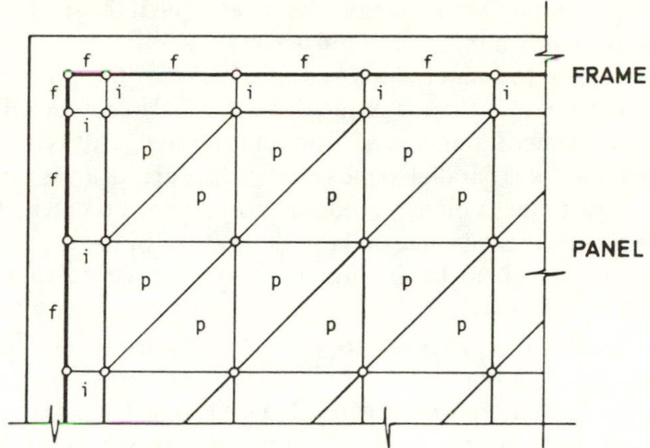


FIG. 5.7. Modelling of finite elements.

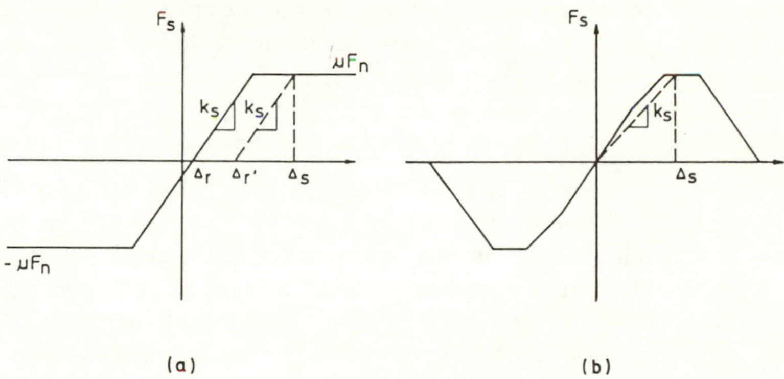


FIG. 5.8. Friction–slip relation at the interface: (a) without connectors; (b) with connectors.

### *5.3.1.2 Panel Element*

In the panels, constant strain, three-noded, triangular plane-stress elements having two degrees of freedom at each node are used.

In tension, the panel material is idealised as a linearly elastic brittle material, assuming that cracks occur when the principal tensile stress exceeds the tensile strength. Bearing in mind that cracks may be closed and reopened again, a cracked element is checked in every iteration step for closure and reopening: a crack is closed or opened up if the stress normal to the crack surface is compressive or tensile respectively.

Before cracking, the material is assumed to be isotropic. After cracking, due to the presence of the crack surface, the material becomes anisotropic. For a cracked element, the Young's modulus normal to the crack surface and the shear modulus parallel to the crack surface are taken as zero. When a crack is closed, the Young's modulus normal to the crack surface is restored to the uncracked value and the shear stress of the crack surface is assumed to be taken up by friction in a way similar to the interface elements.

In compression, the panel material exhibits extensive non-linearity in the stress–strain relation. As the infill panels are two-dimensional, the panel material is under biaxial stresses. However, the biaxial stress state is approximately uniaxial (i.e. one of the principal stresses is much smaller than the other) as can be inferred from the following. When no connectors are provided, the panels act as diagonal struts so that they are subjected to nearly uniaxial stresses. When connectors are provided, the numerous and extensive cracks remove the stresses normal to the crack surfaces so that the stresses approximate to the uniaxial state again. In both cases, strain gauge measurements of the panel stresses have confirmed the uniaxial approximation.

Therefore, the analysis can be simplified with the following assumptions:

(1) The anisotropy is due solely to cracking. Thus for uncracked elements the material is isotropic; and for close-cracked elements the Young's moduli normal and parallel to the crack surface are equal. Hence in every case (whether uncracked, open-cracked or close-cracked) only a single value of Young's modulus is needed.
(2) The Young's modulus varies as a function of the larger compressive principal strain, as in the uniaxial case.
(3) The Poisson's ratio remains constant at any stress level.
(4) The panel is crushed when the principal compressive strain exceeds the ultimate limit. Once the material fails in compression, it can never regain any strength.

*5.3.1.3 Frame Element*

The frame elements are standard prismatic bending elements having three degrees of freedom at each node. The material is idealised as elastoplastic. In each iteration step, the axial force and bending moment are integrated from the normal stress in the section, and the secant stiffness is modified accordingly.

*5.3.1.4 Iterative Procedure*

The iterative procedure with incremental displacement is used. During the iteration process, prescribed displacement in increments is applied to the structure. At each displacement step, the structure is analysed using the secant stiffness at the previous displacement level; and after the stresses are evaluated, the secant stiffness is modified according to the new stress state.

### 5.3.2 Results and Discussions

The results of two typical models chosen from a series of model study are presented here. The two models were identical except that one was without connectors, i.e. non-integral interface, and the other was provided with connectors, i.e. integral interface.

*5.3.2.1 Interface Stresses*

The normal and shear stresses at the structural interface of the third storey (other storeys are similar) as obtained analytically are shown in Fig. 5.9 for the infilled frame without connectors, and in Fig. 5.10 for that with connectors.

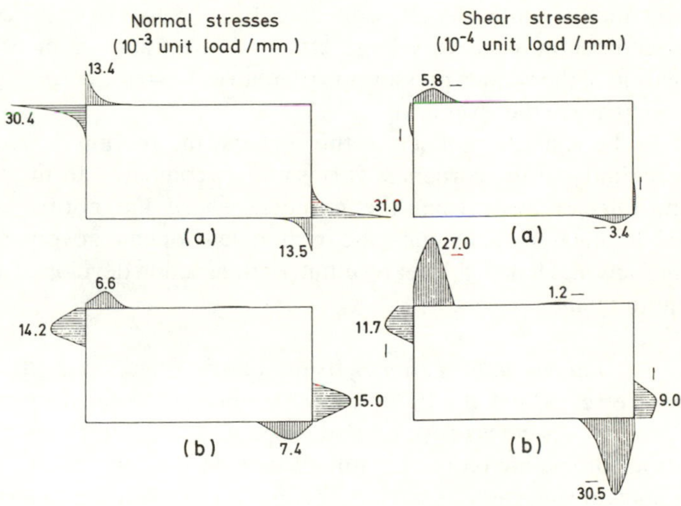


FIG. 5.9. Interface stresses at third storey of model A2: (a) at 4·8 mm top deflection; (b) at 14·4 mm top deflection.

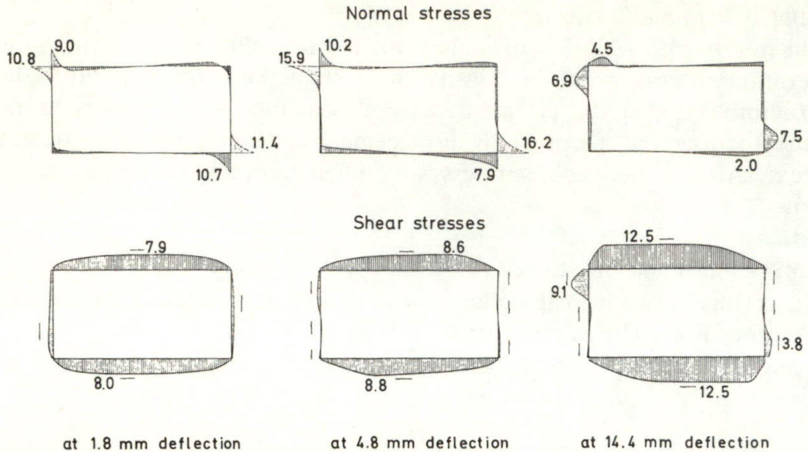


FIG. 5.10. Interface stresses at third storey of model C2III (normal stresses in $10^{-3}$ unit load/mm; shear stresses in $10^{-4}$ unit load/mm).

*5.3.2.1.1 Non-integral infilled frame.* In the elastic stage, the contact pressures are highly concentrated at the compression corners. With larger deflections, when the panel corners have yielded, the contact pressure evens out and the contact length propagates further away from the corners. This phenomenon has a very large effect on the frame moment as the propagation of the contact pressures further away from the corners leads to a large increase in the lever arm.

Due to the restriction of slip at the corners, the friction developed at close proximity to the corners is very small as compared to the limiting friction. Further away from the corners, where the normal stresses diminish to much smaller values, the friction developed is governed by the limiting friction. Hence, the net resultant of the friction developed appears to be quite small.

*5.3.2.1.2 Integral infilled frame.* In the elastic stage, concentration of normal stresses occurs at all four corners, while the shear distribution is fairly uniform. As cracks develop and propagate, the interface stresses at the tensile corners are reduced, while those near the compressive corners are significantly increased. With larger deflections, when the panel corners have yielded, the normal pressures at the compressive corners even out and propagate further away from the corners; and the interface shear capacity is mobilised at the structural interface until the shear connection also fails.

*5.3.2.2 Panel Stresses*

Panel stresses of both models are shown in Fig. 5.11. Since the panel stresses are similar in different storeys, only those of the third storey are shown. It can be seen from the experimental results that the panel stresses were approximately uniaxial—confirming the assumption made in the analysis.

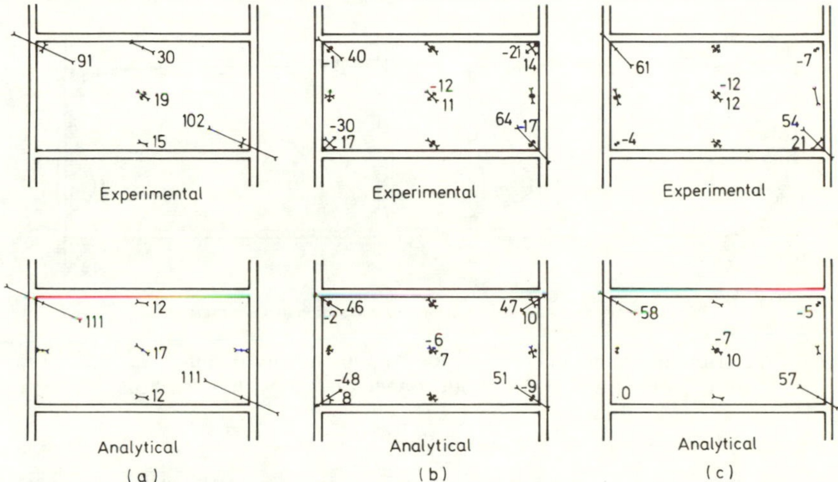


FIG. 5.11. Principal stresses (in $10^{-5}$ load/mm$^2$, positive values are compressive) in third storey panel: (a) without connectors, top deflection 4·8 mm before cracking; (b) with connectors, top deflection 1·8 mm before cracking; (c) with connectors, top deflection 4·8 mm after cracking.

*5.3.2.2.1 Non-integral infilled frame.* When the connectors are not provided, the stress concentrations at the compressive corners are very high. These high stress concentrations led to early crushing of the compressive corners, resulting in low strength of the structure. The high stress concentrations near the corners are the consequence of the small contact areas. As the tensile corners are subjected to very small stresses, only the diagonal region is really effective in the bracing action.

*5.3.2.2.2 Integral infilled frame.* Before cracking appears in the panel, stress concentrations occur at all four corners, showing that the infilled

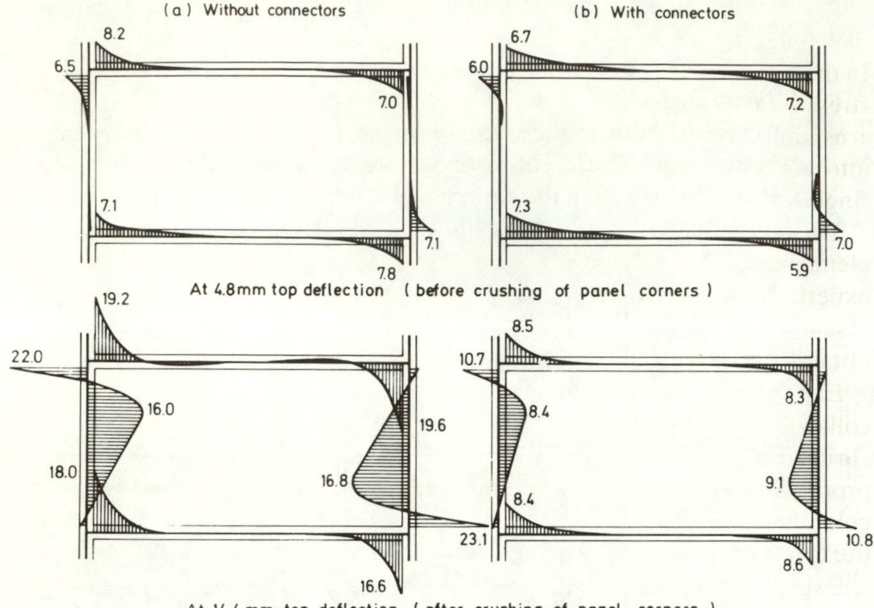


FIG. 5.12. Bending moments (in unit load—mm shown on compression side) in third storey frame: (a) without connectors; (b) with connectors.

panel acts as two diagonal struts, one in tension and the other in compression. As cracks develop, the tensile stresses at the tensile corners are reduced and each panel acts as a single diagonal strut, resulting in slightly higher stress concentrations at the compressive corners. However, part of the shearing load is transmitted from the frame to the panels through the connectors, and the stress concentrations become less serious than that in the model without connectors.

### *5.3.2.3 Frame Moments*

The theoretical bending moments in the frame at different stages of loading are shown in Fig. 5.12. Before crushing of the corners of the panels, frame moments in both models are negligible. At higher load, due to the propagation of the contact areas further away from the corners, the frame moments increase significantly until yielding occurs in the members. Frame moments in the model without connectors are much greater than in that with connectors.

## 5.4 PLASTIC THEORY

In the following, a plastic theory of infilled frames is developed in which the stress redistribution, due to the development of cracks together with the crushing of the infill toward collapse, and the interface conditions are taken into account. The theory is applicable to single-storey and multistorey single bay infilled frames.

The plastic theory is developed with the backing of nonlinear finite element analysis as briefly discussed in the previous section. Numerous experimental results conducted by the author's group and other researchers are used to verify the theory. Good agreements are generally obtained in both the structural strengths and the modes of failure.

Experiments showed that both models A-2 and A-3 failed with the same collapse mechanism. The development of the collapse mechanism at various stages as observed in the tests is shown in Fig. 5.6. No cracks were produced in both models except one or two vertical cracks near the leeward column at large deflection, because the panels were in contact with both the top and bottom beams near the leeward column.

Experiments showed that the C-models failed with the same collapse mechanism. The development of the collapse mechanism at various stages as observed in the tests is also shown in Fig. 5.6.

In developing the plastic theory for infilled frames with different interface conditions, the most important aspect is to recognise the characteristics of the interaction between the infills and the frame due to the interface conditions. The theory for integral infilled frames (C-models) is first developed, in the following, taking into account the shear strength of the connectors. Then, the theory for non-integral infilled frames (A-models) is derived directly from the C-models by setting the shear strength of the connectors to zero.

### 5.4.1 Integral Infilled Frames

#### *5.4.1.1 Single Storey*

The behaviour of integral infilled frames ranging from elastic to collapse modes has been studied both experimentally and using nonlinear finite element analysis. It has been shown that the cracks develop approximately at 45° to the beams (Fig. 5.11), and four possible collapse modes have been identified as shown in Fig. 5.13. In the light of the results obtained from nonlinear finite element analysis and by considering the structure at collapse, whereby the infill is replaced with the simplified interface stresses

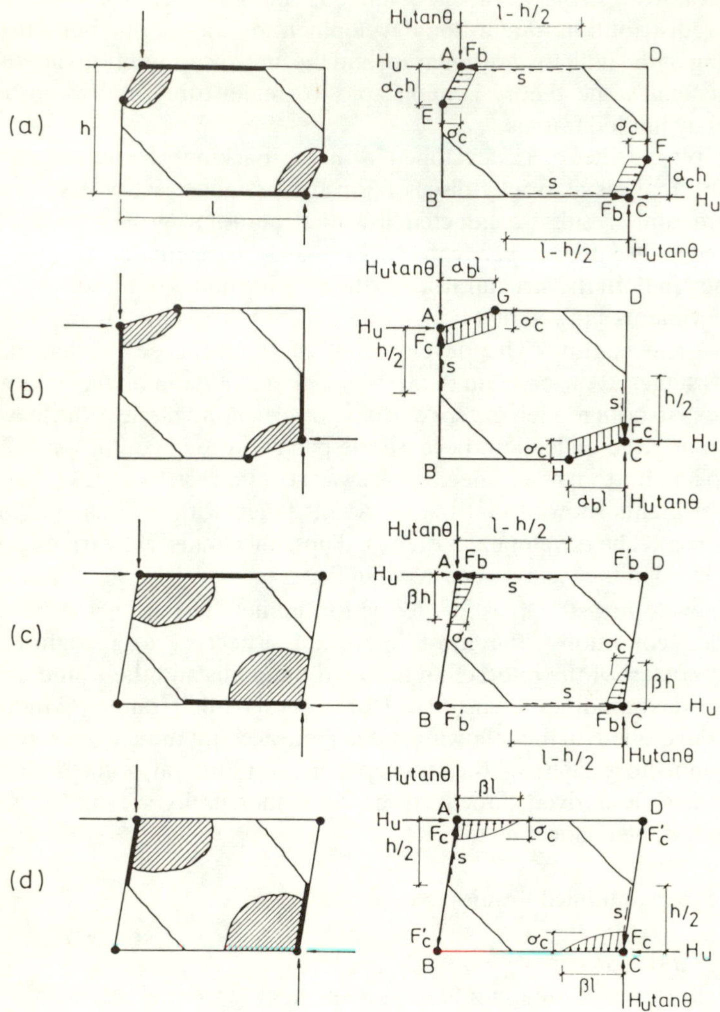


FIG. 5.13. Failure modes of single-storey integral infilled frames. Shaded areas and thick lines show, respectively, crushing regions and yielded connectors; only one pair of contact pressures is shown, the other pair being omitted for clarity.

shown in Fig. 5.13, the collapse shear strengths of all modes can be derived using the virtual work approach as follows.

*5.4.1.1.1 Mode 1.* In collapse mode 1, Fig. 5.13(a), the compressive corner regions of the panels are crushed, plastic hinges form at the loaded corners and at the columns, and the infill–beam shear connections have yielded. Assuming perfect plasticity of the crushing region, the crushing pressures developed between the columns and the panel are taken at the crushing strength of the panel material. Also, assuming perfect plasticity of the interface connection, the shear stresses developed at the interfaces are taken as the shear strength of the connection.

Let the relative horizontal displacement between corners A and C be $\delta$. The work equation is

$$H_u \delta = \frac{(M_{pj} + M_{pc})}{\alpha_c h} \cdot \delta + \frac{1}{2} \sigma_c t \alpha_c h \delta + s\left(l - \frac{h}{2}\right) \cdot \delta \tag{5.4}$$

Eliminating $\delta$ from eqn (5.4) and minimising for $H_u$ yield

$$\alpha_c = \sqrt{\frac{2(M_{pj} + M_{pc})}{\sigma_c t h^2}} \tag{5.5}$$

$$H_u = \sqrt{\frac{2(M_{pj} + M_{pc})}{\sigma_c t h^2}} \sigma_c t h + s\left(l - \frac{h}{2}\right) \tag{5.6}$$

*5.4.1.1.2 Mode 2.* In mode 2, Fig. 5.13(b), the compressive corner regions of the panels are crushed, plastic hinges form at the loaded corners and beams, and the infill–column shear connections have yielded. The same assumptions and similar derivations to those in mode 1 yield

$$\alpha_b = \tan\theta \sqrt{\frac{2(M_{pj} + M_{pb})}{\sigma_c t h^2}} \tag{5.7}$$

$$H_u = \frac{1}{\tan\theta}\left[\sqrt{\frac{2(M_{pj} + M_{pb})}{\sigma_c t h^2}} \sigma_c t h + s\frac{h}{2}\right] \tag{5.8}$$

*5.4.1.1.3 Mode 3.* In mode 3, Fig. 5.13(c), all the plastic hinges are formed at the joints and the infill–beam shear connections have yielded. It has been shown in the previous section that, due to the formation of cracks, the tensile stresses developed at the infill/frame interface are negligible. The contact pressure distribution along the length of compressed interface is assumed parabolic.

Let the relative horizontal displacement between corners A and C be δ. The work equation is

$$H_u\delta = \frac{4M_{pj}}{h}\delta + \sigma_c th\left(\frac{2}{3}\beta - \frac{1}{2}\beta^2\right)\delta + s\left(l - \frac{h}{2}\right)\delta \tag{5.9}$$

Eliminating δ from eqn (5.9) gives

$$H_u = \frac{4M_{pj}}{h} + \sigma_c th\left(\frac{2}{3}\beta - \frac{1}{2}\beta^2\right) + s\left(l - \frac{h}{2}\right) \tag{5.10}$$

With improved bonding provided by the connectors, the contact length of the compressed interface represented by $\beta h$ is at least as long as that in the case of non-integral infilled frames. The overall effect of the variation of $\beta$ on the strength prediction is small, and it is on the safe side to take $\beta$ as 1/3 (see Section 5.4.2.1). Therefore, eqn (5.10) becomes

$$H_u = \frac{4M_{pj}}{h} + \frac{1}{6}\sigma_c th + s\left(l - \frac{h}{2}\right) \tag{5.11}$$

*5.4.1.1.4 Mode 4.* Collapse mode 4, Fig. 5.13(d), is identical to mode 3 if it is turned through a right angle. Therefore, by the same assumptions and similar derivations as that in collapse mode 3, the following equation can be obtained:

$$H_u = \frac{4M_{pj}}{h} + \frac{\sigma_c th}{6\tan^2\theta} + \frac{sh}{2\tan\theta} \tag{5.12}$$

*5.4.1.1.5 Collapse shear.* The actual collapse shear is the minimum of the values derived from all the kinematically possible failure modes. In terms of the relative strength parameters, the collapse shears for various cases are summarised in Table 5.1. The sequence of collapse modes is illustrated in Fig. 5.14 for the case in which the cross-sections of the beams and columns are the same.

### *5.4.1.2 Multistorey (C-models)*

The theory is developed so that multistorey infilled frames can be designed storey by storey. The terms storey shear and collapse shear are defined as

TABLE 5.1
COLLAPSE SHEAR OF SINGLE-STOREY INTEGRAL INFILLED FRAMES (ALL MODES)

| *Collapse shear* | *Beam stronger than column* | *Column stronger than beam* | *Eqn no.* |
|---|---|---|---|
| $H_u = \min$ | $\min\begin{cases} m_c\sigma_c th + s\left(l - \dfrac{h}{2}\right) \\ \dfrac{m_{bc}}{\tan\theta}\sigma_c th + \dfrac{sh}{2\tan\theta} \\ m_c^2\sigma_c th + \dfrac{1}{6}\sigma_c th + s\left(l - \dfrac{h}{2}\right) \\ m_c^2\sigma_c th + \dfrac{\sigma_c th}{6\tan^2\theta} + \dfrac{sh}{2\tan\theta} \end{cases}$ | $\min\begin{cases} m_{bc}\sigma_c th + s\left(l - \dfrac{h}{2}\right) \\ \dfrac{m_b}{\tan\theta}\sigma_c th + \dfrac{sh}{2\tan\theta} \\ m_b^2\sigma_c th + \dfrac{1}{6}\sigma_c th + s\left(l - \dfrac{h}{2}\right) \\ m_b^2\sigma_c th + \dfrac{\sigma_c th}{6\tan^2\theta} + \dfrac{sh}{2\tan\theta} \end{cases}$ | (5.13) |

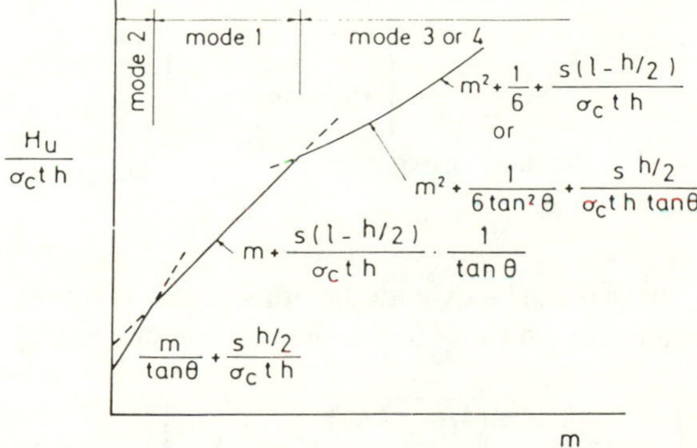


FIG. 5.14. Collapse shear of single-storey integral infilled frames with same section for beams and columns.

follows: the storey shear of a storey is the total lateral shear force acting at and above the storey referred to, whereas the collapse shear of a storey is the property of the storey such that if it is not exceeded by the corresponding storey shear, the structure is safe against collapse.

The collapse modes of multistorey infilled frames are basically the same as those of single-storey infilled frames. However, many different combinations of failure modes are possible. The evaluation of ultimate load and the design of multistorey infilled frames against collapse are inevitably complex and tedious. It is, therefore, proposed that the design of multistorey infilled frames should be carried out storey by storey on the basis of simple design rules which are proposed as follows. If the axial force in the lowest columns is large, its effect on the plastic moment capacity of the cross-section should be taken into account.

*5.4.1.2.1 Mode 1.* The analysis of mode 1 is identical to that of single-storey infilled frames. Thus, the collapse shear $H_u$ for the top storey is the same as that given by eqn (5.6), and for storeys other than the top one, where $M_{pj} = M_{pc}$, it is given by

$$H_u = m_c \sigma_c th + s\left(l - \frac{h}{2}\right) \tag{5.14}$$

*5.4.1.2.2 Mode 2.* A typical mode 2 collapse is shown in Fig. 5.15. By the principle of virtual work, Liauw and Kwan (1983b) have shown that the load factor $\lambda \geq 1$ if the following conditions are satisfied:

$$S_i \leq \frac{1}{\tan \theta_i}\left[m_b \sigma_c th + s\frac{h}{2}\right]_i \tag{5.15}$$

for all storeys.

The above analysis can be extended readily to general cases. Therefore, it can be concluded that if the collapse shear in any storey as given by

$$H_u = \begin{cases} \dfrac{1}{\tan\theta}\left[\sqrt{\dfrac{2(M_{pj}+M_{pb})}{\sigma_c th^2}}\,\sigma_c th + s\dfrac{h}{2}\right] & \text{for the top storey} \\[2ex] \dfrac{1}{\tan\theta}\left[m_b \sigma_c th + s\dfrac{h}{2}\right] & \text{for lower storeys} \end{cases} \tag{5.16}$$

is not exceeded by the storey shear, then the structure is safe against mode 2 collapse.

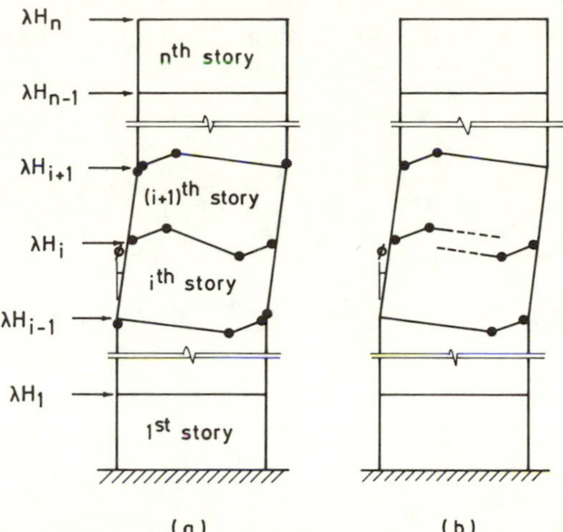


FIG. 5.15. Corner crushing with failure in beams: (a) actual collapse mechanism; (b) approximation for evaluation of internal energy absorption.

*5.4.1.2.3 Modes 3 and 4.* Using the principle of virtual work and following the same procedure as the preceding section, the collapse shear of any storey due to diagonal crushing can be given as shown in Table 5.2. It can be concluded that if, in any storey, the collapse shear as given by eqn (5.17) in the table is not exceeded by the storey shear, then the structure is safe against diagonal crushing.

*5.4.1.2.4 Experimental comparison.* The experimental results of four-storey models are given in Table 5.3, and compared with the theoretical predictions in which account has been taken of the effect of axial force on the plastic moment capacity of the cross-section. The discrepancies between the theoretical and measured strengths are within 11%, with an average of 8%, and the prediction of mode 1 failure is consistent with experimental observations.

*5.4.1.2.5 Simplified design rules.* A structure should be designed against all kinematically possible collapse modes. However, there can be many such modes and it is tedious to evaluate the ultimate load or to check the design against all possible collapse modes. Therefore, simple design rules are preferred. Grouping the results for all collapse modes together, the collapse shear in lower storeys, eqn (5.18), is given in Table 5.4. The

TABLE 5.2
COLLAPSE SHEAR OF MULTISTOREY INTEGRAL INFILLED FRAMES (MODES 3 AND 4)

| Collapse shear | Eqn no. |
|---|---|
| $$H_u = \min \begin{cases} \dfrac{4M_{pc}}{h} + \dfrac{1}{6}\sigma_c th + s\left(l - \dfrac{h}{2}\right) \\ \dfrac{4M_{pc}}{h} + \dfrac{\sigma_c th}{6\tan^2\theta} + \dfrac{sh}{2\tan\theta} \\ \dfrac{2M_{pb}}{h} + \dfrac{1}{6}\sigma_c th + s\left(l - \dfrac{h}{2}\right) \\ \dfrac{2M_{pb}}{h} + \dfrac{\sigma_c th}{6\tan^2\theta} + \dfrac{sh}{2\tan\theta} \end{cases}$$ | (5.17) |

collapse shear in the top storey can be obtained similarly. Each storey of the infilled frame should be designed so that the strength, as calculated using Table 5.4, is greater than the storey shear in each storey of the infilled frames.

### 5.4.2 Non-integral Infilled Frames

#### *5.4.2.1 Single-storey (A-models)*

The failure modes in single-storey non-integral infilled frames depend on panel proportions and relative strengths of the columns, beams and infills. Based on nonlinear finite element analysis and the laboratory observations, three collapse modes have been identified as shown in Fig. 5.16. The actual collapse mode is the one which gives the smallest collapse shear strength. The stress redistribution and the plastic hinges forming the collapse mechanism for the three modes are also shown in Fig. 5.16.

These three modes of failure correspond to the respective modes of failure for the fully integral infilled frames (C-models), except that there are no shear connectors in the case of A-models and the shear strength of connectors does not exist. Therefore, all the previously derived equations

TABLE 5.3

COMPARISON OF STRENGTH; FOUR-STOREY MODELS C

| | | Theoretical strength (kN) | | | | | | | Discrepancy (%) | | |
|---|---|---|---|---|---|---|---|---|---|---|---|
| | | Plastic theory | | | | | | | | | |
| *Specimen* (1) | *l/h* (2) | *Mode 1* | *Mode 2* | *Mode 3* | *Mode 4* | *Minimum* (3) | *Finite element analysis* (4) | *Measured strength (kN)* (5) | $\frac{(3)-(4)}{(4)}$ (6) | $\frac{(3)-(5)}{(5)}$ (7) | $\frac{(4)-(5)}{(5)}$ (8) |
| C–2–I | 2·0 | 67·9 | 126·2 | 86·6 | 183·9 | 67·9 | 80·1 | 76·2 | −15·2 | −10·9 | +5·1 |
| C–2–II | 2·0 | 67·9 | 126·2 | 86·6 | 183·9 | 67·9 | 76·3 | 72·5 | −11·0 | −6·3 | +5·2 |
| C–2–III | 2·0 | 67·9 | 126·2 | 86·6 | 183·9 | 67·9 | 77·0 | 76·1 | −11·8 | −10·8 | +1·2 |
| C–3–I | 3·0 | 99·0 | 188·3 | 115·6 | 385·1 | 99·0 | 117·5 | 109·2 | −15·7 | −9·3 | +7·6 |
| C–3–II | 3·0 | 99·0 | 188·3 | 115·6 | 385·1 | 99·0 | 111·9 | 106·6 | −11·5 | −7·1 | +5·0 |
| C–3–III | 3·0 | 99·0 | 188·3 | 115·6 | 385·1 | 99·0 | 107·3 | 103·2 | −7·7 | −4·1 | +4·0 |
| | | | | | | | | Average = | −12·2 | −8·1 | +4·7 |

TABLE 5.4
COLLAPSE SHEAR IN LOWER STOREYS OF INTEGRAL INFILLED FRAMES (ALL MODES)

| *Collapse shear* | *Eqn no.* |
|---|---|
| $$H_u/\sigma_c th = \min \begin{cases} m_c + \dfrac{s\left(l - \dfrac{h}{2}\right)}{\sigma_c th} \\ \dfrac{m_b}{\tan\theta} + \dfrac{s}{2\sigma_c t \cdot \tan\theta} \\ m_c^2 + \dfrac{1}{6} + \dfrac{s\left(l - \dfrac{h}{2}\right)}{\sigma_c th} \\ m_c^2 + \dfrac{1}{6\tan^2\theta} + \dfrac{s}{2\sigma_c t \cdot \tan\theta} \\ \dfrac{1}{2}m_b^2 + \dfrac{1}{6} + \dfrac{s\left(l - \dfrac{h}{2}\right)}{\sigma_c th} \\ \dfrac{1}{2}m_b^2 + \dfrac{1}{6\tan^2\theta} + \dfrac{s}{2\sigma_c t \cdot \tan\theta} \end{cases}$$ | (5.18) |

for C-models can be used for A-models if the shear strength, $s$, is set to zero. Setting the shear terms to zeros in eqn (5.13), Table 5.1, the collapse shear $H_u$ for A-models can be given by eqn (5.19), as shown in Table 5.5. Subsequently, the collapse shear can be expressed in terms of the strength parameters, $m_c$, $m_b$ and $m_{bc}$, similar to that shown in Table 5.1 but with the shear terms set to zero.

In the model tests of Barua and Mallick (1977), all specimens had span greater than height and beams stronger than columns. In the model tests of Mallick and Severn (1967), all the specimens had span greater than height

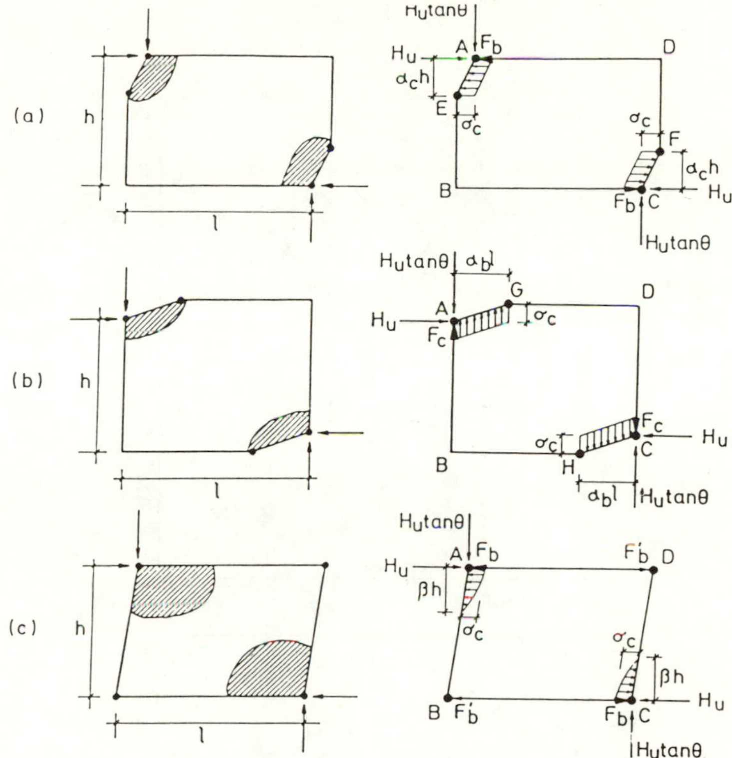


FIG. 5.16. Failure modes of single-storey non-integral infilled frames. Shaded areas show crushing regions; only one pair of contact pressures is shown, the other pair being omitted for clarity.

and beam strength equal to column strength. In the tests of Mainstone (1971), the specimens had span equal to height and beams equal to columns. Thus, for all these specimens

$$\frac{H_u}{\sigma_c t h} = \min \begin{cases} m_c \\ m_c^2 + 1/6 \end{cases} \qquad (5.20)$$

Their experimental results are compared with the theoretical predictions in Fig. 5.17(a), (b) and (c). The values of $\beta$ range from about 1/3 to 1/2, and it is accurate enough to take $\beta$ as 1/3, which is on the conservative side.

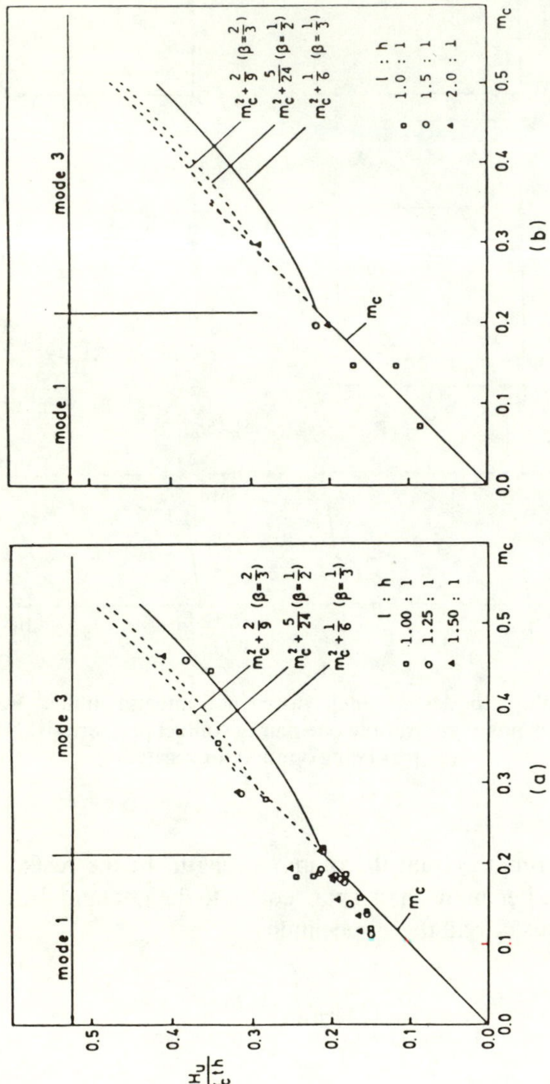

mode 1
mode 3
$m_c$
$m_c^2 + \frac{2}{9}$ $(\beta = \frac{2}{3})$
$m_c^2 + \frac{5}{24}$ $(\beta = \frac{1}{2})$
$m_c^2 + \frac{1}{6}$ $(\beta = \frac{1}{3})$
l : h
1.00 : 1
1.25 : 1
1.50 : 1
1.0 : 1
1.5 : 1
2.0 : 1
$\frac{H_u}{\sigma_c t h}$
(a)
(b)

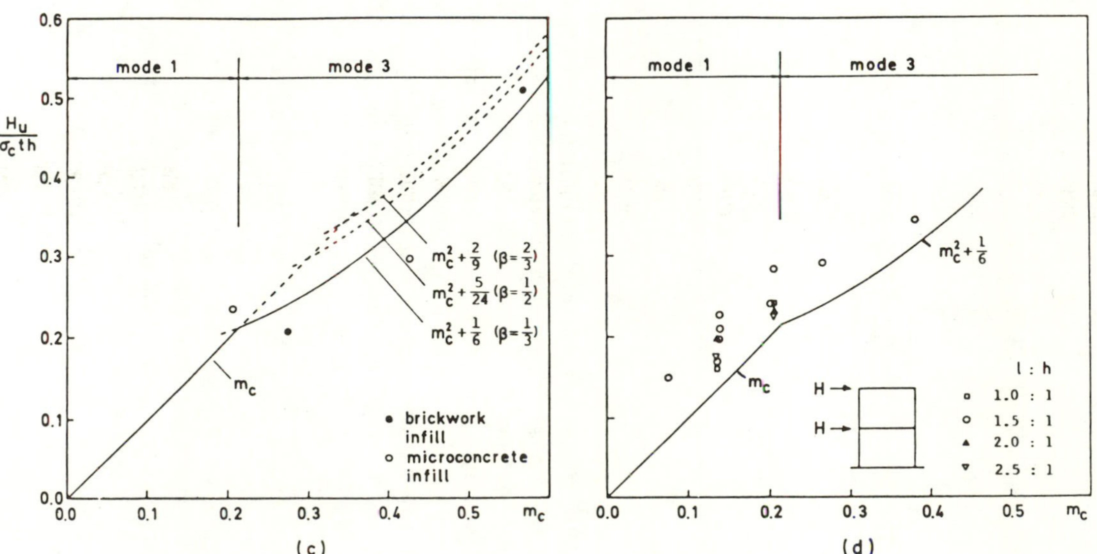


FIG. 5.17. Comparisons of strength predictions: (a) with Barua and Mallick's (1977) experimental results (single-storey models): (b) with Mallick and Severn's (1967) experimental results (single-storey models): (c) with Mainstone's experimental results (single-storey square models); and (d) with Stafford Smith's (1967) experimental results (two-storey models).

TABLE 5.5
COLLAPSE SHEAR OF SINGLE-STOREY NON-INTEGRAL INFILLED FRAMES (ALL MODES)

| Collapse shear | Eqn no. |
|---|---|
| $$H_u/\sigma_c th = \min \begin{cases} \sqrt{\dfrac{2(M_{pj}+M_{pc})}{\sigma_c th^2}} \\ \dfrac{1}{\tan\theta}\sqrt{\dfrac{2(M_{pj}+M_{pb})}{\sigma_c th^2}} \\ \dfrac{4M_{pj}}{\sigma_c th^2}+\dfrac{1}{6} \end{cases}$$ | (5.19) |

*5.4.2.2 Multistorey (A-models)*
The collapse shear of multistorey non-integral infilled frames is obtained directly from that of integral infilled frames by setting the shear strength, $s$, to zero and noting that tan $\theta<1$ when $l>$h. The corresponding eqns (5.21)–(5.23) for the three modes of failure are given in Table 5.6. If the collapse shear in any storey, as given above, is not exceeded by the storey shear, then the structure is safe against the diagonal crushing mode. If the axial force in the lowest column is large, its effect on the plastic moment capacity of the cross-section should be taken into account.

*5.4.2.2.1 Experimental comparison.* The experimental results of four-storey models A are given in Table 5.7 and compared with the theoretical predictions in which the effect of axial force on plastic moment capacity of the cross-section has been taken into account. The discrepancies between the measured strengths and the analytical values are reasonable.

In the model tests of Stafford Smith (1967), the specimens of two-storey models were subjected to equal lateral load, Fig. 5.17(d). All specimens, except one, had span greater than height and beams stronger than columns. The results for all failure modes can be condensed to

$$\frac{2H}{\sigma_c th} = \min \begin{cases} m_c \\ m_c^2 + 1/6 \end{cases} \tag{5.24}$$

TABLE 5.6
COLLAPSE SHEAR OF MULTISTOREY NON-INTEGRAL INFILLED FRAMES

| $H_u/\sigma_c th =$ | | |
|---|---|---|
| *Mode 1, corner crushing with column failure (1)* | *Mode 2, corner crushing with beam failure (2)* | *Mode 3, diagonal crushing (3)* |
| $\begin{cases} \sqrt{\dfrac{2(M_{pj}+M_{pc})}{\sigma_c th^2}} & \text{for top storey} \\ m_c & \text{for lower storeys} \end{cases}$ | $\begin{cases} \dfrac{1}{\tan\theta}\sqrt{\dfrac{2(M_{pj}+M_{pb})}{\sigma_c th^2}} & \text{for top storey} \\ \dfrac{1}{\tan\theta} m_b & \text{for lower storeys} \end{cases}$ | $\min \begin{cases} m_c^2 + \dfrac{1}{6} \\ \dfrac{1}{2} m_b^2 + \dfrac{1}{6} \end{cases}$ |
| Eqn (5.21) | Eqn (5.22) | Eqn (5.23) |

The theoretical prediction based on the above equation is compared with Stafford Smith's experimental results in Fig. 5.17(d), and good agreement is obtained.

*5.4.2.2.2 Simplified design rules.* From eqn (5.18) (Table 5.4), by eliminating the shear terms and noting that tan $\theta<1$ when $l>h$, the collapse shear in the lower storeys can be given by eqn (5.25), as shown in Table 5.8. The collapse shear in the top storey can be obtained similarly. The design rule is then directed to provide adequate collapse shear as calculated from Table 5.8, to be greater than the storey shear in each storey.

### 5.4.3 Design Examples with Design Charts

The collapse modes of the infilled frames have been previously identified in Figs 5.13 and 5.16. Of these, only two (modes 1 and 3) are basic collapse modes, with the other two (modes 2 and 4) being complementary collapse modes in the sense that they actually represent mirror images of the basic collapse modes. The actual collapse mode is the one which leads to the smallest value of collapse shear.

TABLE 5.7

COMPARISON OF STRENGTH, FOUR-STOREY MODELS A

| Specimen (1) | $l/h$ (2) | Theoretical strength (kN) | | | | | Measured strength (kN) (5) | Discrepancy (%) | | |
|---|---|---|---|---|---|---|---|---|---|---|
| | | Plastic theory | | | | Finite element analysis (4) | | $\frac{(3)-(4)}{(4)}$ (6) | $\frac{(3)-(5)}{(5)}$ (7) | $\frac{(4)-(5)}{(5)}$ (8) |
| | | Mode 1 | Mode 2 | Mode 3 | Minimum (3) | | | | | |
| A2 | 2·0 | 42·2 | 97·8 | 43·9 | 42·2 ($\mu = 0$) | 41·5 ($\mu = 0$) | – | +1·7 | | |
| | | | | | – | 45·4 ($\mu = 0{\cdot}4$) | 48·1 ($\mu = 0{\cdot}42$) | | −12·3 | −5·6 |
| A3 | 3·0 | 46·2 | 145·8 | 44·1 | 44·1 ($\mu = 0$) | 41·9 ($\mu = 0$) | – | +5·3 | | |
| | | | | | – | 46·2 ($\mu = 0{\cdot}4$) | 48·3 ($\mu = 0{\cdot}42$) | | −8·7 | −4·3 |
| | | | | | | | Average = | +3·5 | −10·5 | −5·0 |

TABLE 5.8
COLLAPSE SHEAR IN LOWER STOREYS OF NON-INTEGRAL INFILLED FRAMES (ALL MODES)

| *Collapse shear* | *Eqn no.* |
|---|---|
| $H_u/\sigma_c th = \min \begin{cases} m_c \\ \dfrac{m_b}{\tan\theta} \\ m_c^2 + \dfrac{1}{6} \\ \dfrac{1}{2}m_b^2 + \dfrac{1}{6} \end{cases}$ | (5.25) |

Three relative strength parameters—for the columns, the beams and the frame/panel connection—are defined respectively as

$$m_c = \sqrt{4M_{pc}/\sigma_c th^2} \tag{5.26}$$

$$m_b = \sqrt{4M_{pb}/\sigma_c th^2} \tag{5.27}$$

$$q = s/\sigma_c t \tag{5.28}$$

In terms of these non-dimensional relative strength parameters, and based on the equations given in Table 5.4 for integral infilled frames and in Table 5.6 for non-integral infilled frames, the design charts in the form shown in Figs 5.18 and 5.19 can be prepared. The use of the charts is illustrated in the following examples.

### *5.4.3.1 Example 1—Integral Infilled Frames*

The steel frame with concrete infill, as shown in Fig. 5.20, is to be designed. Connectors are to be welded to the frame to provide a shear strength of 1·0 MPa, which gives a $q$-value of $1{\cdot}0/10{\cdot}0 = 0{\cdot}10$. The appropriate load factor has already been allowed for in the design load.

The required strength, in non-dimensional form, is given by

$$\frac{H_u}{\sigma_c th} = \frac{1400 \times 10^3}{10 \times 200 \times 3500} = 0{\cdot}20 \tag{5.29}$$

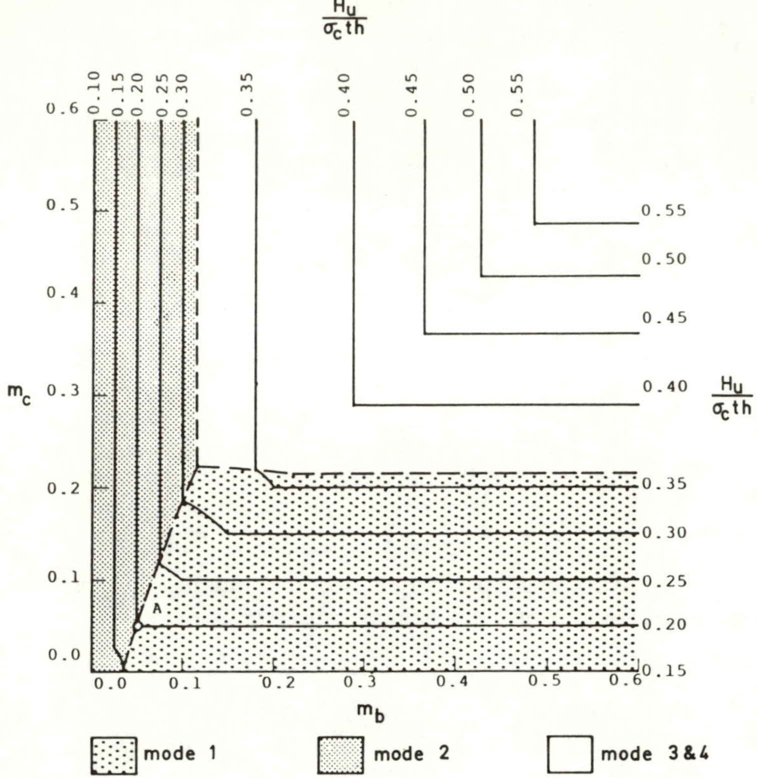


FIG. 5.18. Design chart for integral infilled frames with $l/h = 2$ and $q = 0{\cdot}1$.

From Fig. 5.18, the contour line with $H_u/\sigma_c th = 0{\cdot}20$ can be read out. Point A is chosen, giving $m_b = 0{\cdot}05$ and $m_c = 0{\cdot}05$. The required bending strength of the frame is thus given by

$$M_{pb} = 0{\cdot}05^2 \times 10 \times 200 \times 3500^2/4 = 15{\cdot}3 \text{ kN m} \tag{5.30}$$

$$M_{pc} = 0{\cdot}05^2 \times 10 \times 200 \times 3500^2/4 = 15{\cdot}3 \text{ kN m} \tag{5.31}$$

The frame shear is evaluated as follows:

$$\text{Beam shear} = H_u \tan\theta - s(h/2)$$

$$= 1400 \times \frac{1}{2} - \frac{1{\cdot}0}{1000} \times 200 \times \frac{3500}{2}$$

$$= 350 \text{ kN} \tag{5.32}$$

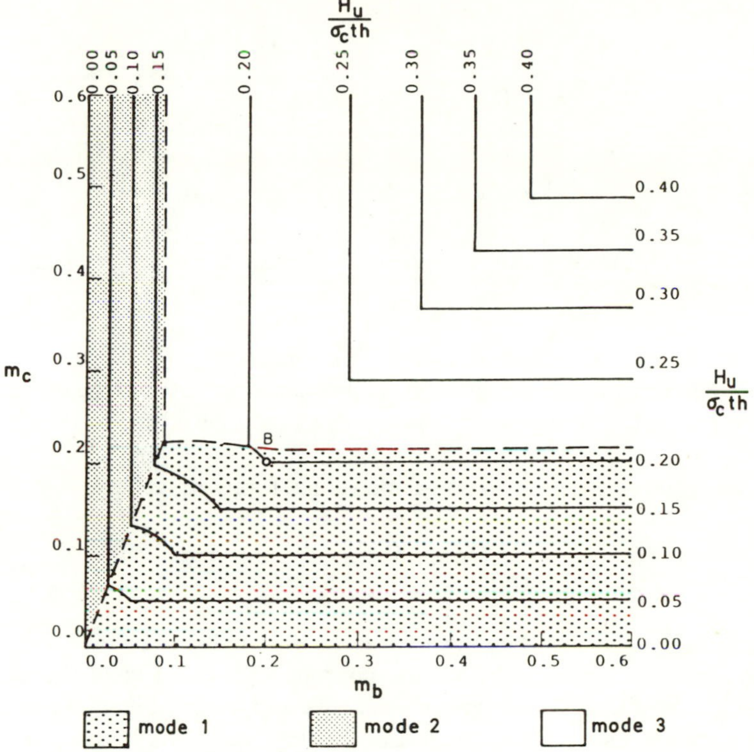


FIG. 5.19. Design chart for non-integral infilled frames ($q = 0$) with $l/h = 2$.

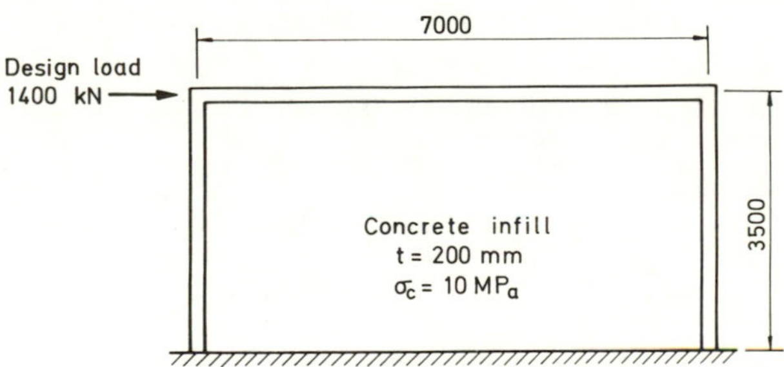


FIG. 5.20. Parameters (mm) of design examples.

Column shear $= H_u - s[l - (h/2)]$

$$= 1400 - \frac{1{\cdot}0}{1000} \times 200 \times \left(7000 - \frac{3500}{2}\right)$$

$$= 350 \text{ kN} \qquad (5.33)$$

Given the required bending and shear strengths, the frame can be designed accordingly.

*5.4.3.2 Example 2—Non-integral Infilled Frames*

The steel frame with concrete infill, as shown in Fig. 5.20, is to be designed. The appropriate load factor has already been allowed for in the design load.

The required strength, in non-dimensional form, is given by

$$\frac{H_u}{\sigma_c th} = \frac{1400 \times 10^3}{10 \times 200 \times 3500} = 0{\cdot}20 \qquad (5.34)$$

From Fig. 5.19 the contour line with $H_u/\sigma_c th = 0{\cdot}20$ can be read out. Any combination of $m_b$ and $m_c$ on the contour line would yield the design strength required. The designer can freely select the combination which, in the particular case, leads to the most economic design. In this case, point B is chosen, which gives $m_b = 0{\cdot}20$ and $m_c = 0{\cdot}20$. The required bending strength of the frame is thus given by

$$M_{pb} = 0{\cdot}20^2 \times 10 \times 200 \times 3500^2/4(10^6) = 245 \text{ kN m} \qquad (5.35)$$

$$M_{pc} = 0{\cdot}20^2 \times 10 \times 200 \times 3500^2/4(10^6) = 245 \text{ kN m} \qquad (5.36)$$

In the corner crushing mode, the joints at the tensile corners are subjected to no shear. Hence

Beam shear $= H_u \tan\theta = 1400 \times 1/2 = 700$ kN (5.37)

Column shear $= H_u = 1400$ kN (5.38)

Given the required bending shear strengths in the beam and the columns, the frame can then be designed accordingly. It is obvious that the frame will have much larger sections than that required in Example 1, in which shear connectors are used. However, the frame in Example 2 will be reasonably modest when it is compared to the bare frame without any infill, which will require a section having 1225 kN m plastic moment to withstand the same horizontal load.

## 5.5 CONCLUSIONS

A general understanding of the behaviour, strength and failure mode of a laterally loaded single-bay steel frame with concrete infill, with or without shear connectors at the infill/frame interface, has been achieved through many years of experimental and analytical studies. A nonlinear finite element analysis has yielded important results beyond the elastic stage, particularly with regard to the different stress distributions at the interfaces of non-integral and fully integral infilled frames.

In the light of these results, a plastic theory with the upper bound approach has been proposed for the predictions of strength and failure mode of an infilled frame with or without shear connectors. For a non-integral infilled frame, the collapse shear is governed by the strengths of the frame and the infill. For a fully integral infilled frame, the collapse shear is governed by the strengths of the infilled frame and the interface connection. Three major collapse modes have been identified: (1) corner crushing with failure in columns and infill/beam connections; (2) corner crushing with failure in beams and infill/column connections; and (3) diagonal crushing with two variations for a fully integral infilled frame: (a) failure in infill/beam connections; and (b) failure in infill/column connections.

Comparison of the analytical predictions with experimental results shows generally good agreement in terms of the ultimate strengths and the failure modes of the structures. Simplified design rules for multistorey infilled frames are recommended in the manner that a structure be designed storey by storey so that the collapse shear of each storey of the structure is greater than the storey shear force including the appropriate load factor.

## REFERENCES

BARUA, H. K. and MALLICK, S. K. (1977) Behaviour of mortar infilled steel frames under lateral load, *Building and Environment*, **12**, 263–72.

BENJAMIN, J. R. and WILLIAMS, H. A. (1957) The behaviour of one-storey reinforced concrete shear walls, *Proc. ASCE*, **83**(ST5).

BERTERO, V. and BROKKEN, S. (1983) Infills in seismic resistant building, *Proc. ASCE*, **109**(ST6), 1337–61.

COULL, A. (1968) Discussion on 'The behaviour of infilled frames under static loading', *Proc. Instn Civ. Engrs*, **41**, Sept., 205–22.

DAWE, J. L. and MCBRIDE, R. T. (1985) Experimental investigation of the shear resistance of masonry panels in steel frames, *Proc. 7th Internat. Brick Masonry Conf.*, Melbourne, Feb., 791–801.

DAWE, J. L. and YOUNG, T. C. (1985) An investigation of factors influencing the behaviour of masonry infill in steel frames subjected to in-plane shear, *Proc. 7th Internat. Brick Masonry Conf.*, Melbourne, Feb., pp. 803–14.

DHANASEKAR, M., PAGE, A. W. and KLEEMAN, P. W. (1985) The behaviour of brick masonry under biaxial stress with particular reference to infilled frames, *Proc. 7th Internat. Brick Masonry Conf.*, Melbourne, Feb., 815–24.

HOLMES, M. (1961) Steel frames with brickwork and concrete infilling, *Proc. Instn Civ. Engrs*, **19**, Aug., 473–8.

HOLMES, M. (1963) Combined loading on infilled frames, *Proc. Instn Civ. Engrs*, **25**, May, 31–8.

KADIR, M. R. and HENDRY, A. W. (1975) The behaviour of brickwork infilled frames under racking load, *Proc. 5th Internat. Symp. on Load Bearing Brickwork*, British Ceramics Research Association, London.

KAHN, L. F. and HANSON, R. D. (1979) Infilled walls for earthquake strengthening, *Proc. ASCE*, **105**(ST2), 283–96.

KING, G. J. W. and PANDEY, P. C. (1978) The analysis of infilled frames using finite elements, *Proc. Instn. Civ. Engrs*, Parts 2, **65**, Dec. 749–60.

KLINGER, R. E. and BERTERO, V. V. (1978) Earthquake resistance of infilled frames, *Proc. ASCE*, **104**(ST6), 973–89.

KWAN, K. H. and LIAUW, T. C. (1984) Nonlinear analysis of integral infilled frames, *Engg Structures*, **6**, July, 223–31.

LIAUW, T. C. (1970) Elastic behaviour of infilled frames, *Proc. Instn Civ. Engrs*, **46**, July, 343–49.

LIAUW, T. C. (1972) An approximate method of analysis for infilled frames with or without opening, *Building Science*, **7**, 233–8.

LIAUW, T. C. (1973) The composite characteristics of infilled frames, *Internat. J. Mech. Sciences*, **15**(7), 517–33.

LIAUW, T. C. (1979) Tests on multistorey infilled frames subject to dynamic lateral loading, *ACI J.*, **76**(4), April, 551–64.

LIAUW, T. C and KWAN, K. H. (1982) Non-linear analysis of multistorey infilled frames, *Proc. Instn Civ. Engrs*, Part 2, **73**, June, 441–54.

LIAUW, T. C. and KWAN, K. H. (1983a) Plastic theory of non-integral infilled frames, *Proc. Instn Civ. Engrs*, Part 2, **75**, Sept., 379–96.

LIAUW, T. C. and KWAN, K. H. (1983b) Plastic theory of infilled frames with finite interface shear strength, *Proc. Instn Civ. Engrs*, Part 2, **75**, Dec., 707–23.

LIAUW, T. C. and KWAN, K. H. (1984a) Nonlinear analysis of non-integral infilled frames, *Computers & Structures*, **18**(3), 551–60.

LIAUW, T. C. and KWAN, K. H. (1984b) Plastic design of infilled frames, *Proc. Instn Civ. Engrs*, Part 2, **77**, Sept., 367–77.

LIAUW, T. C. and KWAN, K. H. (1985a) Static and cycle behaviours of multistorey infilled frames with different interface conditions, *J. Sound & Vibration*, **99**(2), 275–83.

LIAUW, T. C. and KWAN, K. H. (1985b) Unified plastic analysis for infilled frames, *Proc. ASCE, J. Struct. Engg*, **111**(7), July, 1427–48.

LIAUW, T. C. and LEE, S. W. (1977) On the behaviour and the analysis of multistorey infilled frames subjected to lateral loading, *Proc. Instn Civ. Engrs*, **63**, Sept., 641–56.

LIAUW, T. C. and LO, C. Q. (1985) Tests on multibay and multistorey infilled frames, *Trans. Instn of Engrs, Australia, Civil Eng.*, **CE27**(2), May, 200–09.

LIAUW, T. C., TIAN, Q. L. and CHEUNG, Y. K. (1986) Dynamic response of infilled frames incorporating a sliding device, *Proc. Instn Civ. Engrs*, Part 2, **81**, March, 55–69.

LO, C. Q. and LIAUW, T. C. (1984) Study of multibay and multistorey infilled frames subjected to lateral load, *Proc. 3rd Internat. Conf. on Tall Buildings*, Hong Kong & Guangzhou, Dec., 138–43.

MAINSTONE, R. J. (1962). Discussion on steel frames with brickwork and concrete infilling, *Proc. Instn Civ. Engrs*, **23**, 94–9.

MAINSTONE, R. J. (1971) On the stiffness and strengths of infilled frames, *Proc. Instn Civ. Engrs*, Suppl. paper 7360S, 57–90.

MALLICK, D. V. and GARG, R. P. (1971) Effect of openings on the lateral stiffness of infilled frames, *Proc. Instn Civ. Engrs*, **49**, June, 193–209.

MALLICK, D. V. and SEVERN, R. T. (1967) The behaviour of infilled frames under static loading, *Proc. Instn Civ. Engrs*, **38**, Dec., 639–56.

MALLICK, D. V. and SEVERN, R. T. (1968) Dynamic characteristics of infilled frames, *Proc. Instn Civ. Engrs*, **39**, Feb., 261–88.

PALSSON, H., GOODNO, B. J., CRAIG, J. I. and WILL, K. M. (1984) Cladding influence on dynamic response of tall buildings, *Earthquake Engg & Struct. Dynamics*, **12**, 215–28.

POLYAKOV, S. V. (1950) Investigation of the strength and of the deformation characteristics of masonry filler walls and facing of framed structures (in Russian), *Construction Ind.* (5).

POLYAKOV, S. V. (1957) On interaction between masonry filler walls and enclosing frame when loaded in plane of wall, *Construction in Seismic Regions*, Moscow, translation in *Earthquake Engineering*, Earthquake Engineering Research Institute, San Francisco, 1960, 36–42.

RIDDINGTON, J. R. and STAFFORD SMITH, B. S. (1977) Analysis of infilled frames subjected to racking with design recommendations, *Struct. Engr*, **55**, June, 263–68.

SACHANSKI, S. (1960) Analysis of the earthquake resistance of frame buildings taking into consideration the carrying capacity of the filling masonry, *Proc. 2nd World Conf. on Earthquake Engg*, 2127–41.

STAFFORD SMITH, B. (1966) Behaviour of square infilled frames, *Proc. ASCE*, **92**(ST1), Feb., 381–403.

STAFFORD SMITH, B. (1967) Methods for predicting the lateral stiffness and strength of multistorey infilled frames, *Building Science*, **2**, 247–57.

STAFFORD SMITH, B. and CARTER, C. (1969) A method of analysis for infilled frames, *Proc. Instn Civ. Engrs*, **44**, Sept., 31–48.

STAFFORD SMITH, B. and RIDDINGTON, J. R. (1978) The design of masonry infilled steel frames for bracing structures, *Struct. Engr*, **56B**(1), March, 1–7.

THOMAS, F. G. (1953) The strength of brickwork, *Struct. Engr*, **31**, 36–45.

WHITNEY, C. S., ANDERSON, B. G. and COHEN, E. (1955) Design of blast resistant construction for atomic explosions, *ACIJ.*, **51**, 655–73.

WOOD, R. H. (1958) The stability of tall buildings, *Proc. Instn Civ. Engrs*, **11**, Sept., 69–102.

WOOD, R. H. (1968) Discussion on 'The behaviour of infilled frames under static loading', *Proc. Instn Civ. Engrs*, **41**, Sept., 205–22.

WOOD, R. H. (1978) Plasticity, composite action and collapse design of unreinforced shear wall panels in frames, *Proc. Instn Civ. Engrs*, Part 2, **65**, June, 381–411.

# *Chapter 6*

# STEEL–CONCRETE COMPOSITE COLUMNS—I

H. SHAKIR-KHALIL

*Department of Civil Engineering, University of Manchester, UK*

## *SUMMARY*

*A steel–concrete composite column is a compression member which may be either a concrete-encased structural steel section or a concrete-filled steel tube. The structural steel and concrete components resist the external loading by interacting together through the interface bond, or in special circumstances by the use of mechanical shear connectors. A great deal of theoretical and experimental work has been carried out on these types of column with a view to establishing their behaviour, characteristics and load-bearing capacities. The results and conclusions of these studies and investigations are summarized and presented here.*

## NOTATION

| | |
|---|---|
| $A_c, A_r, A_s$ | Cross-sectional area of concrete, reinforcement and structural steel section respectively |
| $E_c, E_r, E_s$ | Modulus of elasticity of concrete, reinforcement and structural steel section respectively |
| $E_{co}$ | Initial modulus of elasticity of concrete |
| $f_{cd}, f_{rd}, f_{sd}$ | Design strength of concrete, reinforcement and structural steel section respectively, taken as the respective characteristic strength divided by the material partial safety factor, $\gamma_m$ |
| $\bar{f}_{cd}$ | Design strength of concrete in bending compression |
| $f'_{cd}$ | Enhanced design strength of triaxially contained concrete |

| | |
|---|---|
| $f_{ck}$ | Characteristic strength of concrete, taken as $0{\cdot}67f_{cu}$ |
| $f_{cu}, f_{cy}$ | Characteristic 28-day cube and cylinder strength of concrete respectively |
| $f_{rk}, f_{sk}$ | Characteristic strength of reinforcement and structural steel section respectively |
| $f'_{sd}$ | Reduced design strength of concrete-filled circular steel tubes |
| $I_c, I_r, I_s$ | Second moment of area of uncracked concrete, reinforcement and structural steel section respectively about appropriate axis |
| $K$ | Reduction factor dependent on the slenderness factor $\bar{\lambda}$, the concrete contribution factor $\alpha$ and the ratio $\beta$ of the end moments on the column |
| $K_1$ | Reduction factor dependent on the slenderness factor $\bar{\lambda}$ and the shape of the structural steel section |
| $K_2, K_3$ | Reduction factors used, in conjunction with $K_1$, to evaluate the global $K$-factor |
| $L$ | Length of pin-ended column |
| $L_c$ | Critical length of pin-ended column, given by the column length for which the column Euler load, $N_c$, is equal to its squash load, $N_u$ |
| $M$ | Bending moment in column about appropriate axis |
| $M_u$ | Ultimate moment of resistance of composite section about appropriate axis |
| $N$ | Compression force on column |
| $N_a$ | Axial compression force to cause failure of column |
| $N_{ax}$ | Failure load of column when loaded axially and constrained to bend about its major axis |
| $N_c$ | Euler critical load of column |
| $N_u$ | Squash load of column |
| $N_x, N_y$ | Compression force on column accompanied by bending about the major and minor axis respectively |
| $N_{xy}$ | Compression force on column accompanied by biaxial bending |
| $r$ | Radius of gyration of section about minor axis |
| $\alpha$ | Concrete contribution factor, given by the proportion of the squash load carried by the concrete |
| $\beta$ | Ratio of end moments acting on column |
| $\gamma_{mc}, \gamma_{ms}$ | Material partial safety factor for concrete and structural steel respectively |
| $\lambda$ | Slenderness ratio of column, given by $L/r$ |

$\lambda_c$ Critical slenderness ratio of column given by the slenderness ratio at which the column Euler load, $N_c$, is equal to its squash load, $N_u$

$\bar{\lambda}$ Slenderness factor of column, given by $\lambda/\lambda_c$

## 6.1 INTRODUCTION

It could be argued that steel–concrete composite columns were developed by chance. It was as a means of fire protection for structural steel in buildings that, early in this century, it became widespread practice to encase steel stanchions in concrete. The concrete used was a wet mix of low strength, and it was common practice to ignore the increase in stiffness and strength of the steel column which resulted from the concrete encasement. Some of the earliest tests on composite columns were carried out by Burr (1912), and were followed over the years by more experimental and theoretical studies by other researchers whose work is referred to by Stevens (1965) and McDevitt and Viest (1972). Results of tests on composite columns by Faber (1956), Stevens (1959), Jones and Rizk (1963), Stevens (1965) and others showed that the concrete encasement increases the carrying capacity of the bare steel column, and that savings in the steelwork could be achieved by the use of better-quality concrete and also by allowing for such composite action in the column design.

By definition, columns are structural members which are subjected mainly to axial forces and very little transverse shear. It is for this reason that complete interaction between the structural components of the column, namely the steel and concrete elements, is achieved through direct interface bond. In the case of concrete-filled circular pipes, Virdi and Dowling (1980) showed that bond occurs through the interlocking of concrete in two types of imperfections of steel, namely the surface roughness of steel, and also the variation in the shape of the tube cross-section. In the case of concrete-encased composite columns it is recommended practice that a reinforcement cage should be used to contain the lateral expansion of the concrete, and also to prevent the premature spalling of the concrete encasement, especially the thin concrete cover to the flange of an encased I-section.

When the steel–concrete interface conditions comply with the relevant codes of practice, the interface bond may be assumed to be satisfactory in achieving complete interaction between the steel and concrete. However, it should be mentioned that the use of mechanical connectors may be necessary in special circumstances in which the limiting bond stress is likely to be exceeded, e.g. in the presence of significant transverse shear on the column, and also in the case of dynamic and seismic loading.

It should be noted that in the following sections, the material partial safety factor for structural steel, $\gamma_{ms}$, is taken equal to unity, and that it should be taken by the designer in accordance with the relevant codes and standards.

## 6.2 TYPES OF COMPOSITE COLUMN

A steel–concrete composite column is conventionally a compression member in which the steel element is a structural steel section. There are two main types of composite column, namely the concrete-encased and concrete-filled types, examples of which are shown in Figs 6.1 and 6.2 respectively.

In the case of concrete-encased composite columns (Fig. 6.1) the structural steel component could be either one or more rolled steel sections which may be tied together. In addition to supporting a proportion of the load acting on the column, the concrete encasement enhances the behaviour of the structural steel core by stiffening it, and so making it more effective against both local and overall buckling. The load-bearing concrete encasement performs the additional function of fireproofing the steel core. The main disadvantage of this type of composite column is that it requires a complete formwork. Furthermore, a reinforcement cage is required in order to prevent the concrete cover from spalling at low load levels, especially in the case of eccentrically loaded columns.

The concrete-filled composite columns shown in Fig. 6.2 have the advantage that they require no formwork and no reinforcement. The most common steel sections used are the hollow rectangular and circular tubes. In the latter case, the strength of the concrete is enhanced due to being triaxially contained within the circular tube. The main disadvantage of the concrete-filled composite column is that the concrete does not provide protection to the steel against fire, and an alternative form of fireproofing may be required. Another disadvantage is the difficulty of effecting connections.

A new type of steel–concrete composite column is shown in Fig. 6.3. The steel component of the column consists of two steel channels facing each other and connected together by end and intermediate batten plates. The rectangular core so formed is then filled with plain concrete in which no

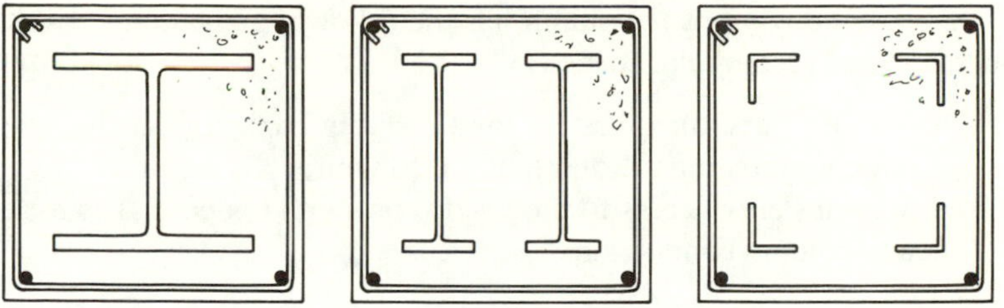

FIG. 6.1. Concrete-encased composite columns.

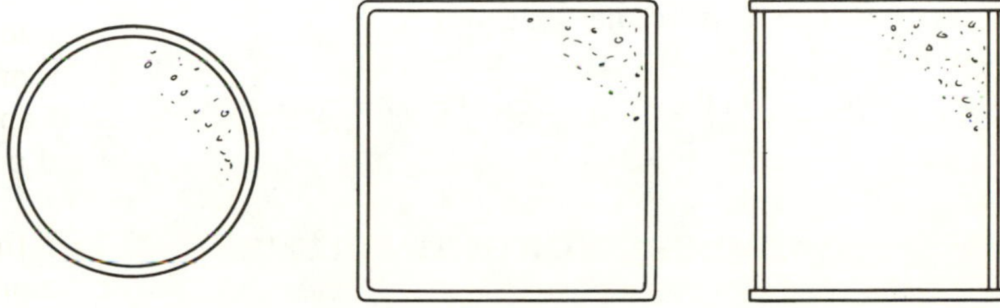

FIG. 6.2. Concrete-filled composite columns.

1. Steel channels.
2. End and intermediate batten plates.
3. Plain concrete core.

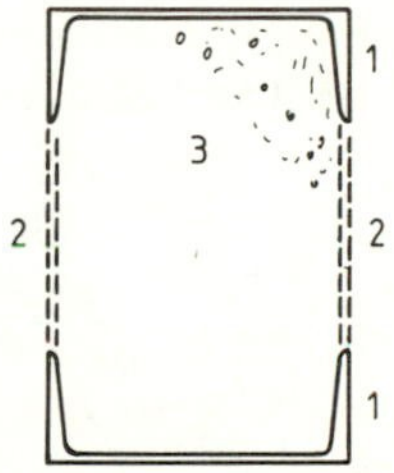

FIG. 6.3. Battened composite column.

reinforcement is used. Results published by Shakir-Khalil and Hunaiti (1985a, 1985b) show that the column behaviour is similar to that of a concrete-filled rectangular composite column. Tests carried out on full-scale columns included both axial and eccentric loading, in which the end eccentricities were as high as the full depth of the column. The tests showed that full composite action was achieved between the steel and concrete components through interface bond, and that mechanical shear connectors are not required.

The battened composite column has the following advantages over the concrete-filled rectangular section:

— it is more versatile since its load-carrying capacity can be altered simply by changing the depth of the column;
— it provides easy access to the inside core, and hence makes for easy beam–column connections.

Compared with concrete-encased composite columns, the battened composite column offers the following advantages:

— it requires shuttering of a simple and cheap form;
— it requires no reinforcement cage;
— it makes very efficient use of the structural steel element since the steel is placed on the outer faces where it is most needed.

## 6.3 SHORT COMPOSITE COLUMNS

The term short column normally refers to a compression member which is capable of attaining its ultimate carrying capacity with no sign of either local or overall buckling. In a short composite column, it is therefore to be expected that the maximum design strengths will develop in both its structural steel and concrete components. Figure 6.4 shows typical stress–

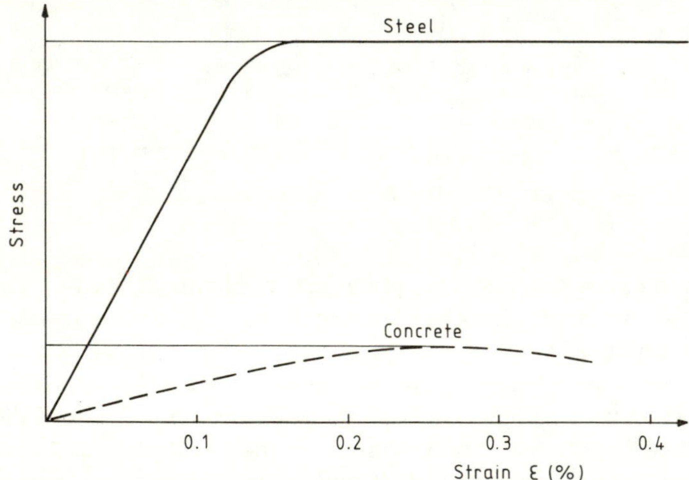


FIG. 6.4. Typical stress–strain curves.

strain curves for commercial structural steel and also for concrete. It can be seen that, when undergoing the same increase in strain, the steel part of a composite section will yield before the concrete reaches its compressive strength.

### 6.3.1 Squash Load, $N_u$

When a perfectly straight short composite column is loaded axially, its structural components undergo the same strain values, as a result of which the steel yields before the attainment of the compressive strength of the concrete. In the case of an encased composite column, the surrounding concrete will restrain the yielded steel and prevent it from buckling. Any further increase in the applied axial load will be supported by the concrete encasement until it finally reaches its compressive strength. This final load level is the ultimate strength of an axially loaded short column, and is normally referred to as the 'squash load', $N_u$. Similarly, in the case of concrete-filled hollow steel sections, when the steel tube reaches its yield strength the local buckling of the steel tube is prevented by the outward push of the contained concrete against the tube wall. As a result of this interaction between the steel and the concrete, the axial load acting on a short concrete-filled hollow section may be increased beyond the load level at which the steel yields. The ultimate load of the column will be reached when the stress in the contained concrete reaches the maximum strength of the concrete.

By definition, the squash load of a concrete-encased composite column is therefore given by:

$$N_u = A_s f_{sd} + A_r f_{rd} + A_c f_{cd} \tag{6.1}$$

The above expression is also applicable to concrete-filled rectangular sections which afford little containment to the concrete core. However, in the case of concrete-filled circular tubes, the concrete containment results in an enhancement of the compressive strength of concrete, and also in the development of hoop stresses in the steel tube which cause a reduction in the yield strength of steel. Gardner and Jacobson (1967), Sen (1969) and Knowles and Park (1969) carried out investigations into the behaviour of concrete-filled circular tubes, as a result of which it was shown that the modified squash load of a concrete-filled circular tube may be given in the following form:

$$N'_u = A_s f'_{sd} + A_c f'_{cd} \tag{6.2}$$

where $f'_{sd}$ is the reduced design strength of steel and $f'_{cd}$ is the enhanced design strength of the triaxially contained concrete.

A term which is useful to evaluate when dealing with the behaviour and design of composite columns is the concrete contribution factor, $\alpha$. This term represents the proportion of the squash load carried by the concrete, and for encased columns is given by:

$$\alpha = A_c f_{cd} / N_u \tag{6.3a}$$

whilst for concrete-filled circular tubes it is given by:

$$\alpha' = A_c f'_{cd} / N'_u \tag{6.3b}$$

### 6.3.2 Ultimate Moment of Resistance, $M_u$

When a short composite column is subjected to pure uniaxial bending, failure will take place on the attainment of its ultimate moment of resistance, $M_u$. At this stage, it is normal practice to assume that the section has become fully plastic, in which case the steel will have reached its design strength, $f_{sd}$, in compression as well as in tension. The concrete, on the other hand, is assumed to have cracked in the tension zone, and reached its design strength in bending compression, $\bar{f}_{cd}$, on the compression side. The calculations are carried out by the use of stress blocks as shown in Fig. 6.5 for concrete-encased and concrete-filled composite columns. The position of the neutral axis in the section is determined by arranging, using a trial and error procedure, that the longitudinal forces due to the stress blocks cancel out, and that the only resultant is a moment which is the ultimate moment of resistance, $M_u$, of the composite section. The hatched stress blocks are added to the stress distribution in the steel section in order to simplify the numerical work involved in calculating the $M_u$ values. Expressions for the ultimate moments of resistance corresponding to the stress blocks shown in Fig. 6.5 are given in both the British Standard BS 5400: Part 5 (1979) and the ECCS recommendations (1981a).

### 6.3.3 Short Column under Eccentric Loading

A short composite column subjected to an eccentric load which causes uniaxial bending could be treated in a way similar to that in the previous section. The only difference is that the position of the neutral axis is so chosen that the stress blocks now reduce to an eccentric force able to counteract the applied external force. An example is given in Fig. 6.6, which shows the stress blocks for concrete in compression and for steel in

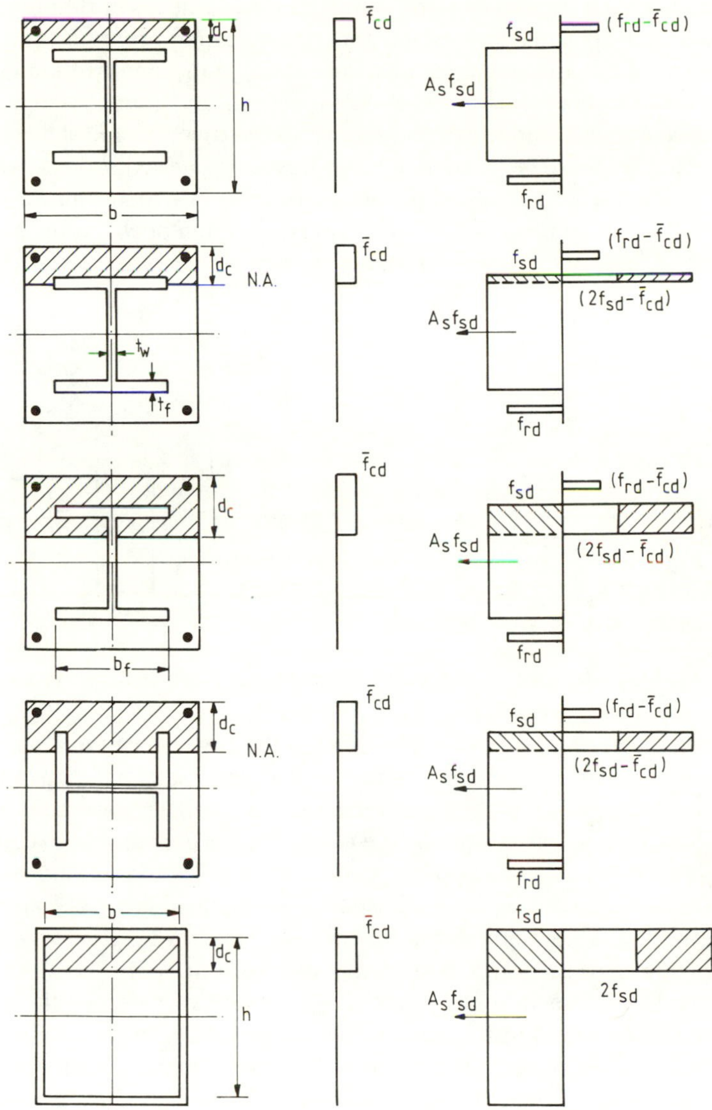


FIG. 6.5. Stress blocks for calculating ultimate moment of resistance $M_u$.

both tension and compression. Adding the diagonally hatched stress blocks to the stress distribution (b) in the steel is a means of simplifying the calculations. It should be noted that the resulting cross-hatched stress blocks in the equivalent stress distribution (c) in the steel section give the full plastic moment in the steel cross-section, and that the sum of the resulting compression forces, $N_c$ and $N_s$, in the concrete and steel sections respectively should be equal to the compression force acting on the section. This procedure also involves trial and error in order to determine the exact position of the neutral axis to maintain equilibrium between the external eccentric force and the internal stress blocks.

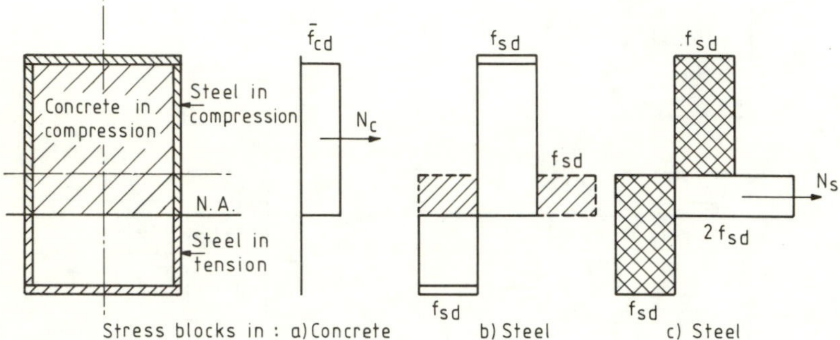


FIG. 6.6. Composite section subjected to eccentric compression.

An indirect approach would be to vary the position of the neutral axis and calculate the resulting sets of the axial force, $N$, and bending moment, $M$, as determined from the stress blocks so obtained. An interaction curve could thus be established giving the combinations of $N$ and $M$ to cause the full 'plasticity' of the composite section, in which the steel and concrete components reach their respective design strengths. Such results could be presented in a non-dimensional form, as given in Fig. 6.7 for a typical composite cross-section. It is clear that combinations of $N$ and $M$ which lie outside the interaction curve are not permissible, whereas those lying within the enclosed area represent loading conditions which are less than the carrying capacity of the composite section. In the latter case, such loading conditions would result in stress distributions of the elastic–plastic type in the section.

As can be seen from Fig. 6.7, when subjected to a small axial load, the composite section is capable of supporting a bending moment in excess of its ultimate moment of resistance, $M_u$. This is similar to the effect a

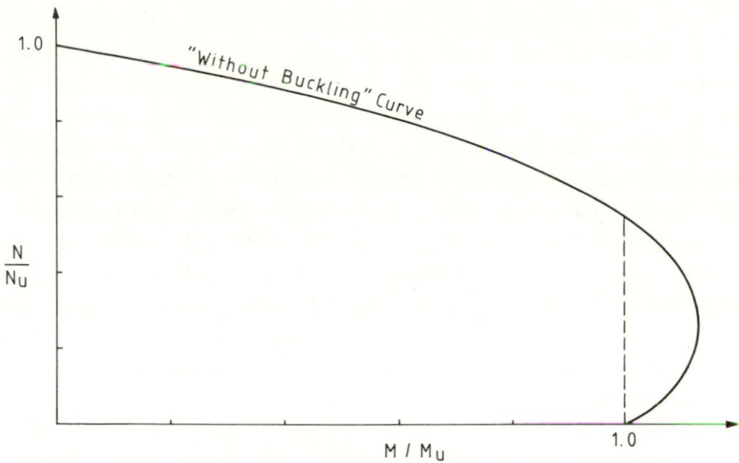


FIG. 6.7. Interaction curve for short column.

prestressing force has on a reinforced concrete section. However, if the axial force and bending moment are independently applied to the composite section, it should be noted that the removal of the axial load would destabilize the section and cause its failure. It is therefore recommended practice in the design of composite columns to consider the cut-off at $M/M_u = 1{\cdot}0$ to form an integral part of the interaction curve.

## 6.4 PIN-ENDED AXIALLY LOADED COLUMNS

### 6.4.1 Introduction

The British Standard BS 449 (1949) recognized the beneficial effects of the concrete encasement on the stiffness of the encased structural steel core. This was allowed for by increasing the radius of gyration of the encased section. Later, the load-bearing function of the encasement was taken into account, and the structural concrete encasement was recognized as contributing directly to the load-carrying capacity of the encased composite column.

The design method in the above British Standard is normally referred to as the 'cased strut' method. The method is empirical, based on little research and very few experimental results. Tests on composite columns by Faber (1956), Stevens (1959), Jones and Rizk (1963) and Stevens (1965) showed the very conservative nature of this method. In some cases the experimental failure load was over six times the predicted design load.

Various theoretical studies and experimental investigations in the behaviour of composite columns have been carried out over the last twenty years. These include work by Bondale (1966), Procter (1967), Basu (1967), Furlong (1968), Basu and Sommerville (1969), Neogi *et al.* (1969), Virdi and Dowling (1973, 1976), Johnson and Smith (1980), Roberts and Yam (1983) and Shakir-Khalil and Hunaiti (1985a, 1985b). Particular reference is made hereafter to the work by Basu and Sommerville (1969) and Virdi and Dowling (1976), the results of which have been incorporated in the British Standard BS 5400: Part 5 (1979), and also in the recommendations by the European Convention for Constructional Steelwork for the design of composite structures, ECCS (1981a).

### 6.4.2 Axially Loaded Steel Columns

Prior to discussing the behaviour of composite columns, it will be helpful to consider steel columns.

The elastic buckling load of a perfect pin-ended steel column is given by the Euler formula

$$N_c = \pi^2 E_s I_s / L^2 \tag{6.4a}$$

$$= \pi^2 E_s A_s / \lambda^2 \tag{6.4b}$$

where $\lambda$ is the slenderness ratio of the column, given by the column length divided by the least radius of gyration of the column section.

Given that the squash load of a steel column is given by

$$N_u = A_s f_{sd} \tag{6.5}$$

the Euler critical load could be expressed as a function of the squash load as follows:

$$K_1 = \frac{N_c}{N_u} = \left(\frac{\pi^2 E_s}{f_{sd}}\right) / \lambda^2 \tag{6.6}$$

The values of $E_s$ and $f_{sd}$ may be taken respectively as 205 kN/mm$^2$ and 265 N/mm$^2$ in accordance with BS 5950: Part 1 (1985) for steel to BS 4360 Grade 43 of thickness not exceeding 40 mm. Adopting these values, the above relationship is shown in Fig. 6.8 as the Euler curve for elastic buckling.

The critical loads of axially loaded columns, as obtained by the above expressions, are unattainable in practice due to the various practical imperfections in pin-ended columns. These imperfections include initial lack of straightness, accidental eccentricity of loading and residual stresses. Such imperfections result in bending moments developing at an early stage of loading in essentially axially loaded columns, and the failure load of a practical steel column can thus be seen to depend not only on its slenderness ratio, $\lambda$, but also on the mechanical properties of its material.

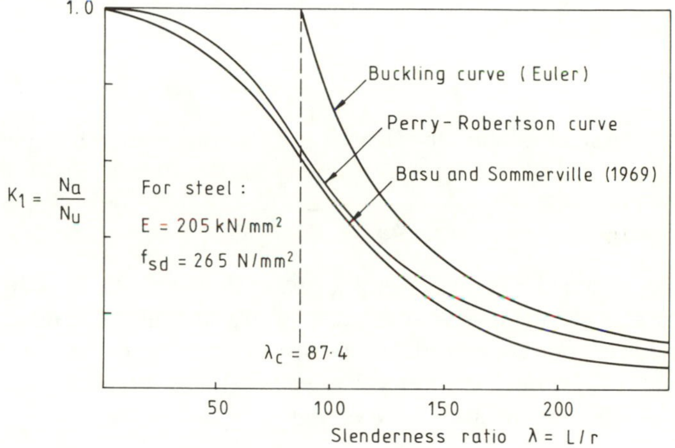


FIG. 6.8. Curves for axially loaded columns.

Consequently, the various design Standards assume an initial imperfection to account for the collective effects of all practical imperfections. The Perry–Robertson formula adopted by the British Standard BS 449 (1969) assumes the initial shape of the column to be a sine-curve with the maximum initial lateral displacement occurring at column mid-height, and the corresponding curve for the theoretical failure load of a pin-ended column is given in Fig. 6.8.

As can be seen from eqn (6.6), several elastic buckling curves would be required in order to represent the design strength range of the various types of steel available. However, this can be avoided if use is made of the concept of a critical slenderness ratio, $\lambda_c$, at which the Euler load, $N_c$, is equal to the squash load, $N_u$. Furthermore, the column slenderness is redefined as the slenderness factor, $\bar{\lambda}$, which is the slenderness ratio, $\lambda$, divided by the critical slenderness ratio, $\lambda_c$. Hence, by definition:

$$\bar{\lambda} = \lambda/\lambda_c \tag{6.7}$$

and

$$\frac{\pi^2 E_s A_s}{\lambda_c^2} = A_s f_{sd} \tag{6.8}$$

Hence

$$\lambda_c^2 = \pi^2 E_s / f_{sd} \tag{6.9}$$

Substituting in eqn (6.6):

$$K_1 = (\lambda_c/\lambda)^2 = (1/\bar{\lambda})^2 \tag{6.10}$$

For a bare steel column, the slenderness function is therefore given by

$$\bar{\lambda} = \frac{\lambda}{\lambda_c} = \frac{L/r}{\pi\sqrt{E_s/f_{sd}}} \tag{6.11}$$

The use of the slenderness factor $\bar{\lambda}$ results in one elastic buckling curve for all grades of steel, and also avoids to a large extent the dependence of the failure curve on the material properties of the column. This can be seen from Fig. 6.9, in which the Perry–Robertson failure curves for $f_{sd}$ = 265, 340 and 430 N/mm$^2$ are seen to be very close to each other.

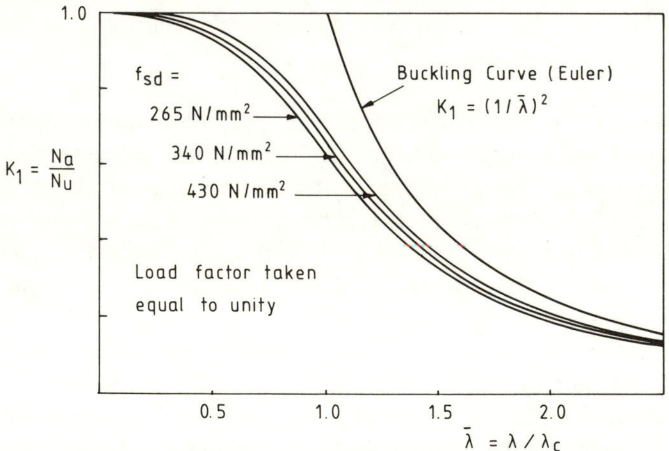


FIG. 6.9. British Standard BS449 (1969) failure curves.

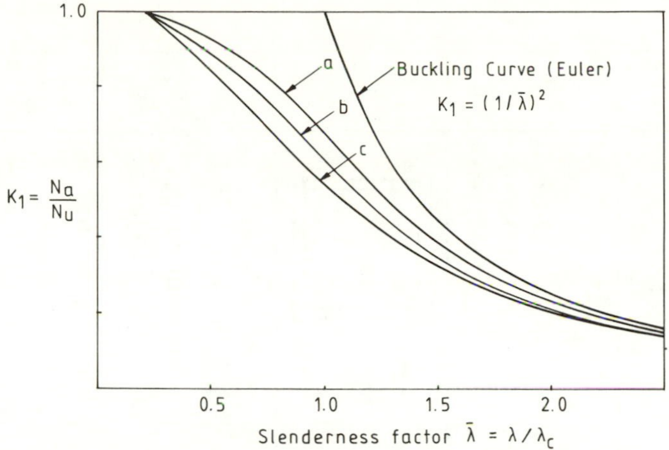


FIG. 6.10. European buckling curves for columns.

Studies by Beer and Shulz (1970) recognized the difference in behaviour between the various shapes of steel cross-section, in particular regarding the initial imperfections and the locked-in stresses due to rolling and welding. The results have been included in the European recommendations for steel construction, ECCS (1981b), and are represented by the three design curves reproduced in Fig. 6.10 which are applicable to all grades of steel. Table 6.1, reproduced from the above reference, enables the designer to select the appropriate design curve for a given column cross-section. A similar design approach is also used in the new British Standard, BS 5950: Part 1 (1985).

### 6.4.3 Axially Loaded Composite Columns

Similar to a steel column, the Euler load, $N_c$, of a composite column can be expressed as a function of the column squash load, $N_u$, as follows:

$$K_1 = \frac{N_c}{N_u} = \frac{\pi^2(E_s I_s + E_r I_r + E_c I_c)/L^2}{(A_s f_{sd} + A_r f_{rd} + A_c f_{cd})} \tag{6.12}$$

in which $E_c$ varies with the stress (or strain) in the concrete.

To present the results of composite columns in a manner similar to that of steel columns, Basu and Sommerville (1969) proposed the use of an equivalent radius of gyration for composite columns. They studied a large

TABLE 6.1
STRUT CURVE SELECTION CHART

| Shape of steel section | | Curve | Shape of steel section | | Curve |
|---|---|---|---|---|---|
| rolled tubes welded tubes (hot finished) | | a | I and H sections stress relieved by heat treatment | Buckling about x-axis | a |
| | | | | Buckling about y-axis | b |
| Welded box sections 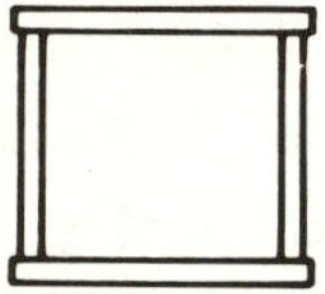 | | b | Welded box sections stress relieved by heat treatment 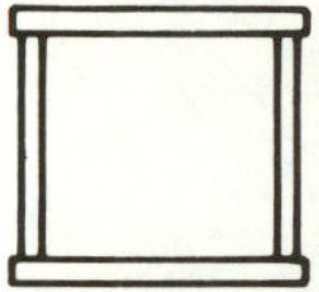 | | a |
| I and H rolled sections | Buckling about x-axis: $h/b > 1.2$ | a | I and H welded sections | Buckling about x-axis: (a) flame cut flanges | b |
| | $h/b \leq 1.2$ | b | | (b) rolled flanges | b |
| | Buckling about y-axis: $h/b > 1.2$ | b | | Buckling about y-axis: (a) flame cut flanges | b |
| | $h/b \leq 1.2$ | c | | (b) rolled flanges. | c |

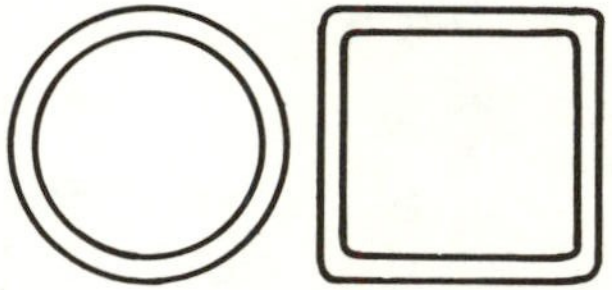

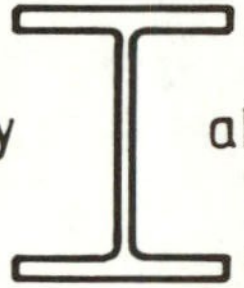

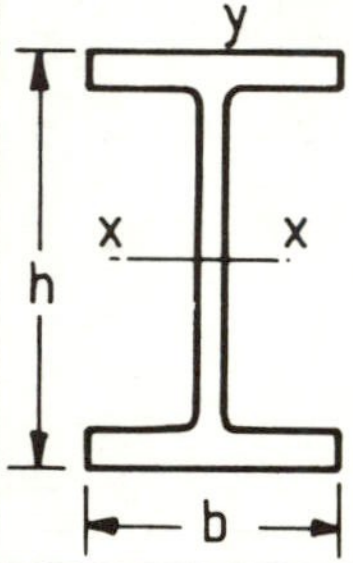

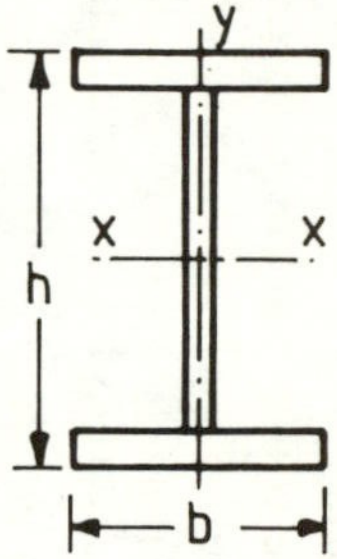

number of axially loaded composite columns and plotted the results so obtained as a curve of $K_1$ versus slenderness ratio. The scatter of the results was in a relatively narrow band which overlapped the Perry–Robertson curve. They chose the $K_1$-curve as the lower bound of the band, and further reduced it in the middle range of $\lambda = 50$–$150$ as a result of which a conservative curve below the Perry–Robertson curve was finally taken, as shown in Fig. 6.8. It should be noted that, due to the conservative choice of the $K_1$-curve, for large slenderness ratios an encased composite column may have a lower carrying capacity than the corresponding bare steel column.

Virdi and Dowling (1976) compared the ultimate loads available from test results on composite columns, and also the theoretical results of a verified computer program, with the design loads obtained by using the $K_1$-curve. These comparisons confirmed the excessively conservative nature of the $K_1$-curve proposed by Basu and Sommerville, and also that the conservativeness increased with the slenderness ratio. In their study, they proposed that use should be made of the slenderness factor, $\bar{\lambda}$, instead of the traditional slenderness ratio $\lambda$. This was in order to develop a unified design method which would allow the steel design curves given in Fig. 6.10 to be used for the design of composite columns. They redefined the slenderness factor, $\bar{\lambda}$, as the ratio of the column length, $L$, to the critical length, $L_c$, where the latter is the column length for which the column Euler load, $N_c$, is equal to its squash load, $N_u$. Hence:

$$\bar{\lambda} = L/L_c \tag{6.13}$$

For a bare steel column of length $L_c$, by definition:

$$\frac{\pi^2 E_s I_s}{L_c^2} = A_s f_{sd} \tag{6.14}$$

Hence

$$L_c = \pi r \sqrt{E_s/f_{sd}} = r\lambda_c \tag{6.15}$$

and

$$\bar{\lambda} = L/L_c = L/r\lambda_c = \lambda/\lambda_c \tag{6.16}$$

which agrees with the original definition of the slenderness factor as given by eqn (6.7).

For a composite column of length $L_c$, by definition, $N_c = N_u$, and hence:

$$\pi^2(E_s I_s + E_c I_c)/L_c^2 = A_s f_{sd} + A_c f_{cd} \qquad (6.17)$$

and

$$L_c = \pi \sqrt{\frac{E_s I_s + E_c I_c}{A_s f_{sd} + A_c f_{cd}}} \qquad (6.18)$$

As can be seen from eqns (6.13) and (6.18), the slenderness factor, $\bar{\lambda}$, of a composite column is measured with a single parameter, namely the critical length $L_c$, which contains the geometric properties of the column cross-section as well as the mechanical properties of the materials used. Having calculated the slenderness factor, $\lambda$, of a composite column, its load-carrying capacity could be determined as a proportion of its squash load by the use of the column design curves in Fig. 6.10. This design approach has been proved satisfactory for the design of axially loaded composite columns both by experimental and theoretical work. Table 6.1 enables the designer to select the appropriate design curve applicable to the structural steel component of the composite column. The merit of the new interpretation by Virdi and Dowling (1976) of the column slenderness factor, $\bar{\lambda}$, is that it is applicable to bare metal as well as to composite columns.

As mentioned earlier, the contained concrete of an axially loaded concrete-filled circular tube exhibits an enhanced strength due to being triaxially contained. However, this is mainly true for columns of short length as the beneficial effect of the containment decreases as the column length increases. Virdi and Dowling (1976) found that the concrete containment has a negligible effect on concrete-filled circular tubes of slenderness factor $\lambda$ larger than 0·7. As the value of $\bar{\lambda} = 1{\cdot}0$ for most practical columns corresponds to a length-to-diameter ratio varying between 24 and 29, they suggested that the triaxial effect be ignored for composite columns of $L/d$ ratios over 25.

### 6.4.4 Elastic Modulus of Concrete, $E_c$

As can be seen from eqn (6.18), the elastic modulus of concrete, $E_c$, is the main material property which would be in dispute when calculating the slenderness factor of a composite column. This is due to the fact that $E_c$ has no unique value, and that it decreases as the strain increases. A number of researchers including Hognestad (1951), Desayi and Krishnan (1964) and Barnard (1964) have proposed equations to represent the concrete stress–

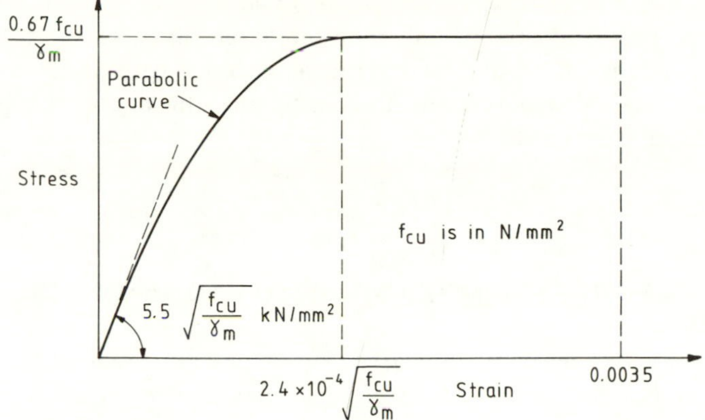


FIG. 6.11. Short-term design stress–strain curve for normal-weight concrete.

strain relationship. The curve recommended by BS 8110: Part 1 (1985) is shown in Fig. 6.11, and is similar to that given by the European CEB/FIP recommendations (1970). The initial elastic modulus of concrete, $E_{co}$, as given by the above recommendations, is given respectively by:

$$E_{co} = 5500\sqrt{f_{cu}/\gamma_{mc}} = 6720\sqrt{f_{ck}/\gamma_{mc}}$$

$$= 6720\sqrt{f_{cd}} \qquad (6.19a)$$

and

$$E_{co} = 1000 f_{cd} \qquad (6.19b)$$

where $f_{cu}$ is the characteristic 28-day cube strength of concrete, and $f_{ck}$ is the characteristic strength of concrete, taken as $0{\cdot}67\,f_{cu}$ in accordance with the British Standard.

British Standard BS 5400: Part 5 (1979) gives the following expression for the elastic modulus of concrete to be used for the purpose of calculating the critical length of a composite column as given by eqn (6.18):

$$E_c = 450 f_{cu} \qquad (6.20)$$

The European recommendations, ECCS (1981a), on the other hand recommend the following:

$$E_c = 600 f_{cy} \qquad (6.21)$$

where $f_{cy}$ is the characteristic 28-day cylinder strength of concrete. However, it is also stated that in order to account for the effects of time-dependent strains of concrete on the carrying capacity of concrete-encased structural steel sections, the elastic modulus of concrete given in eqn (6.21) may be reduced to

$$E_c = 300 f_{cy} \tag{6.22}$$

## 6.5 PIN-ENDED COLUMN WITH END MOMENTS

In Section 6.3.3 an interaction curve (Fig. 6.7) was presented for a short column under the combined effect of an axial force $N$ and a bending moment $M$. In developing the curve, buckling was ignored, and only equilibrium conditions between the external forces and the stress blocks within the section were considered. In the case of a column of a finite length, a series of inelastic analyses would be required in order to develop a similar interaction curve in which buckling considerations are taken into account. Such a curve would be expected to intersect the $N/N_u$ and $M/M_u$ axes at $K_1$ (obtained from Fig. 6.10) and unity respectively, since these represent the limiting loading conditions corresponding to zero bending and no axial force. Figure 6.12 shows a typical curve so obtained .

Basu and Sommerville (1969) studied ~100 column sections for which a wide range of slenderness ratios and eccentricities were considered. Their results showed that the interaction curve of such columns could be approximately represented by the relationship shown in Fig. 6.12 and given by the following expression:

$$\frac{N}{N_u} = K = K_1 - (K_1 - K_2 - 4K_3)\frac{M}{M_u} - 4K_3\left(\frac{M}{M_u}\right)^2 \tag{6.23}$$

where $K_1$ is obtained from Fig. 6.10, and $K_2$ and $K_3$ are variables which are dependent on the ratio of the end moments, $\beta$, the column slenderness and the concrete contributing factor, $\alpha$. This design approach has been used as the basis for the design of composite columns in BS 5400: Part 5 (1979) in which expressions are given for the factors $K_1$, $K_2$ and $K_3$ for concrete-encased and concrete-filled composite columns.

It should be noted that eqn (6.23) is easily applied when the end moments remain constant as the compression force is increased to its failure value. If the actual loading is represented by point P in Fig. 6.12, the

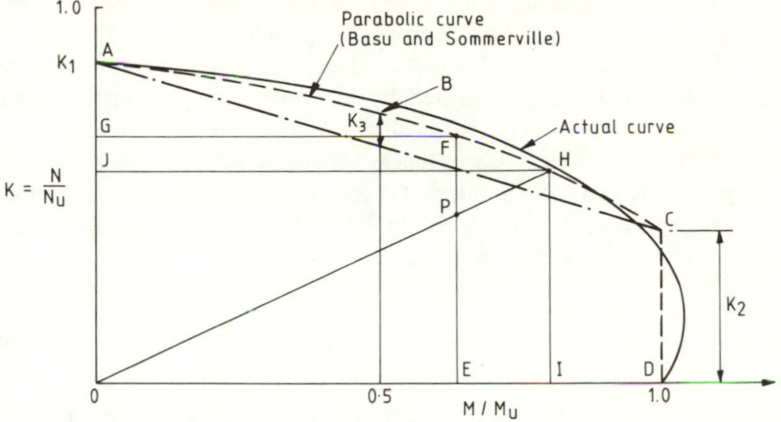


FIG. 6.12. Actual and idealized interaction curves.

failure condition is then represented by point F; E represents the constant moment applied to the column, and G is the corresponding failure load. When the end moments are the result of an eccentrically applied compression force, the value of the failure load is not so easily determined since $M$ and $N$ in eqn (6.23) become interdependent and increase proportionally. The relationship derived by Basu and Sommerville (1969) is represented by line OPH in Fig. 6.12. Point H represents the failure condition, and points I and J represent the failure bending moment and compression force respectively. The corresponding solution for the boundary ABC is given by:

$$\frac{N}{N_u} = K = \frac{-K_4 + \sqrt{K_4^2 + 16K_1K_3(\eta e)^2}}{8K_3(\eta e)^2} \tag{6.24a}$$

where

$$K_4 = 1 + (K_1 - K_2 - 4K_3)\eta e \tag{6.24b}$$

and

$$e = M/N \text{ and } \eta = N_u/M_u \tag{6.24c}$$

Due to the complexity of the calculations involved in the above method, Roberts and Yam (1983) suggested the two simplified methods given below for checking that failure does not occur under a given axial load $N$ and a bending moment $M$.

Under a small axial load, and ignoring the buckling effect, the interaction curve given in Fig. 6.7 may be used to obtain the value of $K_2$ which they proved to be approximately equal to the concrete contribution factor $\alpha$. This could also be easily verified by inspecting the various interaction curves given in ECCS (1981a) both for concrete-encased and concrete-filled composite columns. Using this as a first step in the simplified method, they ignored $K_3$, so that the interaction relationship was represented by the two straight lines AC and CD in Fig. 6.12. The failure-boundary line AC was thus given by the following simple expression:

$$\frac{N}{N_u} = K = K_1 - \frac{M}{M_u}(K_1 - \alpha) \tag{6.25}$$

No mention was made of the limitations of this simplified method, since for columns of practical proportions $K_1$ could be less than the concrete contribution factor $\alpha$, thus leading to incorrect results.

The alternative simplified method suggested by Roberts and Yam (1983) uses the 'without buckling' curve as the design basis. The only assumption they made was that for a given moment, axial loads from the actual curve and the 'without buckling' curve are in a constant ratio, $K_1$. They found that the factored curve so obtained was close to the actual curve, and suggested its use as the failure envelope.

Another simplified design method is given in ECCS (1981a), and is illustrated in Fig. 6.13 which is the 'without buckling' curve for the column section under consideration. When the column has a finite length, it fails under the effect of an axial force $N_a$ represented by point A on the $N/N_u$ axis. Distance AB represents the additional proportion of the ultimate moment of resistance the column section could sustain in addition to the axial load, $N_a$, if the buckling effects were ignored. Length AB could thus be thought of as the bending moment accompanying the axial force on the column, being caused by the practical imperfections in the column. In the absence of an axial force, the moment-carrying capacity of the column section is the ultimate moment of resistance of the section, $M_u$, and is represented by point F on the curve. When subjected to an axial load $N$, represented by point C on the $N/N_u$ axis, the bending moment in the column due to practical imperfections would be proportionately less than that due to the axial force $N_a$. It is assumed that this relationship is linear, being represented by line OB in Fig. 6.13. Consequently, it can be seen that the dashed part of the area enclosed by the failure curve, BEF, represents the additional moment capacity of the column in the presence of

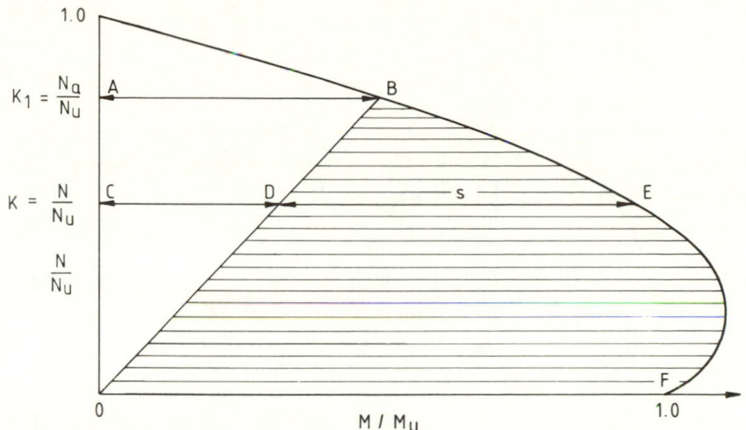


FIG. 6.13. Simplified design method (ECCS, 1981a).

an axial load $N$, and the resulting moment due to the column imperfections. However, the ECCS (1981a) recommends that in the presence of an axial load $N$, the moment-carrying capacity of a column may only be taken as 90% of the values so obtained:

$$M = 0{\cdot}9sM_u \tag{6.26}$$

It should be mentioned that a composite column has two interaction curves which correspond to uniaxial bending about the minor ($y$) and major ($x$) axes, the curves intersecting the $N/N_u$ axis at values of $K_{1y}$ and $K_{1x}$ respectively. For the major axis interaction curve to reach $K_{1x}$, the column would need to be restrained about its minor axis. In the absence of such restraint, both interaction curves would intersect the $N/N_u$ axis at the same value of $K_{1y}$ as a result of the column's natural tendency to buckle about its minor axis. In such case, major axis bending will always be accompanied by bending about the minor axis, as a consequence of which the column should be designed as subjected to biaxial bending.

## 6.6 BIAXIALLY LOADED COLUMNS

Similar to the case of uniaxial bending for which interaction curves can be constructed, it is theoretically possible to develop interaction surfaces for the case of biaxial bending. However, no simple method is available for

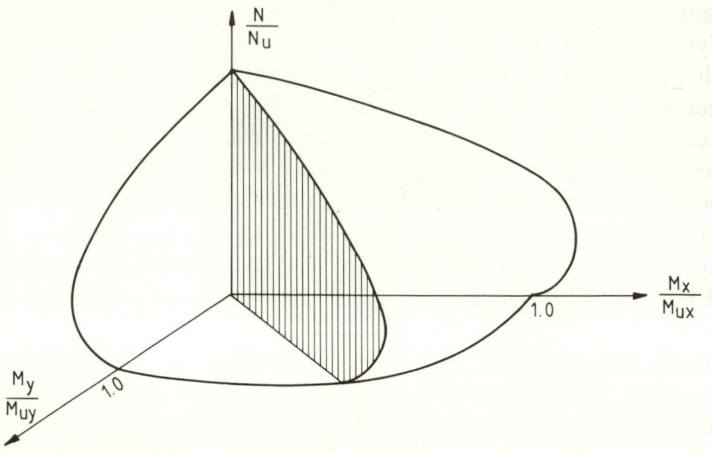


FIG. 6.14. Interaction surface for biaxially loaded column.

this purpose, and failure is defined by an interaction surface of the type shown in Fig. 6.14.

Owing to the complexity of the analysis and behaviour of biaxially loaded composite columns, very little theoretical and experimental work is available in this field. In the case of short reinforced concrete columns, Bresler (1960) proposed the following empirical formula which expresses the biaxial strength of the column in terms of both its uniaxial strengths and its squash load:

$$\frac{1}{N_{xy}} = \frac{1}{N_x} + \frac{1}{N_y} - \frac{1}{N_u} \tag{6.27}$$

Similarly, Basu and Sommerville (1969) suggested the following formula for the design of a composite column of finite length:

$$\frac{1}{N_{xy}} = \frac{1}{N_x} + \frac{1}{N_y} - \frac{1}{N_{ax}} \tag{6.28}$$

where $N_{ax}$ is the failure load of the column when loaded axially and constrained to bend about its major ($x$) axis.

Virdi and Dowling (1973) developed a method for computing the biaxial strength of a composite column, and also tested nine columns in which the column slenderness and the end eccentricities were the only parameters systematically varied. They compared the experimental results with their analytical method, which is too complex for use in design, and also with the predictions of eqn (6.28). The experimental results were found to be in close agreement with the theoretical values for short columns. For long columns, however, the test results were, on average, about 50% higher than both the theoretical and empirical predictions, and in one case nearly twice as high.

Despite its conservative nature, especially in the design of long columns, eqn (6.28) is the recommended method for the design of biaxially loaded composite columns both by the British Standard BS 5400: Part 5 (1979) and the ECCS (1981a). It should be noted that $N_x$ and $N_y$ are calculated on the basis of a constant moment as opposed to the constant eccentricity concept originally proposed by Basu and Sommerville (1969). This modification, proposed by Virdi and Dowling (1976), permits the use of the higher axial load represented by point F in Fig. 6.12, as opposed to that given by point H, when point P represents the actual loading condition of the column.

## 6.7 END RESTRAINTS AND EFFECTIVE LENGTH

The methods described in Sections 6.4 and 6.5 are only applicable to isolated pin-ended columns, which are rarely encountered in practice. In multi-storey structures, columns are connected at their ends to beams and also to other columns, and the load-carrying capacity of the column is consequently affected by the construction details. Such structures are normally braced, since bracing is the cheapest form of resisting transverse forces, and suffer little sway. The columns of braced frames may thus be designed in accordance with the previous sections provided that the beams are designed as simply supported and the beam–column connections are 'simple'.

For rigid-jointed braced frames, the end restraints provided to a column depend on the relative collective stiffness of the structural elements connected at the column ends to that of the column, the latter being a reduced value of the actual column stiffness due to the presence of the column compression force. Wood (1974) has provided curves for the design of columns in elastically designed rigid-jointed frames. The method is based on the concept of an effective length of the column, which allows for the end restraints by taking a reduced column length in the design. The column is then designed as a pin-ended column of length equal to the effective length so obtained, and subjected to the same end force and bending moments as the actual column.

External columns in rigid-jointed frames are normally checked for the case of loading in which all beams above the column support the full imposed load. This case of loading results in the maximum compression force and bending moment in the column. The internal columns, however, require to be checked for the two loading cases, that which gives the maximum bending moment and that which gives the maximum compression force as obtained from pattern loading of the beams. In the majority of cases, the former loading condition governs the design of internal columns except perhaps for columns in the upper floors. In the design of a rigid-jointed frame by the ultimate load method, plastic hinges will form at the ends of beams supporting the imposed load, and the calculations of the column end rotations will become highly complicated due to the interaction between beams and columns. The solution could be simplified by the use of the limited frame recommended by BS 5950: Part 1 (1985). Such an idealized frame consists of the column under consideration and the members connected to its ends, in which the far ends of the connected members are assumed to be fixed. However, the analysis is particularly complicated in the case of composite structures by the incomplete knowledge of the moment–curvature characteristics of the beam–column connections, and also by lack of information on the deteriorated stiffness of a composite column under the effect of a compression force. Johnson (1975) has outlined an interim design method in which he made the conservative assumption that beams carrying factored dead load have plastic hinges at their ends.

Little information is available in the behaviour of rigid-jointed composite structures, and there is ample scope for research in this field.

## 6.8 BEAM–COLUMN CONNECTIONS

Composite columns are normally used in conjunction with composite beams, and only connections in such composite structures are considered here. As with beam–column connections in structural steelwork, there are a number of ways in which connections could be detailed in composite construction. Figure 6.15 shows typical beam–column connections where the floor slab could be either reinforced concrete or composite with profiled steel sheeting. Figures 6.15a and 6.15c show simple joints in composite construction in which only web and bottom flange cleats are used to connect the steel beam to the column. The details of the connections are such that relative rotation is allowed between beam and

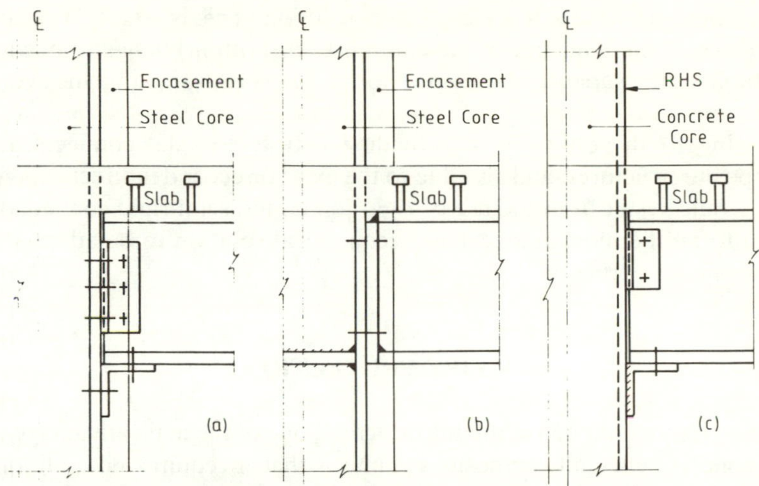


FIG. 6.15. Typical connections in composite construction.

column by the use of black bolts in clearance holes, and also by ensuring that the longitudinal reinforcement in the slab is only that required for crack control. It should be noted that this type of joint might be difficult to achieve when the profile of the steel sheeting of the composite slab runs parallel to the beam.

The connections shown in Figs 6.15b and 6.15c are suitable when semi-rigid connections are required, provided that the bolts in the bottom flange cleat are precision bolts or of the friction grip type, and that the cleat is big enough to accommodate the number of bolts required. The rest of the bolts in the two connections could be black bolts (preferably grade 8.8) in clearance holes. It should be noted that the tension in the top flange is primarily taken past the columns through the slab reinforcement, and that the force in the compression flange in Fig. 6.15b is transmitted to the column by direct bearing. These two types of connection have been investigated by Johnson and Law (1981) and Owens and Echeta (1981) respectively. The main disadvantage of the connection in Fig. 6.15b, in which a beam end plate is used, is the costly high standard of fit-up required in manufacturing the beam due to the practical tolerances in structural steel sections.

Rigid beam–column connections can also be detailed in a manner similar to the connections shown in Fig. 6.15. However, the use of rigid connections would transfer bending from midspan of the beam to its ends, and thus would result in a more expensive solution. This is largely as a consequence of the tendency of the compression (bottom) flange to become unstable at the support whereas the top flange is restrained by the slab at midspan.

Owing to the difficulty of providing a truly 'simple' connection in composite structures, and also due to the extra direct and indirect expense that would result from the use of rigid joints, the semi-rigid connections seem to be the most practical and economical solution to be adopted in composite construction.

## 6.9 FIRE PROTECTION

As has been mentioned in the Introduction, one of the main advantages of the concrete-encased composite column is that it requires no additional protection against fire.

The steel section of the concrete-filled column is, on the other hand, exposed and requires to be protected. However, it has been shown by CCERTA (1980) that the presence of the concrete core can enhance the fire resistance of such a column by acting as a 'heat sink', and that the fire resistance could be further improved by the use of internal reinforcement. Such reinforcement becomes necessary in the case of unprotected concrete-filled composite columns subjected to high bending moments.

Previous research has been summarized by Elliott (1981), and it has been shown by tests that unprotected concrete-filled columns possess a minimum of 30 min fire life, which increases to 1 h for 300 mm sections. It should be noted that, due to the presence of moisture in the contained concrete, holes should be provided both at the top and bottom of such columns in order to allow any steam to escape and thus avoid excessive damage to the column under fire conditions.

Fire protection could be provided by an active protection system in which fire sprinklers are used. Alternatively, when the minimum period of fire resistance required by the Building Regulations, summarized by Elliott (1980), exceeds that of the unprotected concrete-filled composite columns, additional fire protection could be provided in the form of passive protection by insulating the steel from the fire. The various types of insulating methods available are generally grouped under the headings of intumescent coatings, sprayed, boarded and preformed systems. The spray-on system is the least expensive, but owing to its relatively rough finish and unattractive appearance, it is only suitable for structural steel elements which are concealed.

## 6.10 CONCLUSIONS

Information available on the behaviour of composite columns has been summarized in the previous sections. This includes both theoretical and experimental investigations, as well as design recommendations in accordance with the British Standards and the European recommendations. These design methods resulted from extensive theoretical and experimental studies in the behaviour of axially loaded composite columns, and also of columns subjected to compression and uniaxial bending. Due to the complexity of the behaviour of biaxially loaded columns, these are only treated by an empirical method which yields safe values of their load-bearing capacity. Beam–column connections have also been briefly reviewed since the type of connection used affects the column as well as the overall behaviour of the composite structure.

It may be concluded that the behaviour of isolated columns has been adequately investigated. However, further research is required in the field of constructional details and joint behaviour of composite structures. Additional work in this area would lead to a better understanding of this relatively new type of construction, and also to its wider acceptance and use in the building industry.

## ACKNOWLEDGEMENTS

The author is grateful to his colleague Mr R. Taylor for his comments on the draft of this work. He also acknowledges with thanks the help received from Dr Yasser M. Hunaiti in the preparation of the figures.

## REFERENCES

BARNARD, P. R. (1964) Researches into the complete stress–strain curve for concrete. *Mag. Conc. Res.*, **16**(49), Dec., 203–10.

BASU, A. K. (1967) Computation of failure loads of composite columns. *Proc. Instn Civ. Engrs*, **36** (March), 557–78.

BASU, A. K. and SOMMERVILLE, W. (1969) Derivation of formulae for the design of rectangular composite columns. *Proc. Instn Civ. Engrs*, Supplementary vol., Paper 7206S, 233–80.

BEER, H. and SCHULZ, G. (1970) The theoretical basis of the new column curves of the European Convention for Constructional Steelwork. *Construction Métallique* (3), Sept. (in French).

BONDALE, D. S. (1966) Column theory with special reference to composite columns. *Consulting Engr*, **30**, July, 72–7, Aug., 43–8, Sept., 68–70.

BRESLER, B. (1960) Design criteria for reinforced concrete columns under axial load and biaxial bending. *ACI J.*, **57**, Nov., 481–90.

BRITISH STANDARDS INSTITUTION (1949, 1969) *The use of structural steel in building*. BS 449: Part 2.

BRITISH STANDARDS INSTITUTION (1979) *Steel, concrete and composite bridges. Code of practice for design of composite bridges*. BS 5400: Part 5.

BRITISH STANDARDS INSTITUTION (1985) *Structural use of steelwork in building. Code of practice for design in simple and continuous construction: hot rolled sections*. BS 5950: Part 1.

BRITISH STANDARDS INSTITUTION (1985) *Structural use of concrete. Code of practice for design and construction*. BS 8110: Part 1.

BURR, W. H. (1912) Composite columns of concrete and steel. *Proc. Instn Civ. Engrs*, **188**, 114–26.

CCERTA (Commission des Communautés Européennes Recherche Technique Acier) (1980) Determination of the fire life of concrete filled tubular sections. Rapport final (15B-80/10) par Cométube, Paris (in French).

CEB/FIP (Comité Européen du Béton and Fédération Internationale de la Précontrainte) (1970) *International recommendations for the design and construction of concrete structures. Principles and recommendations*.

DESAYI, P. and KRISHNAN, S. (1964) Equation for the stress-strain curve for concrete. *ACI J., Proc.*, **61**(3), March.

ECCS (European Convention for Constructional Steelwork) (1981a) *European recommendations for composite structures*. The Construction Press.

ECCS (European Convention for Constructional Steelwork) (1981b) *European recommendations for steel construction*. The Construction Press.

ELLIOTT, D. A. (1980) *Fire and steel construction*. The Building Regulations 1976. Application to steelwork of Part E: Section I: Structural fire precautions. Constrado.

ELLIOTT, D. A. (1981) *Fire and steel construction*. Protection of structural steelwork. Constrado.

FABER, O. (1956) Savings to be effected by the more rational design of cased stanchions as a result of recent full size tests. *Struct. Engr*, **34**(3), March, 88–109.

FURLONG, R. W. (1968) Design of steel-encased concrete beam–columns. *J. Struct. Div., Proc. ASCE*, Jan. 267–81.

GARDNER, N. J. and JACOBSON, E. R. (1967) Structural behaviour of concrete filled steel tubes. *ACI J.*, July, 404–13.

HOGNESTAD, E. (1951) A study of combined bending and axial load in reinforced concrete members. Univ. of Illinois, Bulletin No. 399, Nov.

JOHNSON, R. P. (1975) *Composite structures of steel and concrete*, Vol. 1, *Beams, columns, frames and applications in buildings*. Constrado Monograph, Granada.

JOHNSON, R. P. and LAW, C. L. C. (1981) Semi-rigid joints for composite frames. *Joints in structural steelwork*, Proc. Conf. Middlesbrough, UK, April, 3.3–3.19.

JOHNSON, R. P. and SMITH, D. G. E. (1980) A simple design method for composite columns. *Struct. Engr*, **58A**(3), March, 85–93.

JONES, R. and RIZK, A. A. (1963) An investigation on the behaviour of encased steel columns under load. *Struct. Engr*, **41**(1), Jan., 21–33.

KNOWLES, R. B. and PARK, R. (1969) Strength of concrete filled steel tubular columns. *J. Struct. Div., Proc. ASCE*, Dec., 2565–87.

McDEVITT, C. F. and VIEST, I. M. (1972) (a) Interaction of different materials. Prelim Report, 55–79. (b) A survey of using steel in combination with other materials. Final Report, 101–17. *Ninth IABSE Congress*, Amsterdam.

NEOGI, P. K., SEN, H. K. and CHAPMAN, J. C. (1969) Concrete-filled tubular steel columns under eccentric loading. *Struct. Engr*, **47**(5), May, 187–95.

OWENS, G. W. and ECHETA, C. B. (1981) A semi-rigid design method for composite frames. *Joints in structural steelwork*, Proc. Conf., Middlesbrough, UK, April, 3.20–3.38.

PROCTER, A. N. (1967) Full size tests facilitate derivation of reliable design methods. *Consulting Engr*, Aug., 54–60.

ROBERTS, E. H. and YAM, L. C. P. (1983) Some recent methods for the design of steel, reinforced concrete and composite steel–concrete columns in the UK. *ACI J.*, March–April, 139–49.

SEN, H. K. (1969) Triaxial effect in concrete-filled tubular steel columns. PhD Thesis, University of London.

SHAKIR-KHALIL, H. and HUNAITI, Y. M. (1985a) Behaviour of battened composite columns. *Applied Solid Mechanics*, Proc. Conf., Strathclyde Univ., UK, March, 415–33; publ. Elsevier Applied Science.

SHAKIR-KHALIL, H. and HUNAITI, Y. M. (1985b) Battened composite columns. *Steel in buildings*, IABSE Symposium Report, Luxembourg, Sept., **48**, 325–33.

STEVENS, R. F. (1959) Encased steel stanchions and BS 449. *Engineering*, Oct., 376–7.

STEVENS, R. F. (1965) Encased stanchions. *Struct. Engr*, **43**(2), Feb., 59–66.

VIRDI, K. S. and DOWLING, P. J. (1973) The ultimate strength of composite columns in biaxial bending. *Proc. Instn Civ. Engrs*, Part 2, **55**, March, 251–72.

VIRDI, K. S. and DOWLING, P. J. (1976) A unified design method for composite columns. IABSE Symposium, **36II**, 165–84.

VIRDI, K. S. and DOWLING, P. J. (1980) Bond strength in concrete-filled steel tubes, IABSE Symposium, Periodica 3/1980, Aug., pp. 125–39; Proceedings P–33/80.

WOOD, R. H. (1974) Effective lengths of columns in multi-storey buildings. *Struct. Engr*, **52**, July, 235–44; Aug., 295–302; Sept., 341–6.

# *Chapter 7*

# STEEL–CONCRETE COMPOSITE COLUMNS—II

R. W. Furlong

*Department of Civil Engineering, University of Texas, Austin, Texas, USA*

## *SUMMARY*

*A composite column is any concrete column reinforced with steel other than reinforcing bars, and can be regarded as a natural development of the original reinforced concrete column. Since rigid or semi-rigid framing is a characteristic of most systems that contain structural concrete, this chapter treats composite columns as if each were in fact a beam–column. Methods for evaluating cross-section thrust and flexural strength are described. Biaxially eccentric thrust is discussed. Shear strength estimates are suggested, and techniques for the rational analysis of slenderness effects are described. Example calculations for strength and for slenderness are included both for an encased steel shape and for a concrete-filled steel tube. A brief discussion of connections and auxiliary reinforcement concludes the chapter.*

## NOTATION

$A_c$ Cross-sectional area of concrete
$A_r$ Cross-sectional area of reinforcing steel
$A_s$ Cross-sectional area of structural steel shape or tube
$A_{vs}$ Area of shear reinforcement at each tie bar space
$A_w$ Area of the web of a steel shape or sides of steel tube
$b_w$ Width of cross-section effective for shear strength
$c_r$ Cover distance from center of bar to edge of cross-section

| | |
|---|---|
| $E_c$ | Material stiffness of concrete |
| $EI$ | Flexural stiffness of a cross-section |
| $E_s$ | Material stiffness of steel |
| $f_{cd}$ | Design strength of concrete |
| $f_{ck}$ | Characteristic strength of concrete from cylinder tests |
| $f_{cu}$ | Characteristic strength of concrete from cube tests |
| $f_{rd}$ | Design strength of reinforcing bars |
| $f_{rk}$ | Characteristic strength of reinforcing bars |
| $f_{sd}$ | Design strength of structural steel shape or tube |
| $f_{sk}$ | Characteristic strength of structural steel shape or tube |
| $h_1$ | Width of cross-section perpendicular to plane of bending |
| $h_2$ | Thickness of cross-section in plane of bending |
| $I_c$ | Moment of inertia for concrete in cross-section |
| $I_r$ | Moment of inertia for reinforcing bars in cross-section |
| $I_s$ | Moment of inertia for structural shapes or tube in cross-section |
| $kl$ | Effective length of column |
| $kl_e$ | Length of column below which elastic buckling does not occur |
| $M$ | Limit moment on a cross-section |
| $M_u$ | Ultimate moment capacity with zero thrust |
| $M_x$ | Moment acting about the major axis |
| $M_{xu}$ | Moment capacity with thrust $N_x$ acting |
| $M_{xy}$ | Moment capacity with biaxially eccentric thrust $N_{xy}$ acting |
| $M_y$ | Moment acting about the minor axis |
| $M_{yu}$ | Moment capacity with thrust $N_y$ acting |
| $N$ | Thrust (axial force) |
| $N_{cr}$ | Critical buckling force on slender column |
| $N_u$ | Squash thrust capacity of short column |
| $N_x$ | Limit thrust eccentric about major axis |
| $N_{xy}$ | Limit thrust eccentric about both major and minor axes |
| $N_y$ | Limit thrust eccentric about minor axis |
| $V_c$ | Shear strength of reinforced concrete portion of composite member |
| $V_s$ | Shear strength of structural steel portion of composite member |
| $V_u$ | Combined total shear strength of composite member |
| $\beta$ | $M_{xu}/M_{yu}$ = ratio between flexural strength about major axis and flexural strength about minor axis |
| $\gamma$ | Material reliability safety factor |
| $\epsilon_{cu}$ | Limit strain at which concrete fibers spall |
| $\phi$ | Coefficient, not greater than 0·9, for reducing apparent biaxial bending strength from that indicated by elliptical flexural strength contours on a thrust/moment interaction surface |

## 7.1 INTRODUCTION

Concrete compression members built with structural steel shapes, pipe, or tubes in addition to bar reinforcement are called composite columns. In a generic sense, *all* reinforced concrete columns are composite members, but common terminology classifies as composite only those elements that include structural shapes, pipe, or tubing in addition to concrete. North American terminology at one time included the term 'combination' column to indicate a structural steel or cast iron core encased in concrete but without any additional binding bars or ties. That type of column is no longer permitted by the Building Code of the American Concrete Institute (ACI–318 1983), but under the general definition used for this chapter it would be classified as a composite column.

More than a century ago, pioneers in reinforced concrete technology used concrete with structural angles, pipe, bars, and built-up systems of either steel or cast iron (Emperger, 1907; Burr, 1908; Withey, 1909; Talbot and Lord, 1912; Mensch, 1917). The pioneers had some confidence in the reliability of the iron or steel elements, and they knew that encasing the metal core in concrete provided protection against corrosion from chemicals as well as insulation from any rapid temperature rise in the presence of fire. Any enhancement of stiffness or strength from concrete was frequently ignored. The early engineers who suggested some load capacity for the concrete tended to undervalue the contribution of the concrete. The early investigators recognized that the strength of short columns was the sum of the strengths from each of the component parts. They knew that strength was reduced as slenderness increased, but no systematic studies of flexure plus axial load or even studies of the flexural stiffness for composite columns were reported.

Rational procedures for evaluating slenderness effects in compression members developed rapidly after the introduction of electronic computers. Influences from end connections, misalignment, residual stresses, and overall beam–column behavior were modeled analytically for correlations with laboratory observations. Steel column studies, concentrated on isolated, axially loaded slender members, led to the ECCS column curves mentioned in Section 6.4.3 (Chapter 6). Investigations for concrete columns generally considered the compression member as a part of a monolithic frame. Some composite columns can function much the same as a concrete column. Other composite columns will behave almost the same as steel columns. Structural behavior of composite columns will involve some combinations of steel column and concrete column characteristics that are dependent upon cross-section configurations and upon framing conditions.

## 7.2 TYPES OF COMPOSITE COLUMN

In addition to the basic forms indicated in Figs 6.1–6.3 (Chapter 6), cross-sections for representative types of concrete-encased composite columns are shown in Fig. 7.1 in this chapter. In each of the cross-sections, one or more structural shapes are encased with concrete, and the concrete is further reinforced with longitudinal and transverse tie bars. The variety

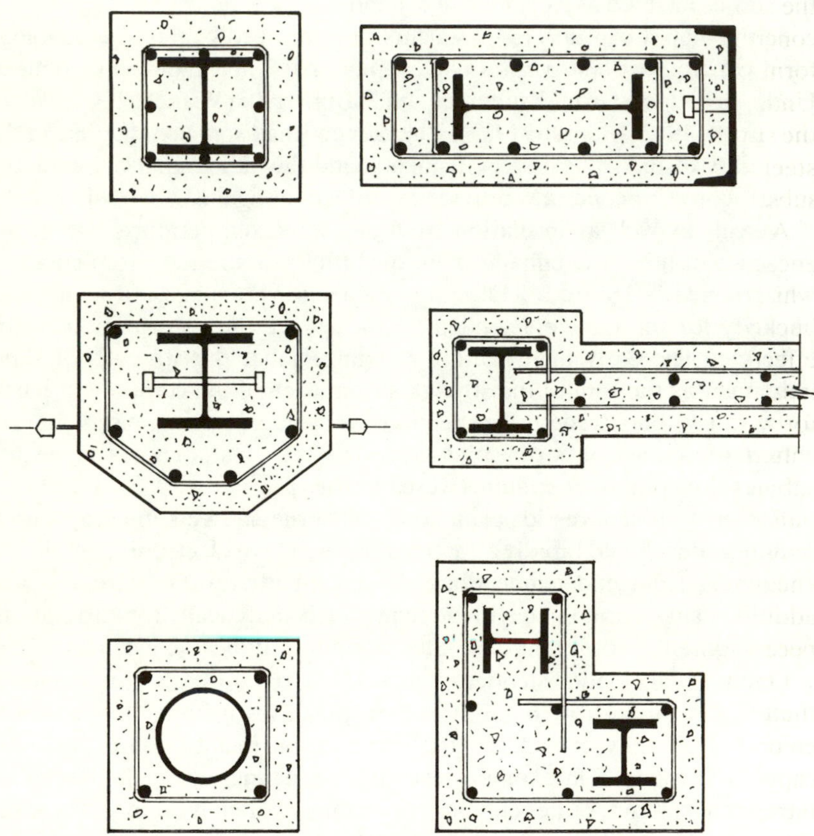

FIG. 7.1. Concrete-encased column cross-sections.

of cross-section shapes also illustrates the potential to use almost any exterior form, in the same manner that any reinforced concrete can be formed. For several of the applications suggested in Fig. 7.1, the structural steel components of the composite columns or walls were erected as a part of a steel erection structure. Composite concrete floor slabs and the concrete encasements of columns were added after the steel structure had been erected several floors ahead of the concrete work.

The composite columns are stiffer than non-composite steel columns that would support the same loads, and composite columns are stronger than reinforced concrete columns of the same size. Supplemental bar reinforcement for the encasement can be made strong enough to prohibit extensive loss of concrete under severe overload. With careful attention to the supplemental reinforcement and to the connections to floor members, concrete-encased steel composite columns have become the preferred form for many seismic-resistant structures in Japan (Wakabayashi, 1984). Under severe flexural overload the concrete encasement cracks, modifying the stiffness and providing effective damping to vibratory movements. The steel cores provide shear capacity and a ductile, tough resistance to subsequent cycles of overload.

A composite column form that is even tougher than the concrete-encased steel-shape column is the concrete-filled steel tube, examples of which are shown in Fig. 6.4. Concrete inside the tube prohibits all inward buckling modes of the steel tube wall, and the tube wall in turn provides effective lateral confinement to the concrete inside the tube. Therefore, local failures cannot occur until material strength limits are reached both for the steel and for the concrete. Concrete-filled steel tube composite columns have been used for earthquake-resistant structures, bridge piers subject to impact from traffic, and for parking level columns in high-rise buildings. Heat transfer from the tube wall into the concrete core provides only a few minutes of delay to the rise in the temperature of the steel encasement during a severe fire. Concrete-filled steel tubes require additional fire-resistant insulation if fire protection of the structure is necessary.

On the basis of strength alone, the composite column is more expensive than a plain steel-shape column of the same capacity, and the composite column is more expensive than the reinforced concrete column of the same capacity. Under some conditions of required structural performance (strength, stiffness, or ductility), construction scheduling, and corrosion or thermal protection, the composite column may offer economic advantages over the plain steel column or the reinforced concrete column.

## 7.3 STRENGTH OF CROSS-SECTIONS

Longitudinal forces can be applied to composite columns by:

- direct attachments to the structural steel or tube;
- bearing against concrete;
- forces in longitudinal bars;
- bearing against steel;
- combinations of these mechanisms.

At a longitudinal distance (perhaps three thicknesses of the cross-section) away from local points at which forces have been applied, it is reasonable and accurate to assume that strains will vary linearly with the distance from a neutral axis of zero strain. If the stress/strain characteristics of steel and concrete can be defined, all the stresses and consequent forces on a composite cross-section can be related to any specified plane of strains that is assumed to exist across the composite section.

Stress/strain relationships for steel shapes, tubes, and reinforcing bars reflect unique values of stress for any specific value of strain. Some representative stress/strain curves for steel are shown in Fig. 7.2. Steel bars, shapes or tubes with a nominal yield stress less than 450 MPa (65 ksi) exhibit a yield plateau region in which stress remains almost constant as strains are increased. Stresses higher than the nominal yield stress can be reached only after strains have become so large that corresponding deformations would be undesirable.

Stress/strain curves for typical concrete cylinder compression specimens 150 mm diameter and 300 mm long (6 in. diameter and 12 in. long) are displayed in Fig. 7.3. The index value for concrete strength is the maximum stress $f_{ck}$ indicated by the stress/strain curve. European standard cube specimens indicate characteristic $f_{cu}$ strength values about 20% greater than those from cylinder specimens. Concrete has so little tensile strength that it is totally discounted and ignored in strength estimates for composite cross-sections. The stiffness (slope of the stress/strain curve) increases with the characteristic strength of concrete. Higher strength concretes have an effective initial modulus of elasticity that increases roughly in proportion to the second or third root of the compressive strength and density of the concrete.

Methods and equations for estimating the strength of short composite columns were suggested in Chapter 6. For composite columns connected to floor members with shear plate or shear angle connectors that transmit negligible moment into the column, the columns may be designed as if each

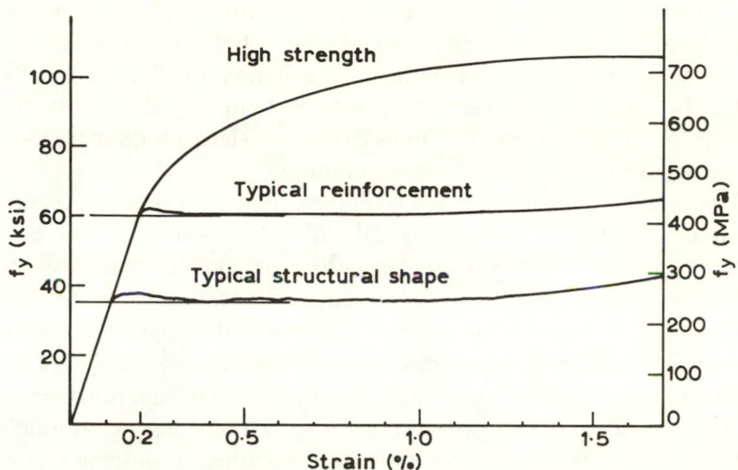


FIG. 7.2. Typical stress–strain curves for steel (Salmon and Johnson, 1980).

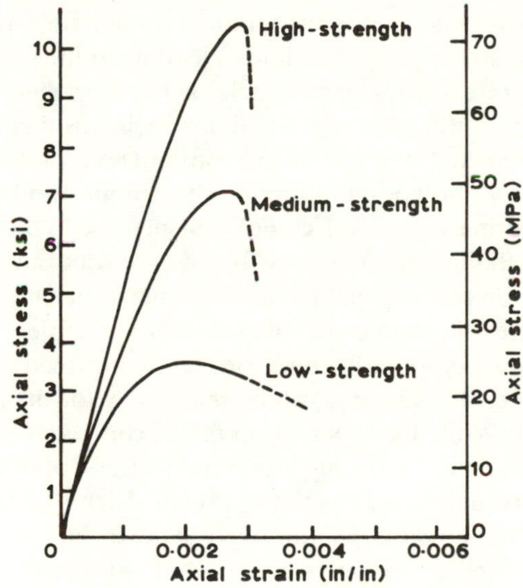


FIG. 7.3. Typical stress–strain curves for concrete (Nilson, 1985).

were loaded only by longitudinal forces without flexural force. The squash load $N_u$ is an adequate measure of strength for short columns, and the squash load serves as an upper limit strength that is reduced according to the slenderness of long, axially loaded columns, as discussed in Section 6.4.2. For many structural applications, columns must resist bending as well as longitudinal force. Columns that are designed for flexure as well as thrust are commonly called beam–columns.

The case has been made that all columns are to some extent eccentrically loaded, and that consequently all columns should be perceived as beam–columns. Most concrete-encased composite structural shapes are connected to floor members through the concrete encasement such that the beam-to-column joints behave as monolithic parts of continuous frames. The required design moments for such columns must be derived from an analysis of the rigid frame. Most building design practices permit the use of elastic analysis procedures for rigid frame design, although limit strength concepts serve as the basis for evaluating cross-section strength. The description of beam–column capacity is most readily presented in the form of thrust/moment interaction diagrams, as suggested in Fig. 6.7. The calculation of coordinates for such diagrams is here discussed in more detail.

At any strain value, the stress/strain response of concrete reflects a unique value of stress only if the time of loading to the strain value is held constant. Concrete specimens to which loads are applied very slowly will appear to be less stiff and less strong than specimens that are loaded very quickly. Concrete that is subjected to a constant level of stress will continue to deform under the constant stress. The amount and the rate of the continuing deformation, called 'creep', will increase as the level of stress is increased, but the rate of increase will decrease with the passage of time. The phenomenon of creep makes virtually impossible any precise analysis of a state of stress in composite columns that are subject to variable and changing amounts of load. A good correlation between changes in stress conditions and changes in load can be made only for the instant after the load is applied. While load remains in place, concrete tends to creep out from under pressure, reducing the actual stress of the concrete and increasing the actual stresses on the steel components of cross-sections.

Although it is impossible quantitatively to determine accurately the state of stress in composite columns for every load stage, it is possible to determine analytically the limit state of strength for steel–concrete composite cross-sections if material properties, $f_{sk}$ for steel and $f_{ck}$ or $f_{cu}$ for concrete, are known. The limit state of strength for composite cross-sections is signalled when some concrete fibers begin to crush and spall. When the strength of concrete disintegrates, steel will be forced to yield, and resistance to load cannot be restored in such circumstances of inelastic instability. A convenient definition of a failure state has been considered to be the load stage at which any concrete fiber reaches a critical limit strain value, commonly set as 0·30%, although any value from 0·25% to 0·40% could produce reasonable results. For the unique condition of uniaxially concentric compression load, the stress/strain curves of Fig. 7.2 suggest that at a strain near 0·2% concrete begins to crush and spall before higher strains are reached. Steel would yield also after concrete fails. Consequently, some design regulations for composite columns limit the yield strength of the structural steel to 380 to 450 MPa (55 to 65 ksi), the stress at which steel strains reach 0·2%. The 'squash' load $N_u$ (eqn (6.27)), with $f_{cd}$ taken to be 0·85 times the cylinder strength or about 0·67 times the cube strength, exists when the neutral axis is located an infinite distance from the cross-section. The same 'squash' load would be indicated when the neutral axis is located at a distance of only 5 or 6 times the member thickness from the section. The existence of a neutral axis signifies the presence of bending, and strength analysis for various locations of the neutral axis will produce coordinates for the thrust/moment interaction diagram.

As the neutral axis is located closer to the centroid of the section, some of the section elements will be strained at values associated with stresses less than the yield or failure stresses. In the presence of a significant strain gradient (neutral axis not more than one column thickness away from the section centroid) there will be some plasticity in the concrete. Edge fibers will not crush and spall immediately after the nominal $f_c$ or $f_{cu}$ stress is reached, as adjacent fibers at lower strains can help resist the maximum stress nearer the edge. The maximum edge strain $\epsilon_{cu}$ that is used analytically as an index of the failure limit state can be taken at values between 0·25% and 0·40%, the specific value having small influence on the analytic magnitude of limit loads $N_x$ and $M_u$. A convenient value of $\epsilon_{cu} = 0{\cdot}3\%$ has been advocated by many investigators, and that value has been used for some of the examples of this chapter. Detailed procedures for the strain-compatible limit strength analysis of reinforced concrete cross-sections are given in all reference textbooks (Ferguson, 1976; Nawy, 1984; Nilson and Winter, 1985; Wang and Salmon, 1983; Johnson, 1975). The same procedures can be used for composite steel–concrete cross-sections. The procedures are tedious to apply, not well suited to design application, but useful as a basis for design tables or computer programs.

Design applications can be made easily if design tables or computer software were available for indicating strength limits for similar configurations of cross-sections. Design applications can also be made easier similarly by the use of simplified, though less accurate, calculation procedures. A simplified force–moment strength interaction function has been recommended by an American Structural Specifications Liaison Committee (Task Group 20, SSLC, 1979) as:

$$N_u = A_s f_{sd} + A_r f_{rd} + 0{\cdot}85 A_c f_{cd} \tag{7.1}$$

$$(N_x/N_u)^2 + M_x/M_u = 1 \tag{7.2}$$

An equation for estimating $M_u$ is given as the sum of yield moment on the structural steel, plus an estimate of the moment capacity for longitudinal reinforcement, plus the moment capacity of a reinforced concrete cross-section with a depth of $0{\cdot}5h_2$, a width of $h_1$, and reinforced by the web (or sides of a tube) of the structural steel:

$$M_u = S_s f_{sd} + A_r f_{rd}(h_2 - 2c_r)/3 + A_w f_{sd}(h_2/2 - A_w f_{sd}/1{\cdot}7 f_{cd} h_1) \tag{7.3}$$

A more general procedure for estimating points on a $N_x$ versus $M_x$ interaction diagram was suggested by Roik and Bergmann (1984). The procedure uses plastic analysis of the cross-section in order to determine a value for $M_x$, and subsequent use of limit forces on the section to find two additional points on the interaction diagram. The assumed plastic stress value for concrete should be taken as 60% of the cube strength or 75% of the cylinder strength of the concrete. There will be no axial force when $M_u$ occurs. Thus, the neutral axis of symmetric shapes can be located by equating compression concrete strength above the neutral axis to the strength of the steel nearest the section centroid (usually part of the web of the steel shape plus some reinforcing bars near the centroid).

The plastic analysis procedure is illustrated with the examples displayed in Figs 7.4, 7.5 and 7.7. The SSLC procedure is used also for examples in Fig. 7.4. Figure 7.4 includes interaction functions determined for the SSLC, plastic analysis, and strain-compatible (ACI–318, 1983) analysis. The concrete stress/strain functions that were used for different strain-compatible analyses are shown in Fig. 7.4c. The ACI rectangular stress block uses a critical failure strain of 0·3%, and the parabola–rectangle stress function used had a critical strain value of 0·4%. The interaction curves show very little influence from the specific details of the concrete stress function assumed at failure of the concrete. The steel stress/strain functions were considered to be elastic-plastic in both tension and compression, with a modulus of elasticity equal to 200 000 MPa (29 000 ksi).

If the results from the strain-compatible analyses were accepted as the 'true' representation of strength, it is apparent that both the plastic analysis and the SSLC simplified procedures produce results that are acceptable for most design applications.

## 7.4 BIAXIAL BENDING

Composite columns for bridge piers, and composite columns for some applications in buildings, must be designed to resist flexural forces that are not restricted to bending about a principal axis of the column cross-section. As indicated in the foregoing discussion, analytic procedures already validated by comparison with test results can be used to compute for any value of thrust, $N_x$ or $N_y$, the flexural capacity to resist moment about a principal axis of symmetric cross-sections. The basic concepts for integration of stresses consistent with strains at the limit state of strength can be applied to any cross-section with any orientation of the zero strain neutral axis. The process has been called tedious and impractical without computer software, even when applied to conditions of major axis flexural strains. It is even more cumbersome when applied to analysis for biaxial bending, but it serves as a useful tool for general investigations that have been used to describe the interaction surfaces of typical cross-sections.

The design of symmetric cross-sections to resist biaxial bending forces can be simplified by the use of analytic approximations of the three-dimensional interaction surface. The serviceability design of steel shapes has used a simple superposition of elastic stresses caused by thrust, bending about the major axis, and bending about the minor axis to obtain a maximum stress that is kept below a serviceability elastic limit stress. For any level of axial stress due to thrust, the plane of constant axial stress intercepts the interaction surface along the straight line between allowable bending stresses for the moment about each principal axis alone. A serviceability limit stress interaction surface is indicated by the dashed lines of Fig. 7.6. Analytic and test data for plain steel shapes (Santathadaporn and Chen, 1973) and for reinforced concrete (Pannell, 1963; Ramamurthy, 1966; Furlong, 1979) indicate that the limit strength interaction surface is not a plane, but that the surface is more accurately described by elliptical contours at constant thrust. Roberts and Yam (1983) suggest the use of an elliptic contour interaction surface for composite columns:

$$(M_x/M_{xu})^2 + (M_y/M_{yu})^2 < 1{\cdot}0\phi \qquad (7.4)$$

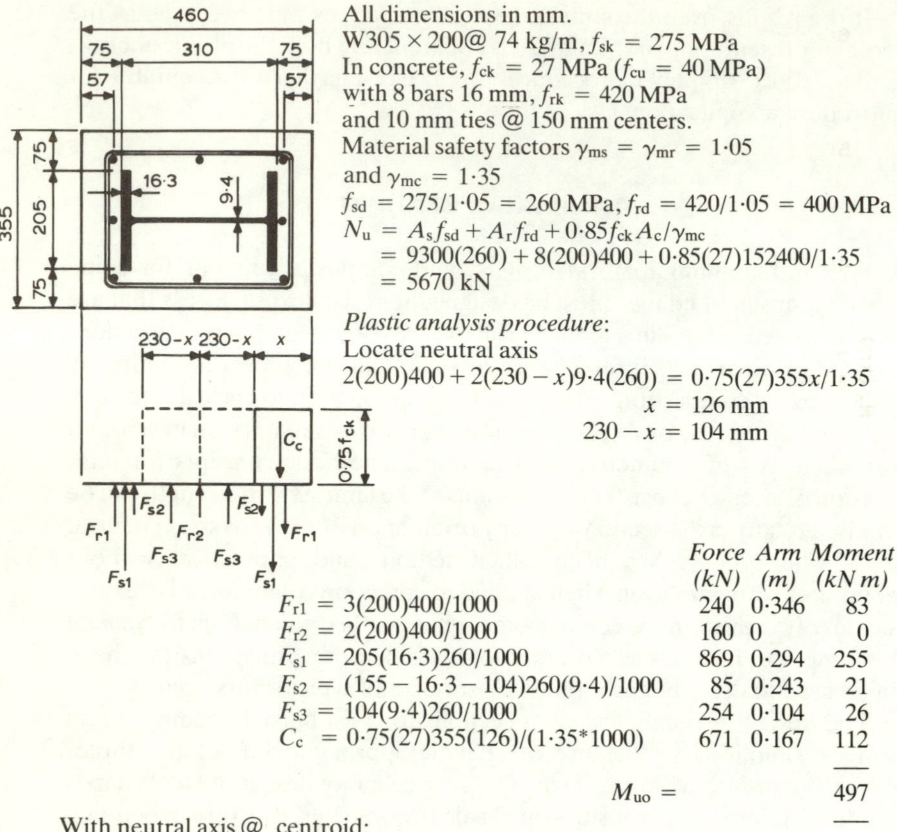


All dimensions in mm.
W305 × 200@ 74 kg/m, $f_{sk} = 275$ MPa
In concrete, $f_{ck} = 27$ MPa ($f_{cu} = 40$ MPa)
with 8 bars 16 mm, $f_{rk} = 420$ MPa
and 10 mm ties @ 150 mm centers.
Material safety factors $\gamma_{ms} = \gamma_{mr} = 1{\cdot}05$
and $\gamma_{mc} = 1{\cdot}35$

$$f_{sd} = 275/1{\cdot}05 = 260 \text{ MPa}, f_{rd} = 420/1{\cdot}05 = 400 \text{ MPa}$$

$$\begin{aligned} N_u &= A_s f_{sd} + A_r f_{rd} + 0{\cdot}85 f_{ck} A_c/\gamma_{mc} \\ &= 9300(260) + 8(200)400 + 0{\cdot}85(27)152400/1{\cdot}35 \\ &= 5670 \text{ kN} \end{aligned}$$

*Plastic analysis procedure*:
Locate neutral axis

$$\begin{aligned} 2(200)400 + 2(230 - x)9{\cdot}4(260) &= 0{\cdot}75(27)355x/1{\cdot}35 \\ x &= 126 \text{ mm} \\ 230 - x &= 104 \text{ mm} \end{aligned}$$

| | *Force (kN)* | *Arm (m)* | *Moment (kN m)* |
|---|---|---|---|
| $F_{r1} = 3(200)400/1000$ | 240 | 0·346 | 83 |
| $F_{r2} = 2(200)400/1000$ | 160 | 0 | 0 |
| $F_{s1} = 205(16{\cdot}3)260/1000$ | 869 | 0·294 | 255 |
| $F_{s2} = (155 - 16{\cdot}3 - 104)260(9{\cdot}4)/1000$ | 85 | 0·243 | 21 |
| $F_{s3} = 104(9{\cdot}4)260/1000$ | 254 | 0·104 | 26 |
| $C_c = 0{\cdot}75(27)355(126)/(1{\cdot}35{*}1000)$ | 671 | 0·167 | 112 |
| $M_{uo} =$ | | | 497 |

With neutral axis @ centroid:

$$N_x = F_{s3} + 0{\cdot}75(27)355(104)/1000{*}1{\cdot}35 = 808 \text{ kN}$$

$$M_u = M_{uo} + \tfrac{1}{2}104(N_x + F_{s3})/1000 = 552 \text{ kN m}$$

With neutral axis @−104:

$$N_x = 2(808) = 1616 \text{ kN}$$

$$M_u = M_{uo} = 497 \text{ kN m}$$

*SSLC equation procedure*:

$$\begin{aligned} M_{uo} &= [S_s f_{sd} + A_r f_{rd}(h_2 - 2c_r)/3 + A_w f_{sd}(\tfrac{1}{2}h_2 - A_w f_{sd}\gamma_{mc}/1{\cdot}7 f_{ck} h_1)]10^{-6} \\ &= [(1{\cdot}04\text{E}6)260 + 8(200)400(460 - 114)/3 + 9{\cdot}4(277)260 \\ &\quad \{\tfrac{1}{2}460 - 9{\cdot}4(277)260(1{\cdot}35)/1{\cdot}7(27)355\}]10^{-6} \\ &= 462 \text{ kN m} \end{aligned}$$

Use graphical construction of parabola to $N_u$.

FIG. 7.4a. Longitudinal strength of concrete-encased steel shape.

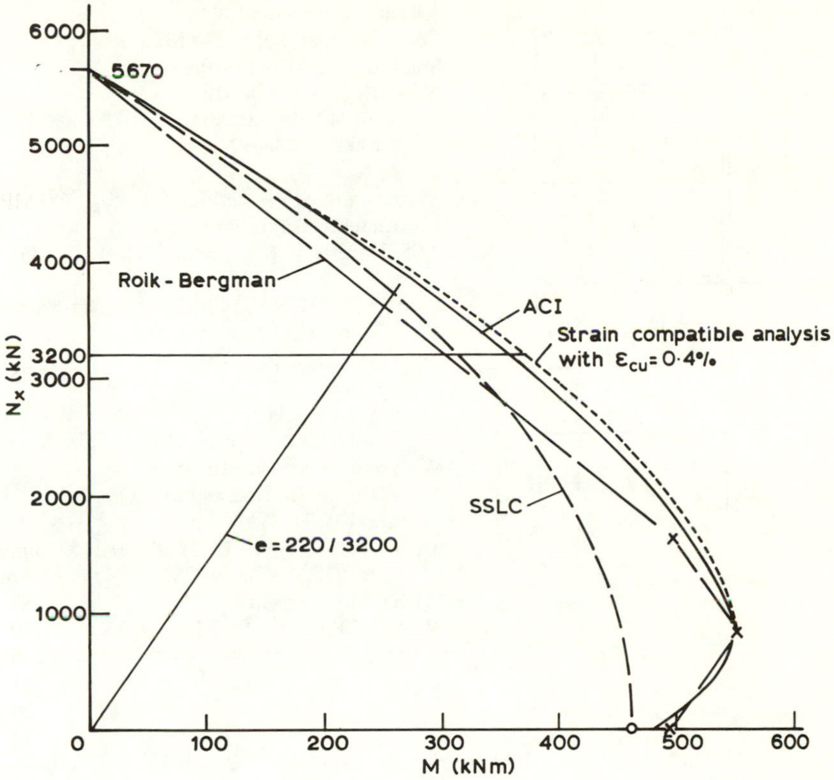

6000
5670
5000
4000
Roik - Bergman
ACI
Strain compatible analysis
with $\varepsilon_{cu} = 0{\cdot}4\%$
3200
3000
$N_x$ (kN)
2000
SSLC
e = 220 / 3200
1000
0
100
200
300
400
500
600
M (kNm)

FIG. 7.4b. Longitudinal strength of concrete-encased steel shape.

$f_{cd}$
Probable real curve
$0{\cdot}85 f_{cd}$
ACI stress block
Assumed for
calculations
with $\varepsilon_{cu} = 0{\cdot}4\%$
0·1
0·2
0·3
0·4
Strain (%)

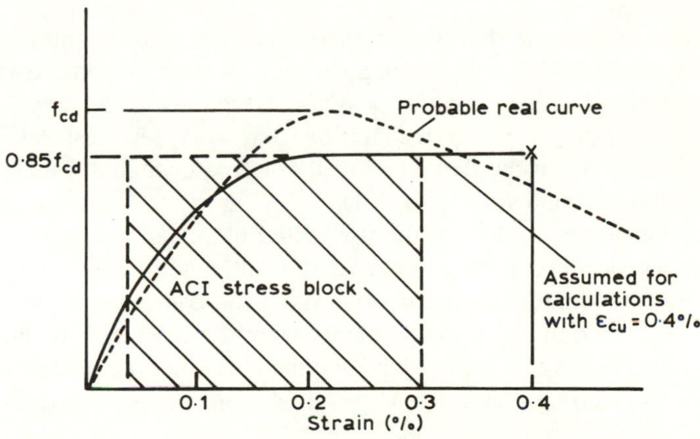

FIG. 7.4c. Longitudinal strength of concrete-encased steel shape.

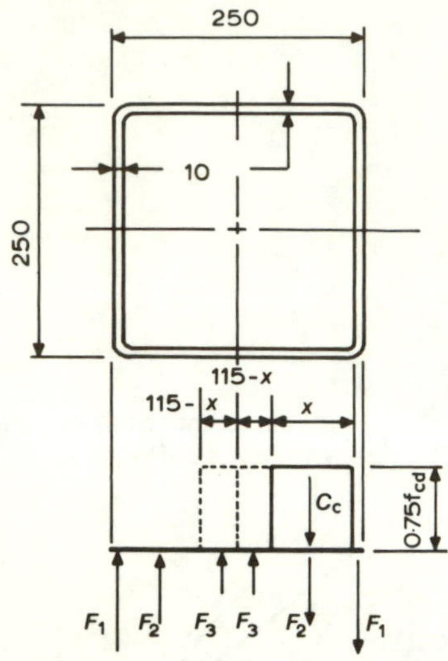


All dimensions in mm.
Concrete filled $f_{cd} = 27$ MPa
Steel tube $f_{sd} = 300$ MPa
$N_u = A_s f_{sd} + 0{\cdot}85 f_{cd} A_c$
$= [4(240)10(300) + 0{\cdot}85(27)230^2]/1000 = 4094$ kN

*Plastic analysis procedure*
Locate neutral axis with $N_x = 0$
$0{\cdot}75(27)230x = (115 - x)2(20)300$
$x = 82{\cdot}85$
$C_c = 0{\cdot}75(27)230(82{\cdot}85)/1000 = 385{\cdot}8$ kN
$F_1 = 300(10)250/1000 = 720{\cdot}0$
$F_2 = 300(20)82{\cdot}85/1000 = 497{\cdot}1$
$F_3 = 300(20)(115 - x)/1000 = 192{\cdot}9$
$M_u = [385{\cdot}8(111 - \frac{1}{2}x) + 720(250 - 10) + 497{\cdot}1(115 - \frac{1}{2}x)2]/1000 = 275{\cdot}3$ kN m
With neutral axis at centroid:
$N_x = F_3 + 0{\cdot}75(27)230(115 - x)/1000 = 342{\cdot}6$ kN
$M_x = [(115 - x)F_3 + 0{\cdot}75(27)230(115 - x)^2/2000]/1000 + M_u = 283{\cdot}9$ kN m
With neutral axis at $-(115 - x)$:
$N_x = 2(342{\cdot}6) = 685{\cdot}2$ kN
$M_x = M_u = 275{\cdot}3$ kN m

*SSLC equation procedure*
$M_u = S_s f_{sd} + A_w f_{sd}(\frac{1}{2}h - A_w f_{sd}/1{\cdot}7 f_{cd} b)$
$= [738600(300) + \frac{1}{2}(230)20(300)(115 - 230*300/1{\cdot}7*27*230)]10^{-6} = 256{\cdot}0$ kN m
Graphically construct parabola from $N_u$ to $M_u$.

FIG. 7.5a. Longitudinal strength of concrete-filled steel tube.

Analytic data indicate that the interaction surface contours may reflect a concave 'dip' for planes of bending through or near diagonal corners of cross-sections. In recognition of possible reductions in capacity for such planes of bending, and in recognition that very few test results are available, it is suggested that a factor of not greater than 0·9 should yield safe strength estimates with eqn (7.4).

For applications in design, the use of the elliptical interaction surface suggests a procedure that uses a modified resultant moment assumed to act in uniaxial bending about the major axis of the composite cross-section. Given a required thrust $N$ and moments $M_x$ and $M_y$, let it be assumed that the $x$ axis is the major axis. Let it be assumed also that the ratio $\beta$ between zero thrust bending capacities $M_{uy}$ and $M_{ux}$ can be estimated. A cross-

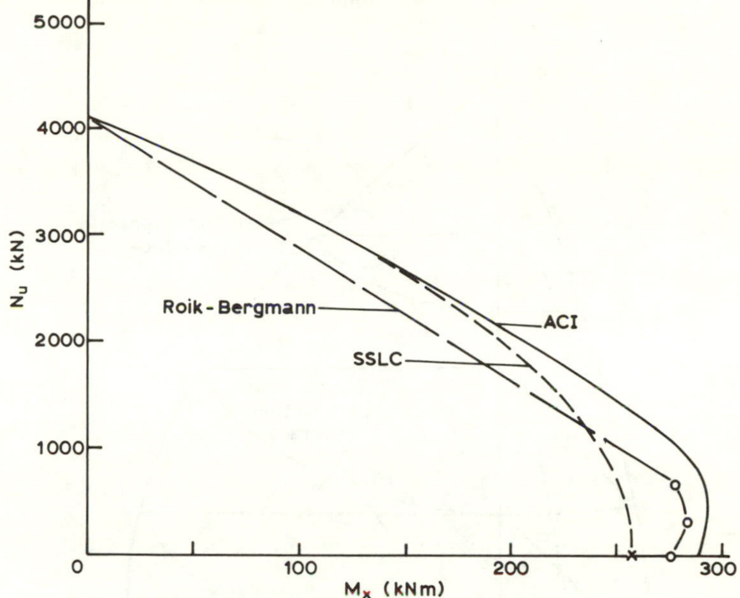


FIG. 7.5b. Longitudinal strength of concrete-filled steel tube.

section can be selected on the basis of strength to support an effective resultant moment $M_{xy}$ assumed to act only about the major axis:

$$M_{xy} = 1{\cdot}15\sqrt{M_x^2 + M_y^2} \tag{7.5}$$

The 1·15 factor is suggested to accommodate a $\phi$ factor of the order of 0·9 in eqn (7.4), against which the selected cross-section should be checked.

The reciprocal thrust relationship, eqn (6.27), is a very convenient and easily remembered expression for estimating skew bending capacity $N_{xy}$, but it tends to give results significantly lower than the actual cross-section strength. It does offer a lower bound value for checking the strength of a trial section.

The use of eqns (7.4) and (7.5) will be demonstrated with applications for the column cross-section analyzed for principal axis bending capacities in Fig. 7.4. Let it be assumed that a cross-section similar to that of Fig. 7.4 is required to resist limit loads $N_{xy} = 3200$ kN, $M_x = 220$ kN m, and $M_y = 70$ kN m. The $\beta$ factor for eqn (7.5) can be estimated approximately

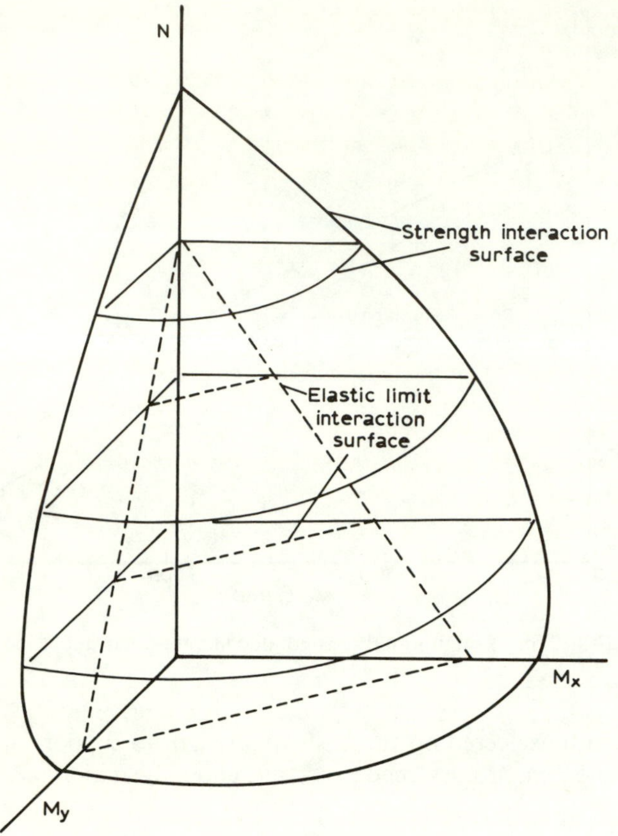


FIG. 7.6. Biaxial bending interaction surfaces.

simply as the ratio between the lengths of the longer and the shorter sides, $\beta = 460/355 = 1{\cdot}30$. Then $M_{xy} = 1{\cdot}15\sqrt{220^2 + (1{\cdot}30 \times 70)^2} = 275$ kN m.

In Fig. 7.4, the SSLC and plastic analysis graphs indicate that for $N = 3200$ kN, a moment capacity greater than 275 kN m exists; therefore the cross-section should be adequate. The cross-section capacity will be checked using eqn 7.4 with the value of $M_{xu}$ taken as 315 kN m from Fig. 7.4 and $M_{yu} = 209$ from Fig. 7.7b when $N_{xy} = 3200$ kN. A value $\phi = 0{\cdot}9$ in eqn 7.4 gives:

$$0{\cdot}9 = \sqrt{(220/315)^2 + (M_y/230)^2}$$

for which $M_y = 100$ kN m, which is larger than the required amount of 70 kN m.

The cross-section strength can be checked also with the more conservative eqn (6.27). A value $N_x = 3770$ kN can be read from Fig. 7.4 for the eccentricity 220/3200 = 0·069 m. A value $N_y = 4240$ can similarly be read from Fig. 7.7b for the eccentricity 70/3200 = 0·022 m. The amount of $N_u$ was computed to be 5670 kN in Fig. 7.4. These values in eqn (6.27) give:

$$1/N_{xy} = 1/3770 + 1/4240 - 1/5670$$

for which $N_{xy} = 3080$ kN.

The required value for $N_{xy}$ is greater than 3080, and the cross-section is not adequately strong according to eqn (6.27). If that relationship were to govern the design, it is suggested that a larger steel shape be used, or else the longitudinal reinforcement be made larger. If the elliptical contour criterion of eqn (7.4) were accepted, the cross-section could be considered adequate.

## 7.5 SHEAR STRENGTH

The shear strength of composite columns can be estimated as the sum of shear capacity for the structural steel alone plus the shear capacity of the reinforced concrete in the cross-section. Plain concrete without reinforcement will have no shear strength after either flexural or shear cracking, unless it is confined inside relatively short lengths of steel pipe or tubing. Structural shapes and tubing resist transverse shear through the stiffness of plates (webs, flanges, or walls) parallel to the shear forces. The limit shear stress of steel plates is taken as $f_{sd}/\sqrt{3}$, which is approximately equal to $0{\cdot}6f_{sd}$. Steel plates encased in concrete will not buckle elastically in shear, and the full material shear capacity can be used for strength estimates. Steel walls for concrete-filled tubes or boxes could be thin enough to buckle in shear, and in the absence of laboratory data to reveal actual composite response, the shear strength of empty steel or cold formed metal tubes should be accepted as the shear capacity of composite columns made with thin-walled steel tubing or boxes filled with concrete. Suzuki and Kato (1984) indicate that in relatively short concrete-filled tubes or boxes, the confined concrete can act as a diagonal compression strut together with tension field action of the steel side walls to resist limit shears that are significantly greater than the shear capacity of the steel sidewalls alone.

The truss action of reinforcing tie bars, $V_c$, can be estimated simply as the tensile strength of the tie bar multiplied by the number of tie bars in a length equal to one thickness of a cross-section. The total amount of $V_c$ is limited by the compression strut strength of the concrete, which can be estimated as the stress limit, $0{\cdot}15f_{cd}$, times the concrete area outside the edges of the structural shape. Concrete within the perimeter of the steel shape cannot react with the ties as part of the shear-resistant truss. Shear strength for composite columns can be computed from:

$$V_u = V_s + V_c \tag{7.6}$$

with

$$V_s = 0{\cdot}4A_{vs}f_{sd} \tag{7.6a}$$

and

$$V_c = A_{vr}f_{rd}h/s < 0{\cdot}15f_{cd}b_w h \tag{7.6b}$$

Shear strength estimates for the cross-sections of Fig. 7.4 are demonstrated in Fig. 7.8. The typical encased steel H column with nominal longitudinal and transverse tie reinforcement will have more capacity to resist shears in the plane of the flanges than in the plane of the web. The shear strength available from concrete and reinforcement will be less than that from the structural shape when concrete cover around the shape is thick enough only to provide minimum cover for the reinforcement around the structural shape.

Earthquake-resistant columns must be designed for shear capacities large enough to resist the largest shear that flexural strength at the ends can deliver to the column, generally equal to twice the largest moment strength (under any axial force) divided by the length of the column (Suzuki and Kato, 1984; ACI–318, 1983). Most concrete-encased composite columns of structural shape meet that requirement readily for shears in the plane of the flanges, as the flexural strength in that plane is relatively smaller than the flexural strength in the plane of the web, and flanges often possess shear strength greater than that of the web. The example cross-section of Fig. 7.4 was found to have a maximum flexural strength of 552 kN m (4885 in.-k) and a shear capacity of 494 kN (111 k) in the plane of the web. The minimum length acceptable for that section in a seismic-resistant structure would be $2M_x/V_u = 2 \times 552/494 = 2{\cdot}23$ m (87·8 in.). If the length of the column were shorter, the shear strength would have to be increased. Ties of larger size or closer spacing would provide more shear strength, but the calculations in Fig. 7.8 indicate that the upper limit of reinforced concrete shear contribution would be 207 kN (46·55 k), regardless of the tie size with closer spacing. The comparable minimum length for the concrete-filled tube of Fig. 7.5 is $2M_x/V_u = 2 \times 278\,000/600 = 460$ mm (18 in.).

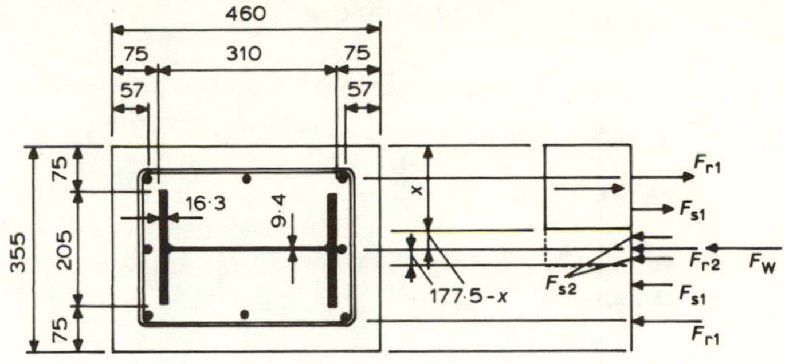


Material safety factors
$\gamma_{ms} = \gamma_{mr} = 1{\cdot}05,\ \gamma_{mc} = 1{\cdot}35$

All dimensions in mm.
W305 × 200@74 kg/m, $f_{sk} = 275$ MPa
In concrete, $f_{ck} = 27$ MPa with 6 bars 16 mm,
$f_{rk} = 420$ MPa and 10 mm ties@150 mm centers.
$f_{sd} = 275/1{\cdot}05 = 260$ MPa,
$f_{rd} = 420/1{\cdot}05 = 400$ MPa
$N_u = 5670$ as for Fig. 7.4.

*Plastic analysis procedure* Bending about the weak axis
Locate neutral axis for zero thrust:

$$(310 - 32{\cdot}6)9{\cdot}4(260) + 2(200)400 + 32{\cdot}6(177{\cdot}5 - x)260 = 0{\cdot}75(27)460x/1{\cdot}35$$

$$x = 152{\cdot}3 \text{ mm}$$

$$177{\cdot}5 - x = 25{\cdot}2 \text{ mm}$$

| | Force (kN) | Arm (mm) | Moment (kN m) |
|---|---|---|---|
| $F_{r1} = 3(200)400/1000$ | 240 | 205 | 49 |
| $F_{r2} + F_w = 277{\cdot}4(9{\cdot}4)260 + 2(200)400/1000$ | 838 | 0 | 0 |
| $F_{s1} = (102{\cdot}5 - 25{\cdot}2)16{\cdot}3(2)260/1000$ | 655 | 127·7 | 84 |
| $F_{s2} = 25{\cdot}2(16{\cdot}3)2(260)/1000$ | 214 | 25·2 | 5 |
| $C_c = 0{\cdot}75(27)460(152{\cdot}3)/1350$ | 1051 | 101·35 | 107 |

$$M_{yu} = 245 \text{ kN m}$$

With neutral axis at centroid:
$N_y = F_{s2} + 0{\cdot}75(27)460(25{\cdot}2)/(1{\cdot}35{*}1000) = 388$ kN
$M_y = 245 + \frac{1}{2}25{\cdot}2(388 + 107)/1000 = 253$ kN m
With neutral axis at 25·2 mm below centroid:
$N_y = 2(388) = 776$ kN $\quad M_y = M_{yu} = 245$ kN m

*SSLC equation procedure*

$$M_{yu} = [228\,000(260) + 8(200)400(355 - 114)/3 + 9{\cdot}4(310)260 \{177{\cdot}5 - 9{\cdot}4(310)260(1{\cdot}35)/(1{\cdot}7{*}27{*}460)\}]/1000 = 209 \text{ kN m}$$

FIG. 7.7a. Weak axis beam–column strength of concrete-encased steel shape.

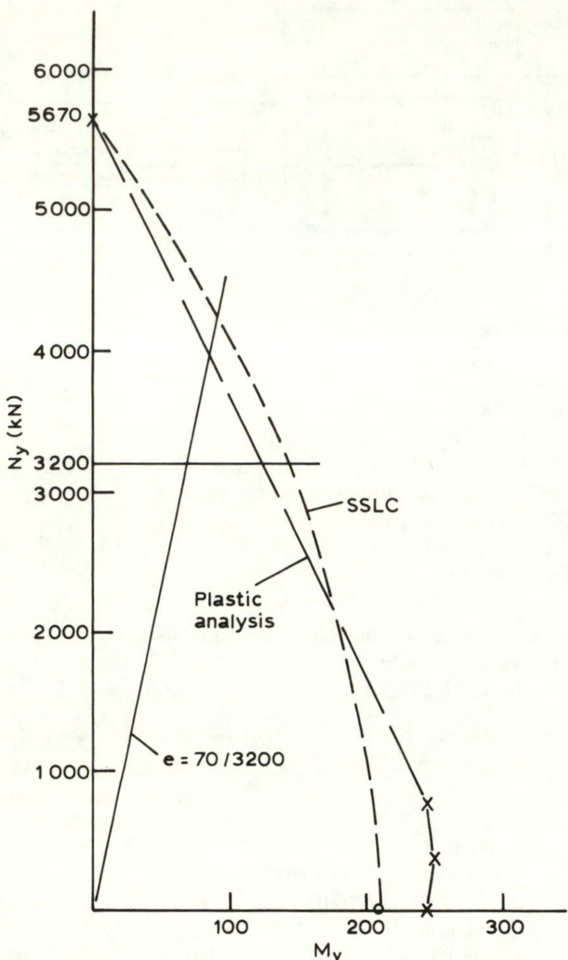


FIG. 7.7b. Weak axis beam–column strength of concrete-encased steel shape.

## 7.6 SLENDERNESS OF COMPOSITE COLUMNS

Rational evaluations of slenderness require knowledge of flexural stiffness, $EI$, for cross-sections. Due to the creep phenomenon under sustained

*Concrete-encased steel shape* (Fig. 7.4)
Shears in the plane of the web:
Limit $V_c = 0{\cdot}15 f_c' b_w h = 0{\cdot}15(3{\cdot}6)2(2{\cdot}96)18 = 57{\cdot}5$ k
$V_{sx} = 0{\cdot}4 A_{vs} f_{sd} = 0{\cdot}4(12{\cdot}19)0{\cdot}37(42) = 75{\cdot}8$ k
$V_{cx} = A_{vr} f_{rd} h/s = 2(0{\cdot}11)60(18)/6 = \underline{39{\cdot}6}$ 57·5 OK

$$V_{ux} = 115{\cdot}4 \text{ k}$$

Shears in the plane of the flanges:
Limit $V_c = 0{\cdot}15 f_c' b_w h = 0{\cdot}15(3{\cdot}6)2(2{\cdot}905)14 = 43{\cdot}9$ k
$V_{sy} = 0{\cdot}4 A_{vs} f_{sd} = 0{\cdot}4(2)8{\cdot}08(0{\cdot}64)42 = 173{\cdot}8$ k
$V_{cy} = A_{vr} f_{rd} h/s = 2(0{\cdot}11)60(14)/6 = \underline{30{\cdot}8}$ 43·9 OK

$$V_{uy} = 204{\cdot}6 \text{ k}$$

*Concrete filled steel tube* (Fig. 7.5)
Assign no shear strength to the unreinforced concrete inside the tube.
$V_u = 0{\cdot}4 A_{vs} f_{sd} = 0{\cdot}4(2)250(10)300/1000 = 600$ kN

FIG. 7.8. Calculations for shear strength.

stress, concrete does not have a constant value for its modulus of elasticity. However, structural steel provides most of the flexural stiffness for concrete-filled tubes or boxes, and for large encased shapes with bending about the major axis of the shape. As indicated in the discussions of buckling loads in Section 6.4.3, the stiffness $EI$ of cross-sections under initial applications of service loads (relatively low stress values) can be taken as the sum of flexural stiffnesses for all component parts of the section:

$$EI = E_s(I_s + I_r) + E_c I_c \tag{7.7}$$

The value of $I_c$ will diminish after the concrete cracks in flexural tension, and the effective value of $E_c$ will be reduced from long-time loading at high levels of compression stress.

The axial force $N_{cr}$ at which composite columns should be expected to buckle elastically can be computed as for any slender compression member:

$$N_{cr} = \pi^2 EI/(kl)^2 < 0{\cdot}5 N_u \tag{7.8}$$

Composite columns short enough that their effective length is low enough to produce values of $N_{cr}$ greater than $0{\cdot}5N_u$ are not truly ‘slender’ columns.

They are columns of intermediate slenderness. The critical axial load capacity of intermediate slenderness columns involves some inelastic instability combined with limits to the material strength of the components of the column. Use of Perry-Robertson failure curves or the ECCS criteria for various types of steel shape, as presented in Chapter 6, is certainly valid. The uncertainties associated with concrete stiffness as a part of composite columns tend to diminish the significance of precise definitions for the most appropriate column curves applied to composite cross-sections.

A simple intermediate slenderness function that can be applied to composite columns is the parabolic reduction of axial capacity from $N_u$ to $0{\cdot}5N_u$ as the effective length increases from 0 to the effective length $kl_e$ associated with the elastic behavior of slender columns under an axial force of $N_u$. The relationships can be expressed:

$$kl_e = \sqrt{2\pi^2 EI/N_u} \tag{7.9}$$

For $kl$ less than $kl_e$:

$$N_{cr} = N_u[1 - 0{\cdot}5(kl/kl_e)^2], \quad N_{cr} < N_u \tag{7.10}$$

For $kl$ greater than $kl_e$, use eqn (7.8).

An application of eqns (7.9) and (7.10) to the example cross-sections of Fig. 7.4 was used to determine values for the graphs of $N_{cr}$ versus effective length shown in Fig. 7.9. The graphs represent an ideal limit load capacity for each cross-section, as no material reliability or load factor considerations have been included. Design regulations for buildings require that material reliability factors should be applied to the values of $f_c$ and to steel yield stresses in a straightforward manner. The reliability factors that are appropriate for material stiffnesses $E_c$ and $E_s$ are less apparent. Concrete at limit load conditions will exhibit creep, and truly slender composite columns will exhibit some cracking of concrete in tension before buckling occurs. Safe design requires that flexural stiffness should not be underestimated; instead, harmful deformations should be overestimated through the intentional use of reduced effective stiffness values $E_s$ and $E_c$ for safety.

It is recommended that the effective moment of inertia for concrete at limit load be taken as not more than two-thirds that of the nominal outline of the concrete cross-section, and the value used for modulus of elasticity at limit load on concrete be half the initial value of $E_c$. Stiffness reductions for reinforcement and for structural steel need not be as severe as those for

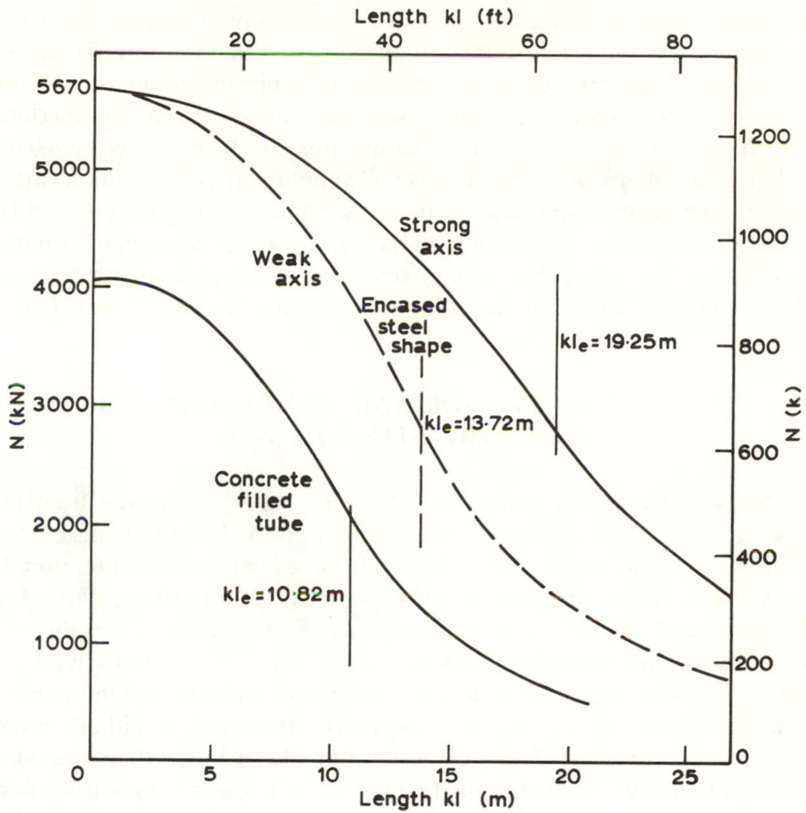


| | *Concrete-encased steel shape* | | |
|---|---|---|---|
| | *strong axis* | *weak axis* | *Concrete-filled steel tube* |
| $I_s$ (mm$^4$) | 161·2E6 | 46·8E6 | 92·3E6 |
| $I_r$(mm$^4$) | 35·9E6 | 17·4E6 | 0 |
| $I_c$ (mm$^4$) | 2680·0E6 | 1650·0E6 | 233·2E6 |
| $EI$(N mm$^2$) | 106·4E12 | 54·1E12 | 24·3E12 |

FIG. 7.9. Column capacity versus effective length.

concrete, but at the limit load condition, 50–75% of the nominal $E_s I_s$ values would be reasonable for design.

In some instances it is obvious that flexural stiffness of structural steel will govern the buckling (slenderness) behavior of composite columns.

Concrete-filled steel tubes are stiffened very little by concrete, to give one example. In other instances it is equally apparent that reinforced concrete behavior will govern buckling response for composite columns, as in the weak axis direction of the encased shape of Fig. 7.4. In the instances for which the steel or the concrete predominates the stiffness, neglecting the other components of cross-sections will simplify eqn (7.7) without introducing significant error to the calculations. For example, good estimates of slenderness effects can be made on the basis of the predominant radius of gyration either for steel or for concrete after an effective length has been determined, without computing $K_1$ factors as indicated in eqn (6.12).

## 7.7 AUXILIARY REINFORCEMENT AND CONNECTIONS TO COMPOSITE COLUMNS

Laboratory studies of composite column strength are derived from test specimens carefully loaded to transmit forces into all components of steel and concrete; or, in some studies, loads have been applied intentionally only to concrete or only to steel (Wakabayashi, 1984; Furlong, 1967). Test loads applied only to concrete can produce local crushing before the total capacity of the composite cross-section is developed; similarly, loads applied only to steel can produce yielding with local deformation before remaining parts of the composite sections participate. In either circumstance, the capacity of the column would be smaller than the ideal formulation suggests with eqns (7.1) and (7.2). Connections from floor members to composite columns must be proportioned with some consideration for the interaction between concrete and steel in the column. The concrete inside filled tubes acquires load from external attachments to the tube only through shear friction at the smooth interface inside the tube. Since the concrete inside is rigidly confined, any deformation of the tube, local or overall, will create pressure against the smooth interface and enhance shear transfer. It may help the local transfer of force to let the attachment plates penetrate through the tube provided the plate within the tube does not obstruct the placement of concrete.

Composite interaction between structural steel inside reinforced concrete encasements can also occur through shear friction. However, as local or overall deformation of the column takes place, transverse pressure at the smooth interfaces between concrete and structural steel is less certain than it would be for filled tubes. Circumferential binding reinforcement certainly helps, but no test data are available to indicate optimum or even minimum requirements. Any appurtenances along smooth faces of the structural shape will improve shear transfer. Studs, bars, or shelf angles welded to the steel shape can be designed to develop specific amounts of shear through bearing against concrete. Recommendations for the shear capacity of studs, bars, or other devices used with composite beams can be applied as well for such devices used with composite columns.

## REFERENCES

AMERICAN CONCRETE INSTITUTE (1983) *Building code requirements for reinforced concrete*. ACI–318.

AISC TASK GROUP 20, Structural Stability Research Council (1979) A specification for the design of steel–concrete composite columns. *Eng. J.*, American Institute of Steel Construction, **16**(4), Chicago, 101–15.

BURR, W. H. (1908) The reinforced concrete work of the McGraw Building. *ASCE Trans.*, **60**(1075), 443 *et seq*.

EMPERGER, F. V. (1907) Welche statische Bedeutung hat die Einbetonierung einer Eisensaule? *Beton Eisen*, Berlin, 172–4.

FERGUSON, P. M. (1976) *Reinforced concrete design*, 4th edn, Wiley & Sons, New York.

FURLONG, R. W. (1967) Strength of steel encased concrete beam columns. *ASCE, J. Struct. Div.*, **93**(ST5), Oct., 119.

FURLONG, R. W. (1979) Concrete columns under biaxially eccentric thrust. *ACI J.*, Oct.

JOHNSON, R. P. (1975) Composite structures of steel and concrete, Vol. 1. Halsted Press, London, 153–77.

MENSCH, J. L. (1917) Tests on columns with cast iron core. *Proc., ACI*, **13**, Detroit, 22 *et seq*.

NAWY, E. (1984) *Reinforced concrete*. Prentice-Hall, NY.

NILSON, A. H. (1985) Design implications of current research on high strength concrete. *High strength concrete*, ACI, SP–87, 108.

NILSON, A. and WINTER, G. (1985) *Design of concrete structures*. 10th edn, McGraw-Hill.

PANNELL, F. N. (1963) Failure surfaces for members in compression and biaxial bending. *ACI J.*, Jan., 129–40.

RAMAMURTHY, L. N. (1966) Investigation of the ultimate strength of square and rectangular columns under biaxially eccentric loads. *Symposium on reinforced concrete columns* (SP13), American Concrete Institute, Detroit, 263–89.

ROBERTS, E. H. and YAM, L. C. P. (1983) Some recent methods for the design of steel, reinforced concrete, and composite steel concrete columns in the UK. *ACI J.*, March–April, 147.

ROIK, K. and BERGMANN, R. (1984) Composite columns—design examples for construction. *Composite and mixed construction*, Proc. US/Japan Joint Seminar, Seattle, July (publ. ASCE, NY), 272–4.

SALMON, C. and JOHNSON, J. (1980) *Steel structures*. 2nd edn, Harper & Row, New York, 41.

SANTATHADAPORN, S. and CHEN, W. F. (1973) Analysis of biaxially loaded steel H columns. *J. Struct. Div., ASCE*, March, 505.

SPRINGFIELD, J. (1975) Design of columns subject to biaxial bending. *Eng. J.*, American Institute of Steel Construction, 3rd Quarter, Chicago, 73–81.

SUZUKI, HIROYUKI and KATO, B. (1984) Shear strength of concrete filled box elements. *Composite and mixed construction*, Proc. US/Japan Joint Seminar, Seattle, July (publ. ASCE, NY), 254–66.

TALBOT, A. N. and LORD, A. R. (1912) Tests of columns: an investigation of the value of concrete as reinforcement for structural steel columns. *Eng. Exp. Sta. Bull.* No. 56, University of Illinois (Urbana).

WAKABAYASHI, M. (1984) Recent developments for composite buildings in Japan. *Composite and mixed construction*, Proc. US/Japan Joint Seminar, Seattle, July (publ. ASCE, NY), 241.

WANG, CHU-KIA and SALMON, C. (1983) *Reinforced concrete design*, 4th Edn, Harper & Row, New York, 416–75.

WITHEY, M. O. (1909) Tests of plain and reinforced concrete columns. Bull. No. 300 (*Engineering Series*, **5**(2)), Univ. Wisconsin (Madison), 143–88.

*Chapter 8*

# DESIGN OF COMPOSITE BRIDGES (CONSISTING OF REINFORCED CONCRETE DECKING ON STEEL GIRDERS)

J. B. KENNEDY

*Department of Civil Engineering, University of Windsor, Ontario, Canada*

and

N. F. GRACE

*Department of Advanced Technology, Giffels Associates, Southfield, Michigan, USA*

*SUMMARY*

*Procedures for the analysis and design of composite concrete deck slab-on-steel girder bridges are presented. Both static and dynamic load effects are considered; the problem of fatigue failure of components such as shear connectors, reinforcing steel, steel girder, steel diaphragm and concrete slab is briefly discussed. Based on a realistic vertical temperature distribution, thermal stress analyses of simply-supported and continuous composite bridges are given. The advantages of prestressing portions of the concrete deck slab in the negative moment regions and in using steel diaphragms, welded to the longitudinal steel girders, are highlighted.*

## NOTATION

| | |
|---|---|
| $A_D$ ($A_G$) | Cross-sectional area of steel diaphragm (girder) |
| $A'_s$ | Area of compression reinforcing steel |
| $A_{sr}$ | Area of longitudinal steel in slab in negative moment region |

| | |
|---|---|
| $A_w$, $A_{tf}$ | Area of web and of top flange of steel girder, respectively |
| $a$ | Depth of compression stress block |
| $2b$ | Width of composite bridge |
| $b_e$ | Effective width of concrete deck slab |
| $D$ | $E_c b_e t^3/12(1-\nu^2)$ |
| $D_1$, $D_2$ | Coupling rigidities arising from Poisson's ratio effect |
| $D_x$ ($D_y$) | Flexural rigidity of bridge per unit width (length) |
| $D_{xy}$ ($D_{yx}$) | Longitudinal (transverse) torsional rigidity of bridge |
| DLA | Dynamic load allowance |
| $d$, $H$ | Diameter and height of shear stud, respectively |
| $d_s$ ($d_d$) | Depth of girder (diaphragm) |
| $d_w$ | Depth of girder web |
| $E_c$ ($E_s$) | Elastic modulus of concrete (steel) |
| $e$ | Clearance distance between diaphragm and concrete deck slab |
| $e_x$ ($e_y$) | Depth of neutral plane from top fibre for bending about $y$ ($x$) axis |
| $F$ | Interface force |
| $F_{sr}$ | Allowable range of stress |
| $F_y$ ($F_s$) | Allowable yield stress in tension (shear) |
| $f'_c$ | 28-day compressive strength of concrete |
| $f'_y$ | Allowable yield stress in compression reinforcing steel |
| $I$ | Impact factor |
| $I_G$ ($I_D$) | Moment of inertia of girder (diaphragm) with respect to its neutral axis |
| $I'_x$ ($I'_y$) | Moment of inertia of girder (diaphragm) with respect to neutral axis of composite section |
| $I_t$ | Moment of inertia of transformed composite slab-on-girder section (transformed to steel) |
| $L$ | Span length of composite bridge |
| $M_R$ | Thermally-induced moment at interior support |
| $M_r$ | Moment of resistance |
| $M_x$ ($M_y$) | Moment due to factored load effects about the $y$ ($x$) axis |
| $m_d$ ($m_{dc}$) | Moment due to specified (superimposed dead) load prior to (after) attainment of 75% $f'_c$ |
| $m_l$ | Moment due to specified live load |
| $N$ | Number of shear connectors |
| $n$ | Modular ratio, $E_s/E_c$ |
| $Q$ | Interface couple |
| $S$ ($S'$) | Spacing of steel girders (diaphragms) |

| | |
|---|---|
| $S_1$ | Shear range per unit length |
| $S_{3t}$ | Section modulus of the transformed composite section referred to the bottom fibre, using modular ratio = $3n$ |
| $S_G$ ($S_t$) | Section modulus of the steel girder (transformed composite section) referred to the bottom fibre |
| $T_0$ | Temperature at casting |
| $T_b$ | Mean seasonal temperature |
| $t$ | Thickness of concrete deck slab |
| $V$ | Shear due to factored load effects |
| $V'$ | Range of shear due to factored live load + impact |
| $V_r$ | Shear resistance of the girder web |
| VTD | Vertical temperature distribution |
| $w$ | Length of weld in a channel connector |
| $x, y, z$ | Rectangular Cartesian coordinates |
| $Z_r$ | Allowable range of shear per connector |
| $\alpha$ ($\theta$) | Torsional (flexural) parameter |
| $\alpha_c$ ($\alpha_s$) | Coefficient of thermal expansion of concrete (steel) |
| $\Delta T$ | Temperature differential between top and bottom of deck slab |
| $\nu$ | Poisson's ratio of concrete |
| $\sigma_{xc}$, $\sigma_{xs}$, $\sigma_{yc}$ | Longitudinal thermal stresses in concrete deck slab, in steel girder, and transverse thermal stresses in concrete deck slab |
| $\phi$ | Performance factor |

## 8.1 INTRODUCTION

A typical composite concrete deck slab-on-steel girder bridge consists of three major structural elements, namely (i) a reinforced concrete slab resting on (ii) steel girders which interact compositely with the slab by means of (iii) mechanical shear connectors; the cross-section in Fig. 8.1 shows the three elements. This type of construction makes use of the excellent compressive property of the concrete and the high tensile strength of the steel. In comparison with non-composite bridges, several advantages accrue from this union of concrete and steel, such as: (i) increased strength and stiffness; (ii) shallower steel girders and hence more economical bridge approaches; (iii) better resistance to lateral (horizontal) loads; (iv) improved dynamic response to loads. These realized economies have been shown to exceed the cost of the required shear connectors.

Caughey and Scott (1929) were the first to investigate the design of composite steel beams; they pointed out the difference in design between shored and unshored construction. The popularity of composite construction led the American Association of State Highway Officials to publish the first specifications for the design of composite bridges; the specifications were updated in 1961 and during that time West Germany included provisions for the design of composite construction in its code of practice, DIN 1078. Past research work on the subject was reviewed by Viest (1974) and then by Johnson (1970). The behaviour of continuous composite bridges has been studied by several investigators such as: Fisher, Daniels and

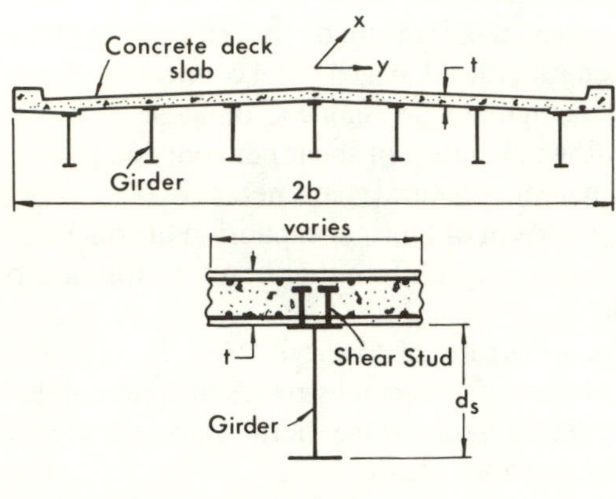


FIG. 8.1.

Slutter (1972); Johnson and Hope-Gill (1976); Botzler and Colville (1979); Salani, Duffield, McBean and Baldwin (1978); Kennedy and Grace (1982, 1983); and Grace and Kennedy (1986). A state-of-the-art report on redundant bridge systems by an ASCE-AASHTO Task Committee on Flexural Members (1985) identifies several aspects in the analysis and design of composite bridges which need further research.

## 8.2 LOADS ON COMPOSITE BRIDGES

A bridge is designed to carry loads specified by the various codes of practice to meet serviceability as well as ultimate limit states. Such limit states are governed by various criteria such as: material yielding; maximum plastic strength; deflections; instability; fatigue; and fracture.

### 8.2.1 Dead Loads

In a preliminary design, dead loads are estimated to be the sum of weights of all bridge components plus any permanent equipment attached to the bridge; snow load is disregarded in this estimate since its presence on the bridge is accompanied by reduction in traffic and its impact effects. This estimate of dead load is refined before the final design calculations are made. Generally, the dead load is assumed to be uniformly distributed provided the bridge cross-section does not change radically. If future addition to the bridge is expected, then an allowance of 10 to 15% is added to the dead load estimates.

### 8.2.2 Live Loads

In order to provide adequate strength in the bridge to carry design vehicle loads as well as future predicted loads, basic units of loads are specified by codes of practice. For example, AASHTO (1983) specifies a two-axle truck loading as H20–44 and a combination truck and trailer as H20–S16–44. An alternate uniform load and a concentrated load is to be used as lane loading whenever it produces a larger moment or shear. Various other countries such as Canada, Great Britain, West Germany, Japan, India, Czechoslovakia, have their own design live loads. Bridge live loads are dynamic in nature; therefore it is necessary to account for their impact or dynamic effects. AASHTO (1983) specifies an impact factor $I$, by which the live load effects are augmented, given by

$$I = \frac{50}{125 + L} \tag{8.1}$$

in which $L$ = loaded length in feet, with $I \leq 0{\cdot}30$. In Canada, the Ontario Highway Bridge Design Code (OHBDC) (1983) specifies a dynamic load allowance (DLA) as a function of the first natural frequency of the bridge; the live load effects are magnified by the DLA, as given in Fig. 8.2. Other forces such as braking force, collision load, and centrifugal force (if the bridge is curved in plan) must also be considered in the design; the details of such loads are found in the various codes of practice. The number of design lanes is determined from the width of the bridge. For Serviceability Limit State Type I (controlled by fatigue or vibration), OHBDC (1983) considers truck loading and not lane loading; whereas for Serviceability Limit State Type II (controlled by permanent deformation or cracking), either lane loading or truck loading is used, whichever produces the maximum load effects; this latter condition also applies to ultimate limit

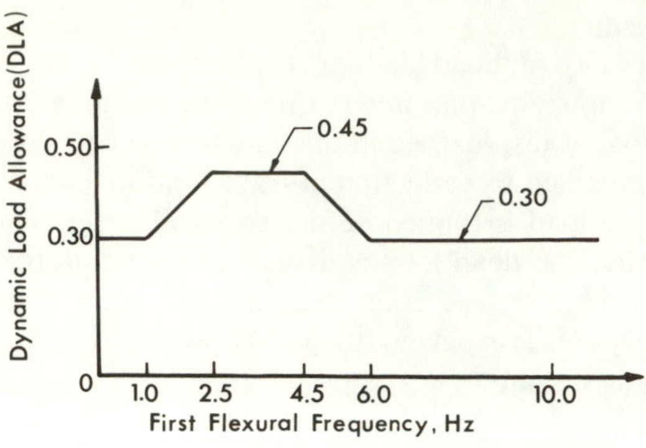


FIG. 8.2.

states, i.e. those pertaining to bridge safety, excluding fatigue. In the case of a multi-lane bridge, the live loads and the DLA are multiplied by a modification factor which is a function of the number of loaded design lanes.

## 8.3 ANALYSIS FOR LOAD EFFECTS

In the analysis of composite bridges for load effects, it is assumed that the contribution of transverse diaphragms to the load-carrying capacity is insignificant and therefore ignored. The bridge is analysed for: (i) longitudinal moment; (ii) transverse moment; and (iii) shear.

### 8.3.1 Longitudinal Bending Moment ($M_x$)

These moments are calculated by considering the flexure of longitudinal strips, each of which is subjected to a fraction of one line of wheel loads of a design truck; the width of the strip centre-to-centre = spacing of the longitudinal steel girders, $S$. The fraction of the wheel loads = $S/D_d$; the quantity $D_d$ represents the maximum longitudinal moment, at a transverse section, due to one line of wheel loads of the design truck, divided by the actual maximum intensity of the longitudinal moment per unit width; $D_d$ has a dimension of length and changes with the number of loaded lanes. AASHTO specifies values of $D_d$ for various structures, independent of any changes in the cross-section or in the aspect ratio of the bridge. However, OHBDC (1983) specifies values of $D_d$ as a function of the orthotropic characteristics of the composite bridge, defined by parameters $\theta$ and $\alpha$ given by

$$\theta = \text{flexural parameter} = \frac{b}{L}(D_x/D_y)^{0\cdot25} \tag{8.2}$$

$$\alpha = \text{torsional parameter} = \frac{D_{xy} + D_{yx} + D_1 + D_2}{2(D_x/D_y)^{0\cdot5}} \tag{8.3}$$

in which $2b$ and $L$ = width and span of the bridge, respectively; $D_x$ ($D_y$) = longitudinal (transverse) flexural rigidity per unit width (length); $D_{xy}$ ($D_{yx}$) = longitudinal (transverse) torsional rigidity per unit width (length); $D_1$ ($D_2$) = coupling rigidity per unit width (length). The above rigidities are estimated by OHBDC (1983) as follows:

$$D_x = E_c \left[\frac{\text{moment of inertia of the transformed composite transverse section}}{\text{spacing } S}\right] \tag{8.4}$$

$$D_y = E_c\left[\frac{t^3}{12(1-\nu^2)}\right] \tag{8.5}$$

$$D_{xy} = \frac{E_c}{2(1+\nu)}\left(\frac{t^3}{6}\right) \tag{8.6}$$

$$\mathrm{D}_{yx} = \frac{E_c}{2(1+\nu)}\left(\frac{t^3}{6}\right) \tag{8.7}$$

$$D_1 = \nu \text{ times (lesser of } D_x \text{ and } D_y) \tag{8.8}$$

$$D_2 = D_1 \tag{8.9}$$

in which $E_c$ = modulus of elasticity of the concrete in the deck slab; $t$ = thickness of deck slab; and $\nu$ = Poisson's ratio of concrete. Based on the values of $\theta$ and $\alpha$, factors $D$ and $C_{f_1}$ are obtained from charts given by OHBDC (1983); thus $D_d$ is calculated for each loading case from

$$D_d = D\left[1 + \frac{\mu C_{f_1}}{100}\right] \tag{8.10}$$

in which $\mu$ = (lane width in metres − 3·3)/0·6, with $\mu \not> 1{\cdot}6$. The calculated $D_d$ is modified for each loading case to account for multi-lane loading and for the DLA; the smallest value of the modified $D_d$ is the governing one to calculate the load fraction $[S/(D_d)_{modified}]$. The design live load moment, $M_x$, for each longitudinal girder is calculated by multiplying the load fraction by the maximum longitudinal factored moment due to one line of truck wheel loads.

### 8.3.2 Transverse Bending Moment ($M_y$)

The transverse moment per unit length, $M_y$, is found by combining the local and global load effects. According to OHBDC (1983), the positive transverse moment $M_y$ due to global load effects can be estimated from

$$M_y = F_d(\text{average } M_x)\sqrt{D_y/D_x} \tag{8.11}$$

in which $F_d = F(1 + \beta C_{ft}/100)$, where the factors $F$ and $C_{ft}$ are obtained from charts in the OHBDC (1983), and $\beta = (2b - 7{\cdot}5)/9$, with minimum and maximum $\beta$ = 0 and 1, respectively. It should be mentioned that AASHTO specifications (1983) do not require the calculation of $M_y$ due to global load effects; however, experience has shown that such a moment can be quite significant and therefore should be considered in the design of the transverse section.

The transverse moment $M_y$ per unit length due to local effects of the design truck can be estimated from a chart in the OHBDC (1983) as a function of girder spacing, $S$. Such a moment can also be derived by assuming the concrete deck slab to be continuous over unyielding supports and subjected to truck wheel loads with specified load dispersion lines; in such cases either plastic or elastic analysis may be performed.

### 8.3.3 Longitudinal Shear ($V$)

The distribution of longitudinal shear is independent of the span and width of the composite bridge. For each loading case on the bridge, the quantity $D_d$ is determined, according to OHBDC (1983), from

$$D_d = D\left[1 + \frac{\mu C_{fs}}{100}\right] \tag{8.12}$$

in which the factors $D$ and $C_{fs}$ are obtained from charts for calculated values of $\alpha$ and $(D_y/D_x)$. The value of $D_d$ is then modified following the procedure in Section 8.3.1. The governing live load shear is determined by multiplying the load fraction $[S/(D_d)_{modified}]$ by the maximum longitudinal shear due to one line of truck wheel loads. This analysis applies also to continuous composite bridges.

### 8.3.4 Deflections

Deflections due to dead load are calculated according to the beam analogy method, i.e. where the bridge structure is treated as a beam. Average deflections due to live load are estimated in the same way; however, maximum deflections due to live load are found by determining the load fraction $[S/(D_d)_{modified}]$ and applying this fraction of the specified load to a longitudinal strip of width $S$, treating the strip as a simple beam in flexure. It should be mentioned that dead load and any long-term loads will produce creep (plastic flow) in the concrete deck slab; it is suggested that such deflections be calculated as though they were elastic, and then a 15% increase be applied to account for creep. The total deflection of the bridge is provided for either by camber of the girders or by varying the thickness of the concrete deck slab. The amount of camber is usually given as a percentage of the computed unfactored dead load deflection, depending on the bridge span. For spans of 15–30 m, 30–75 m, and over 75 m, the percentages are 1·2, 1·15 and 1·10 respectively. For spans less than 15 m the thickness of the concrete deck slab is increased by haunching. In some cases, deflections due to shear may be significant and therefore they should be estimated.

## 8.4 CURRENT DESIGN PRACTICE

A preliminary trial section for the composite bridge is chosen; the capacity of the section is then checked for flexure, shear and deflections, comparing these capacities to those calculated from the load effects; furthermore, proper shear connection between the concrete deck slab and the girders must also be designed to ensure the desired composite action between the two parts.

### 8.4.1 Design for Flexure

The area of the concrete deck slab forming a composite unit with each steel girder must be determined. On the basis of theoretical and test results, the effective width of the slab, $b_e$, is the least value of: (i) one-fourth of the span; (ii) centre-to-centre distance between the steel girders, $S$; (iii) twelve times the least thickness of the concrete deck slab. Furthermore, most

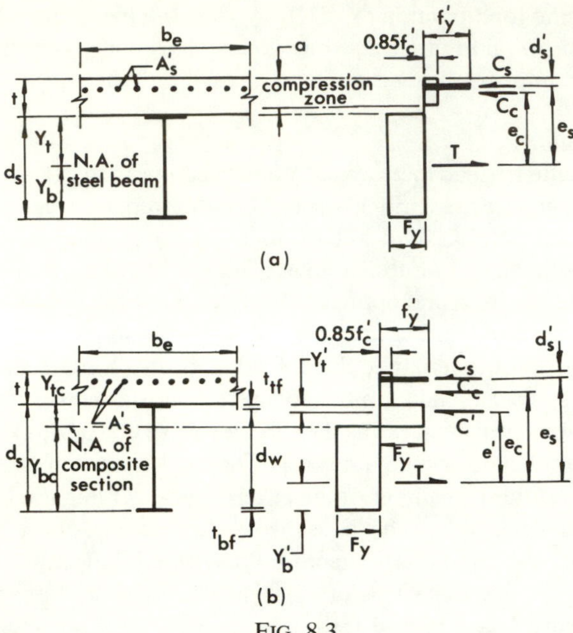


FIG. 8.3.

codes of practice ignore the concrete resistance in tension. On that basis the factored moment of resistance of the composite section can be derived using a fully plastic stress distribution as shown in Fig. 8.3. There are two cases to consider.

### *8.4.1.1 Case I*

Defining the forces $C_c = 0{\cdot}85\phi_c f'_c b_e t$; $C_s = \phi_r A'_s f'_y$; $\Sigma Q_r$ = sum of the factored resistance of shear connectors between the section under consideration and the section of zero moment, then the factored compressive resistance of the concrete deck slab with compression steel is the smallest value of: $(C_c + C_s)$; $\phi A_G F_y$; $\Sigma Q_r$. Here $A_G$ = area of girder section, mm$^2$; $A'_s$ = area of reinforcing steel in concrete deck slab, mm$^2$; $b_e$ and $t$ = effective width and thickness of slab, respectively; $F_y$ and $f'_y$ = specified yield stress of the girder steel and the reinforcing steel in the slab, respectively; $f'_c$ = 28-day compressive strength of concrete; and $\phi$, $\phi_c$ and $\phi_r$ = performance factors. When $(C_c + C_s) \geq \phi A_G F_y$, then the depth of the compression stress block, $a$, is in the slab and is given by

$$a = \frac{\phi A_G F_y - \phi_r A'_s f'_y}{0{\cdot}85\phi_c f'_c b_e} \tag{8.13}$$

The factored moment of resistance becomes, from Fig. 8.3(a),

$$M_r = C_c e_c + C_s e_s \tag{8.14}$$

in which $e_c = Y_t + t - a/2$; $e_s = Y_t + t - d_s'$, where the symbols are explained in Fig. 8.3(a).

If the smallest factored compressive resistance of the slab = $\Sigma Q_r$, the plastic neutral axis, NA, is in the steel girder section. As a result the factored moment of resistance becomes, from Fig. 8.3(b),

$$M_r = (\Sigma Q_r) Y_{tc} + C' e' \tag{8.15}$$

in which $C' = (\phi A_G F_y - \Sigma Q_r)/2$; $e' = d_s - Y_b' - Y_t'$, where the various symbols are explained in Fig. 8.3(b).

*8.4.1.2 Case II*

When the smaller of the two values $(C_c + C_s)$ and $\Sigma Q_r$ is less than $\phi A_G F_y$, the depth $a$ can be taken equal to $t$; then the factored moment of resistance is found from

$$M_r = C_c e_c + C_s e_s + C' e' \tag{8.16}$$

in which $C_c = 0{\cdot}85\phi_c f_c' b_e t$; $C' = [\phi A_G F_y - (C_c + C_s)]/2$; $e_c = d_s + 0{\cdot}5t - Y_b'$; $e_s = d_s + t - d_s' - Y_b'$. The depth $Y_{tc}$ (Fig. 8.3(b)) of the neutral axis of the composite section becomes

$$Y_{tc} = \frac{C'}{\phi A_{tf} F_y} t_{tf}$$

when $C' < \phi A_{tf} F_y$; or

$$Y_{tc} = t_{tf} + \frac{C' - \phi A_{tf} F_y}{\phi A_w F_y} d_w$$

when $C' \geq \phi A_{tf} F_y$; in which $A_{tf}$ and $t_{tf}$ = area and thickness of top steel flange, respectively; and $A_w$ and $d_w$ = area and depth of web of steel girder, respectively.

### 8.4.2 Design for Longitudinal Shear

The web of the steel girder section is designed to carry the total vertical shear on the composite section. Thus the shear resistance of the web, $V_r$, can be expressed as

$$V_r = \phi A_w F_s \tag{8.17}$$

in which $F_s$ = ultimate shear stress = 0·66 $F_y$ for rolled shapes; however, for deep members used as steel girders, it depends on the slenderness ratio of the web.

### 8.4.3 Design of Shear Connectors

In order for the composite section to act as a monolithic unit, mechanical shear connectors are used to resist the horizontal shear force at the interface, and to prevent vertical separation between the slab and the steel girders at any point along the bridge span. Experience has shown that the natural bond and friction between the steel girder and the concrete slab at the interface are easily destroyed by shrinkage, vibration and fatigue loading and therefore they cannot be counted on for composite action. There are several types of shear connector, such as the stud, bent-bar, tee, and channel connector. The stud connector, shown in Fig. 8.1, is by far the most popular in bridge construction; studs are readily welded to the top of the steel girder by means of a special automatic welding gun which dispenses the studs stacked inside and welds them to the steel girder in approximately one second.

Shear connectors are designed against fatigue failure and to develop the ultimate strength of the composite section.

#### *8.4.3.1 Design against Fatigue*

When a bridge is subjected to a moving load, the direction of the horizontal shear force exerted on a shear connector reverses as the load passes the connector; thus a shear connector must be designed for the range of this force, given by the maximum and minimum shears for the location on the bridge being considered. A typical diagram for the range of shear in a simple span bridge is shown in Fig. 8.4; it is observed that the range of shear does not vary much in this case and therefore the spacing of shear connectors can be made uniform. In a continuous composite bridge the

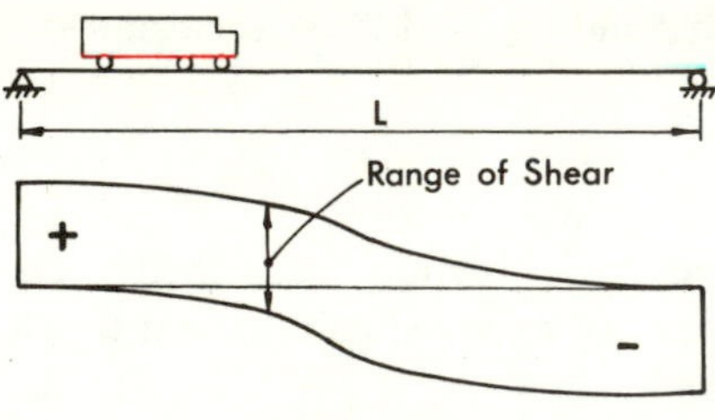


FIG. 8.4.

spacing of connectors can be made uniform from the exterior support to mid-span and from mid-span to the contraflexure point, as well as between this latter point and the intermediate support(s). The shear range per unit length, $S_1$, to be resisted by the shear connectors at the interface is determined as

$$S_1 = \frac{V'Q}{I_t} \tag{8.18}$$

in which $V'$ is the range of shear effects of live load + impact, obtained as the sum of positive and negative shears due to live load plus impact, as discussed in Section 8.3.3; $Q$ = statical moment of area about the NA of the composite section of the transformed area of the concrete deck slab above the interface, or of the area of steel reinforcement embedded in concrete for negative moment; $I_t$ = moment of inertia of the transformed composite section (in terms of steel).

The allowable range of interface shear per connector, $Z_r$, given by OHBDC (1983) is

$$Z_r = F_{sr}\left(\frac{\pi d^2}{4}\right) \tag{8.19}$$

for welded studs with height/diameter ratio $H/d \geq 4$. For channel connectors with 5 mm minimum fillet welds along the heel and toe

$$Z_r = F_{sr}(13{\cdot}5w) \tag{8.20}$$

in which $F_{sr}$ = allowable range of stress, MPa; $Z_r$ is in newtons; $d$ is in mm; and $w$ = length (mm) of welded connection measured in a transverse direction. The value of $F_{sr}$ depends on the number of cycles of maximum stress that a connector is subjected to during its life span, as shown in Table 8.1. The required pitch of shear connectors at any location on the bridge is determined by dividing $\Sigma Z_r$ of all connectors at that location by the

TABLE 8.1
ALLOWABLE STRESS RANGE, $F_{sr}$, MPa

| *Number of cycles* | *100 000* | *500 000* | *2 000 000* | *Over 2 000 000* |
|---|---|---|---|---|
| $F_{sr}$ | 115 | 85 | 65 | 48 |

interface shear per unit length, $S_1$. In continuous bridges the pitch over interior supports can be modified to avoid placing connectors at locations of steep stress gradient, provided that the total required number of connectors remains unchanged. Furthermore, if connectors are not provided in the negative moment (hogging moment) region, an additional number of connectors, $N_c$, is required at each point of contraflexure, given by

$$N_c = 70A_{sr}/Z_r \tag{8.21}$$

in which $A_{sr}$ = area (mm$^2$) of longitudinal steel reinforcement in the concrete deck slab over an interior support per composite member; these connectors should be centred about the point of contraflexure due to dead load and placed within a length = $(1/3)b_e$.

*8.4.3.2 Design for Ultimate Strength Requirements*

The ultimate shear strength of the shear connectors should be at least equal to or greater than the ultimate compressive strength of the concrete deck slab or the ultimate tensile strength of the steel girder, whichever is smaller. At the ultimate limit state, the required number of shear connectors, $N$, is given by

$$N = \frac{P}{0{\cdot}85Q_r} \tag{8.22}$$

in which $Q_r = 0{\cdot}4d^2\sqrt{f'_c E_c}$ for stud connectors with a ratio $H/d \geq 4{\cdot}0$; defining $P_1 = A_G F_y$, $P_2 = 0{\cdot}85 f'_c b_e t + A'_s f'_y$, and $P_3 = A'_s f'_y$, then the value of $P$ to be used in eqn (8.22) is the smaller of $P_1$ and $P_2$, in regions between points of maximum positive moment and points of adjacent end supports. For regions between points of maximum positive moment and points of adjacent maximum negative moments, the value of $P$ = smaller of $(P_1 + P_2)$ and $(P_2 + P_3)$.

The pitch or spacing of connectors can be either uniform or non-uniform along the span of the bridge, provided the total number of connectors, governed by fatigue or ultimate strength requirements, is installed. For good design, maximum and minimum spacings of connectors should be 600 mm and 130 mm, respectively, with a minimum concrete cover of $\geq$50 mm; the connector must extend at least 50 mm into the concrete slab; the ratio $H/d \geq 4$. In the transverse direction, the minimum spacing $\geq$50 mm. It should be noted that, due to thermal stresses in composite bridges, heavy anchorages in the form of additional shear connectors are required near the supports; this is discussed in Section 8.6.3.

## 8.5 ADDITIONAL DESIGN CONSIDERATIONS

### 8.5.1 Influence of Construction Method

The common methods of construction are shored and unshored. For compact steel sections, the condition of yielding of the bottom flange must be avoided at the specified load levels. Thus, if $m_d$ = moment caused by the specified load prior to the attainment of 75% of the required concrete strength, $m_{dc}$ = moment caused by the superimposed dead load after the attainment of 75% of the required concrete strength, and $m_l$ = moment caused by the specified live load, then the following conditions must be satisfied. For shored steel beams,

$$\frac{m_d + m_{dc}}{S_{3t}} + \frac{m_l}{S_t} \leq 0{\cdot}9F_y \tag{8.23}$$

For unshored steel beams,

$$\frac{m_d}{S_G} + \frac{m_{dc}}{S_{3t}} + \frac{m_l}{S_t} \leq 0{\cdot}9F_y \tag{8.24}$$

in which $S_G$ and $S_t$ = elastic section moduli of the steel girder section and the composite section, respectively, referred to the bottom steel flange; and $S_{3t}$ = elastic section modulus of the composite section referred to the bottom steel flange and using a modular ratio of $3n$, where $n = E_s/E_c$; such a ratio accounts for the effect of creep due to dead load.

For non-compact steel sections, the same conditions must be met except that the moments in eqns (8.23) and (8.24) are calculated for the ultimate limit state based on factored loads. It should be mentioned that the ultimate load that can be carried by a given composite cross-section is independent of the method of construction.

### 8.5.2 Design of Negative Moment Section

If an adequate number of shear connectors is provided in the negative moment region of a composite bridge, then the factored moment of resistance of the section can be based on a fully plastic stress distribution for the steel girder section and for the concrete deck slab reinforcing steel (or prestressing steel); such reinforcing (or prestressing) steel must be continuous over the interior support(s). If shear connectors are not provided, then the factored moment of resistance is that of the steel girder section only. For proper effectiveness, the area of the longitudinal steel reinforcement should be not less than 1% of the slab area, with at least 2/3 of this reinforcement placed in the top layer of slab reinforcement and within the effective width of the slab. In the case where shear connectors are not provided in the negative moment region, then such reinforcing steel should be extended into the positive moment regions for anchorage.

### 8.5.3 Design of the Concrete Deck Slab

The concrete deck slab must be designed so that its ultimate moment of resistance is equal or greater than the transverse moment due to load effects, $M_y$, in Section 8.3.2. Based on an assumed depth of slab, $t$, an expression for the ultimate moment of resistance is developed following the usual procedure, depending on whether only tension or both tension and compression steels are present. A minimum concrete cover to the reinforcing steel of 50 mm is recommended, with a placing tolerance for the reinforcement of ±20 mm. Spacing of the reinforcement should be $\leq 3t$ and not greater than 300 mm. The area of shrinkage and temperature reinforcing steel should be at least 0·002 and 0·0018 times the gross concrete area when deformed bars with $f_y < 400$ MPa and with $f_y \geq 400$ MPa respectively; the value of 0·0018 also applies when welded wire fabric is used as reinforcing steel. It should be mentioned that such provision for the temperature reinforcing steel should be checked to see whether it is adequate for the induced thermal stresses discussed in Section 8.6.3. Furthermore, for the serviceability limit state concerning deflections, a differential shrinkage strain in the concrete deck slab of $200 \times 10^{-6}$ can be assumed; creep of concrete reduces the resulting stresses by as much as 60%.

### 8.5.4 Fatigue

As was mentioned in Section 8.4.3.1, the stress range criterion is used to design against fatigue. The allowable stress ranges for the various components of a composite bridge are given in codes of practice. For example, OHBDC (1983) specifies the following values for $2 \times 10^6$ cycles of fatigue loading: 65 MPa for welded shear studs; 90 MPa for welded steel diaphragms to a longitudinal steel girder; and 125 MPa for straight reinforcing steel; these stress ranges depend on the highway classification which is based on the known or projected volume of traffic. It is interesting to note that if the concrete deck slab is prestressed at the interior supports, then fatigue becomes less of a problem since the steel is unlikely to lose its prestress at the specified loads corresponding to the serviceability limit state.

## 8.6 SOME RECENT RESEARCH

The advantages of continuity in the design of composite bridges cannot be fully realized, mainly due to the inevitable transverse cracking of the concrete deck slab in the region of interior supports. Such cracking reduces the stiffness of the bridge, thus requiring deeper steel sections and hence more expensive approaches; it also leads to costly maintenance. Recently the structural response of continuous composite bridges under static and dynamic loads was studied, with an examination of the influence of: (i) prestressing a portion of the concrete deck slab in the vicinity of the interior supports; (ii) steel diaphragms welded to the longitudinal steel girders. Furthermore, simple expressions for the thermal stresses in composite bridges were deduced, using a realistic vertical temperature distribution through the depth of the bridge.

### 8.6.1 The Influence of Prestressing

Tests were carried out by Kennedy and Grace (1982) on two 2-span continuous composite bridge models of 1/10 scale. Bridge models I and II were identical except that a portion of the concrete deck of model II was post-tensioned longitudinally in the region of the intermediate support. The two models were subjected to two-span static loading. In bridge model I the first cracks developed in the vicinity of the intermediate support at a load of 178 kN; in contrast, no cracks were detected in bridge model II up to a load of 356 kN; when the load reached 800 kN, transverse cracks 0·15 mm wide developed in model II. The crack patterns for the two bridge models are shown in Fig. 8.5. This study showed that composite action can be realized in the vicinity of negative moment regions by providing adequate shear connections throughout such regions, coupled with suitable prestressing. Furthermore, such prestressing can eliminate cracking of the concrete deck slab under specified loads, thus increasing substantially the cracking load as well as the stiffness of the bridge. However, no significant increase in the ultimate-load capacity of the bridge due to prestressing should be expected.

FIG. 8.5.

The dynamic response of continuous composite bridges was studied theoretically and experimentally by Grace and Kennedy (1986). A closed-form solution was derived to predict the dynamic response; they examined also the influence of fatigue loading at resonance frequency on shear studs, diaphragms, reinforcing steel in the deck slab, and on the longitudinal steel beams. Two 2-span continuous composite bridges, models III and IV, were tested; a portion of the deck of model IV was post-tensioned as in model II. It is interesting to note that no fatigue cracks were detected at the welded connections between the steel diaphragms and the longitudinal steel beams, even after $10^6$ cycles. This investigation showed that: (i) prestressing enhances the natural frequencies of a composite bridge by eliminating the transverse cracks in the negative moment regions; (ii) complete interaction can be achieved by prestressing the deck slab and by providing an adequate number of shear connectors in such regions; (iii) prestressing can effectively decrease the fatigue stress range and hence increase bridge fatigue life; and (iv) fatigue-induced hair cracks in the secondary members of the bridge have no significant effect on the dynamic response nor on the ultimate-load carrying capacity of the bridge.

### 8.6.2 The Influence of Welded Diaphragms

The contribution of steel I-diaphragms welded to the longitudinal steel girders was examined recently. Analyses by the finite element method and the orthotropic plate theory, as well as test results from six composite bridge models, have shown that the load distribution characteristics of the bridge are greatly enhanced by such diaphragms. Results from these studies show that the orthotropic rigidities of such a bridge can be reliably estimated as follows, taking the effective width of the slab in the longitudinal direction $= 0{\cdot}5\, b_e$ as shown in Fig. 8.6:

$$D_x = \left[D + \frac{b_e t}{1-\nu^2} E_c \left(e_x - \frac{t}{2}\right)^2 + nE_c I'_x\right] / S \qquad (8.25)$$

$$D_y = \left[0{\cdot}5D + \frac{0{\cdot}5 b_e t}{1-\nu^2} E_c \left(e_y - \frac{t}{2}\right)^2 + nE_c I'_y\right] / S' \qquad (8.26)$$

$$D_1 = \nu D'_x = \nu (D_x - nE_c I'_x / S) \qquad (8.27)$$

$$D_2 = \nu D'_y = \nu (D_y - nE_c I'_y / S') \qquad (8.28)$$

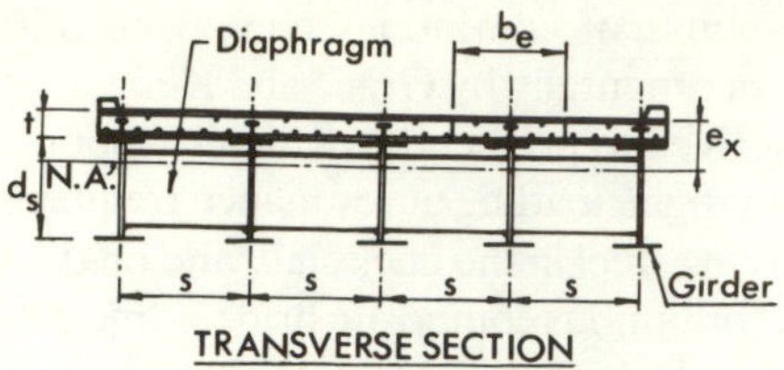


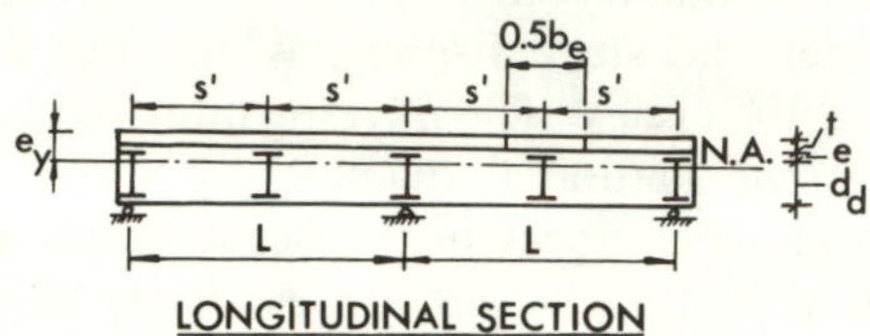


FIG. 8.6.

in which

$$e_x = \frac{nA_G\left(\frac{1}{2}d_s + t\right) + \frac{tb_e}{1-\nu^2}\cdot\frac{t}{2}}{nA_G + \frac{tb_e}{1-\nu^2}};$$

$$e_y = \frac{nA_D\left(\frac{1}{2}d_d + t + e\right) + \frac{tb_e}{2(1-\nu^2)}\cdot\frac{t}{2}}{nA_D + \frac{tb_e}{2(1-\nu^2)}};$$

$I'_x = I_G + A_G\{(1/2)d_s + t - e_x\}^2$; $I'_y = I_D + A_D\{(1/2)d_d + e + t - e_y\}^2$; $D = E_c b_e t^3 / \{12(1-\nu^2)\}$; $I_G$ and $I_D$ = moments of inertia of steel girder section and diaphragm section, respectively, with respect to their individual neutral axes; $A_D$ = cross-sectional area of steel diaphragm; $e$ = clearance between diaphragm and concrete deck slab; and $S'$ = spacing of steel diaphragms.

The above expressions for $D_y$ and $D_2$, given by eqns (8.26) and (8.28) respectively, are based on the assumption that the diaphragms fully interact compositely with the concrete deck slab; however, in practice there will be only partial interaction. To correct for this overestimate of rigidities, $D_y$ and $D_2$ are first calculated on the basis of zero interaction as follows:

$$D_y = (0{\cdot}5D + I_D)/S' \tag{8.29}$$

$$D_2 = \nu(0{\cdot}5D)/S' \tag{8.30}$$

Pending further research, current results have shown that realistic estimates for $D_y$ and $D_2$ can be found by averaging the above expressions. Thus

$$D_y = \tfrac{1}{2}\{D_y \text{ from eqn (8.26)} + D_y \text{ from eqn (8.29)}\} \tag{8.31}$$

$$D_2 = \tfrac{1}{2}\{D_2 \text{ from eqn (8.28)} + D_2 \text{ from eqn (8.30)}\} \tag{8.32}$$

The transverse and longitudinal torsional rigidities per unit width, $D_{xy}$ and $D_{yx}$, can be expressed as

$$D_{xy} = \frac{E_c}{2(1+\nu)} I_{xy} \tag{8.33}$$

$$D_{yx} = \frac{E_c}{2(1+\nu)} I_{yx} \tag{8.34}$$

in which $I_{xy} = I_{yx} = t^3/6$, where the torsional inertias of the steel girder and diaphragm have been ignored.

Results show that an increase in the number of diaphragms beyond a certain limit does not significantly improve the transverse load distribution in a composite bridge subjected to loads below the elastic load limit. However, recent test results obtained by Poggio and Ross (1986) on two bridge models V and VI, one with welded diaphragms and the other with bolted diaphragms, revealed a substantial increase (35–40%) in the ultimate-load carrying capacity due to the welding of the diaphragms; furthermore, deflections were reduced and cracking of the concrete deck slab was much less severe. It was also shown by Kennedy and Grace (1983) that the influence of welded diaphragms becomes more significant for relatively wide bridges and with increasing skew; it is recommended that, for better design, diaphragms in a skew composite bridge should be positioned orthogonally to the longitudinal steel girders. It can be concluded, therefore, that welding the steel I-diaphragms rigidly to the steel girders will enhance the transverse load distribution; it is expected that such design will affect economies as well since it can be used as a means to improve bridge rating and to rehabilitate existing composite bridges. The state-of-the-art report on redundant bridge systems by the ASCE-AASHTO Task Committee on Flexural Members (1985) cited several important contributions relevant to this topic.

### 8.6.3 Thermal Stresses

Thermal stresses are known to cause considerable damage in composite bridges; these stresses can be significant when compared to dead or live load stresses; such stresses also tend to magnify the development of cracks in the concrete deck slab, leading to corrosion of the steel girders, the steel reinforcement and hence to spalling of the concrete, as well as to deterioration of the concrete by allowing the seepage of salt-laden water.

On the basis of a synthesis of previous theoretical and experimental test results on prototype bridges, Kennedy and Soliman (1987) proposed that the most realistic and simple vertical temperature distribution (VTD) through the depth of the composite section is the one-dimensional linear–uniform distribution shown in Fig. 8.7; in this figure, $\Delta T$ = temperature differential between the top and bottom of the concrete deck slab; $T_b$ = mean seasonal temperature or ambient air temperature acquired by the steel girder and found from a map of isotherms for the bridge site; and $T_0$ = temperature at casting of the concrete. It was shown by Hendry and Page (1960) that blacktopping increases the maximum $\Delta T$;

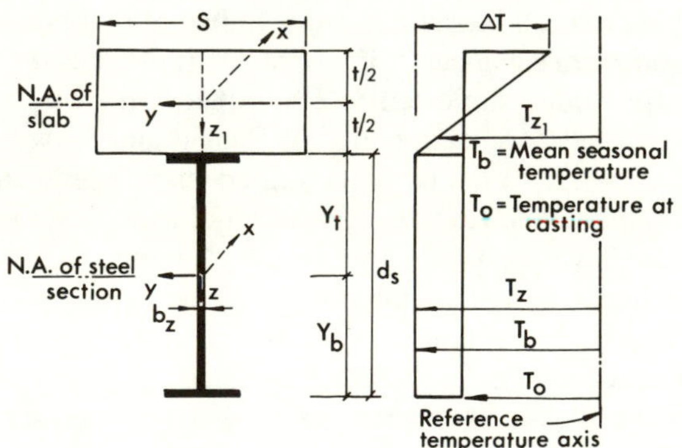


FIG. 8.7.

furthermore, field tests by Zuk (1965) have shown that the presence of full insulation also increases $\Delta T$ by about 25% over that of uninsulated bridges.

Thermal analysis of a composite bridge is based on a number of verified assumptions listed by Soliman and Kennedy (1986). Initially, the slab and girder are assumed to be separated, as in Fig. 8.8, and free to deform independently according to the imposed temperature conditions. From basic thermoelastic theory the correspondingly induced thermal stresses can be readily determined. From these stresses, acting in conjunction with

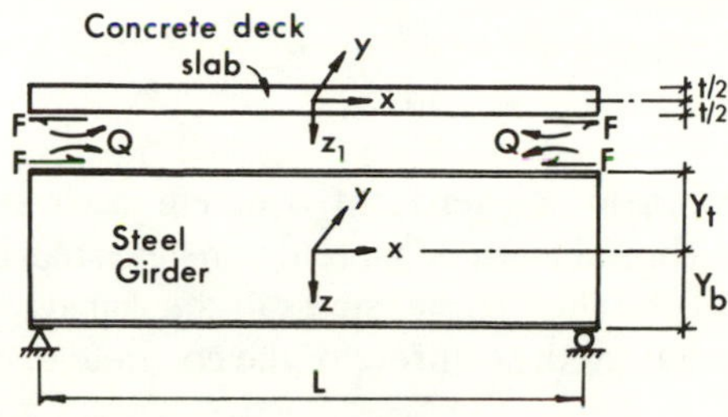


FIG. 8.8.

thermal expansions, the strains are calculated. However, in the actual composite bridge, additional unknown forces in the nature of shears $F$ and couples $Q$ exist at the interface between the slab and the steel girders and close to the ends of the bridge, Fig. 8.8. These interface forces cause the two separated elements to join compositely in such a manner that compatibility of strain and that of radius of curvature at the interface are satisfied. Invoking such conditions of compatibility will yield expressions for $F$ and $Q$. It can be shown that, for the adopted linear–uniform VTD in Fig. 8.7, $Q$ and $F$ are given by

$$Q = (\Delta T)k_1 - (T_b - T_0)k_2 \tag{8.35}$$

$$F = (T_b - T_0)k_3 - Qk_3 \tag{8.36}$$

in which $k_1 = AC_2/(AR - BK)$; $k_2 = KC_1/(AR - BK)$; $k_3 = C_1/A$; and $k_4 = B/A$, where $C_1 = -\alpha_s - (1+\nu)\alpha_c$; $C_2 = -\{(1+\nu)/4\}\alpha_c SE_c E_s I_G t^2$; $A = 1/E_s A_G + Y_t^2/E_s I_G + 4(1-\nu^2)/StE_c$; $B = Y_t/E_s I_G - 6(1-\nu^2)/SE_c t^2$; $K = 16SY_t E_c t^3 - 1{\cdot}5(1-\nu^2)\ E_s I_G t$; $R = 16SE_c t^3 + 3(1-\nu^2)E_s\ I_G$; $\alpha_c$, $\alpha_s$ = coefficients of thermal expansion of concrete and steel girder, respectively; $Y_t$ = distance from top of steel girder to its centroid.

The moment $M_G$ in the steel girder, acting as an individual element, is

$$M_G = FY_t + Q \tag{8.37}$$

The longitudinal thermal stresses in the concrete deck slab and the steel girder, $(\sigma_{xc})_1$ and $(\sigma_{xs})_1$ respectively, reduce to

$$(\sigma_{xc})_1 = \frac{F}{tS} - \frac{Q}{I'}z_1 + \frac{Fa}{I'}z_1 \tag{8.38}$$

$$(\sigma_{xs})_1 = \frac{-F}{tS} + \frac{Q}{I_G}z + \frac{Fd_1}{I_G}z \tag{8.39}$$

in which $I'$ = moment of inertia of concrete deck slab about its own centroid; $z_1, z$ = coordinates of concrete fibre and steel fibre, respectively, as shown in Fig. 8.7. The thermal stress in the transverse $y$ direction in a fibre at $z_1$ distance from the centroid of the concrete deck slab, Fig. 8.7, is given by

$$(\sigma_{yc})_1 = \nu(\sigma_{xc})_1 - \alpha_c E_c(T_{z_1} - T_0) \tag{8.40}$$

The stresses given by eqns (8.38)–(8.40) are for simply supported composite bridges. In the case of continuous composite bridges, the thermally induced curvature is restrained at the intermediate supports, resulting in the development of indeterminate reactions and moments $M_R$; thus additional stresses will develop due to the span continuity of the bridge. The moment diagrams for $M_R$ for different members of spans are shown in Fig. 8.9 in terms of a moment $M_0$ corresponding to the unrestrained curvature and given by

$$M_0 = (-M_G)\frac{I_t}{I_G} \tag{8.41}$$

Thus, the resulting stresses due to continuity are found from

$$(\sigma_{xc})_2 = \frac{M_R}{nI_t}z_1 \tag{8.42}$$

$$(\sigma_{xs})_2 = \frac{M_R}{I_t}z \tag{8.43}$$

The net thermal stresses in continuous composite bridges become

$$\sigma_{xc} = (\sigma_{xc})_1 + (\sigma_{xc})_2 \tag{8.44}$$

$$\sigma_{xs} = (\sigma_{xs})_1 + (\sigma_{xs})_2 \tag{8.45}$$

$$\sigma_{yc} = \nu\sigma_{xc} - \alpha_c E_c (T_{z_1} - T_0) \tag{8.46}$$

In dealing with continuous composite bridges it should be mentioned that the thermal stresses are independent of the span lengths; furthermore, as the number of spans increases the thermally-induced moment $M_R$ tends to converge, as shown in Fig. 8.9.

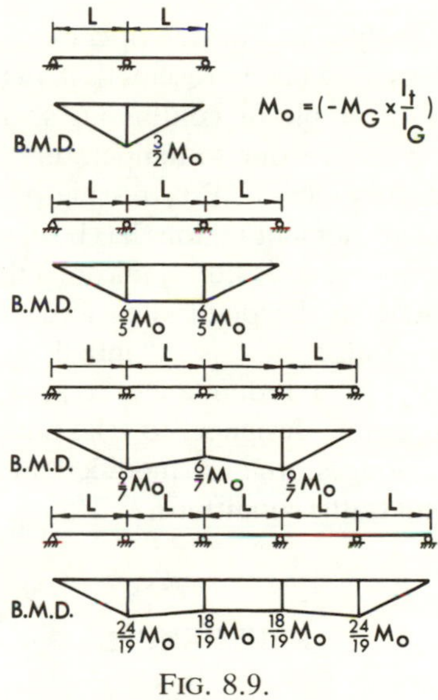


FIG. 8.9.

For proper design, special heavy end anchorages, tieing the concrete deck slab and the steel girders at their interface, must be provided to design for the forces $Q$ and $F$ given by eqns (8.35) and (8.36). It should be noted also that for any thermal stress calculations four cases of temperature conditions, shown in Fig. 8.10, have to be considered; the values of temperature differentials shown are appropriate for certain locations in the USA such as the middle-Atlantic states as well as in southern Ontario; other values may be more appropriate for different locations.

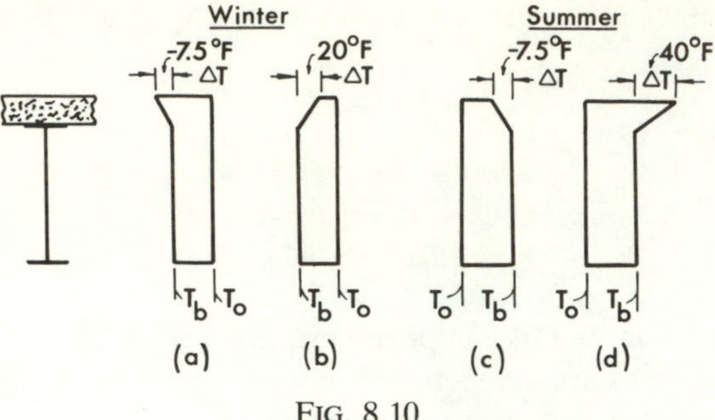


FIG. 8.10.

## 8.7 CONCLUDING REMARKS

Current methods of analysis and design of composite concrete deck slab and steel girder bridges have been presented. A thermal stress analysis of composite bridges with simple or continuous spans is given to aid the designer in estimating stresses due to temperature variations through the depth of the bridge cross-section. Prestressing portions of the concrete deck slab in the negative moment regions has been shown to effect several advantages. Furthermore, results indicate that transverse steel diaphragms, when welded to the longitudinal girders, can improve the load distribution in the bridge and substantially increase its ultimate load-carrying capacity. The redundancy created by the use of such diaphragms will also enable designers to take advantage of the inelastic behaviour of existing bridges, when so modified, in order to improve their load ratings for current traffic conditions.

## REFERENCES

AMERICAN ASSOCIATION OF STATE HIGHWAY AND TRANSPORTATION OFFICIALS (AASHTO) (1983) *Standard specifications for highway bridges*. Washington, DC, USA.

ASCE–AASHTO TASK COMMITTEE ON FLEXURAL MEMBERS (1985) State-of-the-art report on redundant bridge systems. *J. Struct. Eng., ASCE*, **111**(12).

BOTZLER, P. W. and COLVILLE, J. (1979) Continuous composite-bridge model tests. *J. Struct. Div., ASCE*, **105**(ST9), Sept.

CAUGHEY, R. A. and SCOTT, W. B. (1929) A practical method for the design of I-beams haunched in concrete. *Struct. Engr*, London, **7**(8).

EMANUEL, J. H. and HULSEY, J. L. (1978) Temperature in composite bridges. *J. Struct. Div., ASCE*, **104**(ST1), Jan.

FISHER, J. W., DANIELS, J. H. and SLUTTER, R. G. (1972) Continuous composite beams for bridges. *Proc. 9th IABSE Congress*, May.

GRACE, N. F. and KENNEDY, J. B. (1986) Free vibration of a continuous composite bridge. *Proc. Ann. Conf. Can. Soc. Civ. Eng.*, May.

HENDRY, A. W. and PAGE, J. K. (1960) Thermal movements and stresses in concrete slab in relation to tropical conditions. *RILEM Symposium on concrete and r.c. in hot countries*, Haifa, Israel, July.

JOHNSON, R. P. (1970) Research on steel–concrete composite beams. *J. Struct. Div., ASCE*, **96**(ST3), March.

JOHNSON, R. P. and HOPE-GILL, M. (1976) Applicability of simple plastic theory to continuous composite beams. *Proc. Inst. Civ. Engrs*, Part 2, **61**, March.

KENNEDY, J. B. and GRACE, N. F. (1982) Prestressed decks in continuous composite bridges. *J. Struct. Div., ASCE*, **108**(ST11).

KENNEDY, J. B. and GRACE, N. F. (1983) Load distribution in continuous composite bridges. *Can. J. Civ. Eng.*, **10**(3).

KENNEDY, J. B. and SOLIMAN, M. (1987) Temperature distribution in composite bridges. *J. Struct. Eng., ASCE*, **113**(3).

ONTARIO MINISTRY OF TRANSPORTATION AND COMMUNICATIONS (OHBDC) (1983) *Ontario highway bridge design code*. Downsview, Ontario, Canada.

POGGIO, N. V. and ROSS, A. I. (1986) *The effect of diaphragms and their connections on load distribution across simply supported composite bridges*. Internal Report, Dept. of Civil Engineering, University of Windsor, Windsor, Ontario, Canada.

SALANI, H. J., DUFFIELD, R. C., MCBEAN, R. P. and BALDWIN, J. W. (1978) *An investigation of the behavior of a three span composite highway bridge*, Study 71-4, College of Engineering, University of Missouri, Columbia, USA.

SOLIMAN, M. and KENNEDY, J. B. (1986) Simplified method for estimating thermal stresses in composite bridges. *Transportation Research Record No. 1072*, National Research Council, Washington, DC, USA, pp. 23–31.

VIEST, I. M. (1974) Composite steel–concrete construction. *Proc., Struct. Div., ASCE*, **100**(ST5).

ZUK, W. (1965) Thermal behavior of composite bridges—insulated and uninsulated. *Highway Research Record* (T6), National Research Council, Washington, DC, USA.

*Chapter 9*

# COMPOSITE BOX GIRDER BRIDGES

R. GREEN

*Department of Civil Engineering, University of Waterloo, Ontario, Canada*

and

F. A. BRANCO

*Centro de Mecânica e Engenharia Estructurales, Universidade Técnica de Lisboa, Portugal*

*SUMMARY*

*During the construction of composite box girder bridges, the girder section or the individual girders may distort or twist so adding to the construction stresses. This distortion or twist, and the associated stress, may be reduced appreciably with the addition of simple and economical bracing.*

*The action of such bracing within the section of girders in preventing distortion or twist is examined using results obtained from scale models, simple engineering theories and finite strip analyses for both the construction and service loading cases. Interconnecting bracing between girders is also examined. Recommendations for preliminary design based on the response of the girders to various concentrically and eccentrically applied loads are given.*

## NOTATION

$b$ Mid width of box girder

$D_{xs}$ Stiffness or flexural rigidity of stiffened webs

| | |
|---|---|
| $D_{xu}$ | Stiffness or flexural rigidity of unstiffened webs |
| $e$ | Eccentricity of $w_c$ with respect to longitudinal centreline |
| $e_f$ | Eccentricity of formwork load |
| $e_p$ | Eccentricity of finishing machine |
| $e_v$ | Eccentricity of vertical load |
| $e_w$ | Eccentricity of wind load |
| $e_x$ | Eccentricity of uniformly distributed line load |
| $l$ | Member length |
| $L$ | Span |
| $m_c$ | Uniformly distributed torque due to concrete deck |
| $m_f$ | Uniformly distributed torque due to formwork |
| $m_w$ | Uniformly distributed torque due to wind |
| $M_p$ | Concentrated torque due to finishing machine |
| $P$ | Uniform line load, also $W$ |
| $P_c$ | Portion of line load causing member bending |
| $P_D$ | Horizontal component of web reaction, torsional load |
| $P_s$ | Horizontal component of web reaction, concentric load |
| $P_T$ | Portion of line load causing member torsion |
| $r$ | Radius of gyration |
| $R$ | Reaction due to bracing force |
| $S$ | Spacing of girders |
| $T$ | Torque |
| $V$ | Girder reaction due to torque $T$ |
| $w_c$ | Uniformly distributed load due to concrete deck |
| $w_f$ | Uniformly distributed load due to formwork |
| $w_w$ | Uniformly distributed load due to wind |
| $W$ | Uniform line load |
| $u, v, w$ | Displacements |
| $x, y, z$ | Coordinates |

| | |
|---|---|
| $\delta$ | Horizontal deflection of flanges, concentric load |
| $\sigma_B$ | Bending normal stress |
| $\sigma_{DB}$ | Spreading stress due to bending distortion |
| $\sigma_{DW}$ | Distortional warping stress due to torsional distortion |
| $\sigma_u$ | Longitudinal bending stress with unstiffened webs |
| $\sigma_s$ | Longitudinal bending stress with stiffened webs |
| $\sigma_T$ | Total stress |
| $\sigma_W$ | Torsional warping stress due to torsional loading |

## 9.1 INTRODUCTION

### 9.1.1 Background

Composite steel–concrete box girder bridges have been widely used for intermediate span bridge structures in the last few years. Completed composite box girders are very stable, with a three-sided steel box and a concrete deck on the top. However, during construction the girder may have a flexible open section leading to considerable distortion or twist due to construction loads (Reis and Branco, 1980). The behaviour of the open section girders is generally not well understood and difficulties with changes in geometry or excessive rotation, before and during the placement of the deck concrete, have been reported (United States Steel, 1978).

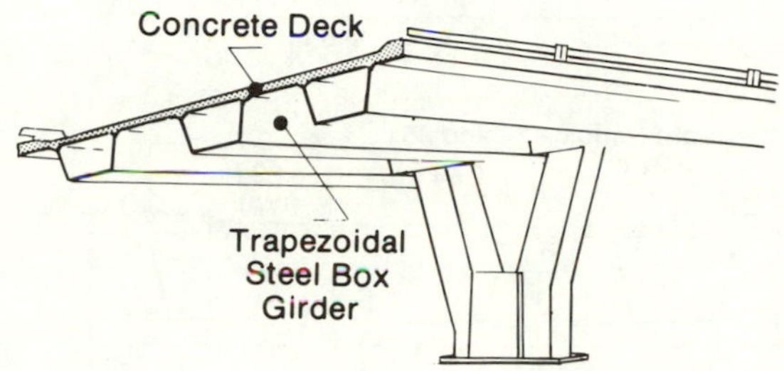


FIG. 9.1. Completed structure.

The completed structure (Fig. 9.1) is aesthetically pleasing and consists of steel box girders made composite with a concrete deck slab. Economies in box girder design are possible in comparison with similar I-beam and plate girder structures because the completed composite girder section has a higher torsional stiffness, and consequently a greater lateral distribution of live load, than I-beam structures having a similar flexural strength. Additional economies are also possible in the fabrication and erection of box girders, as compared to similar I-beam structures, by virtue of the elimination of much wind and transverse bracing.

A variety of cross-sectional geometries for the completed box structure are possible, with the number of girders varying from two to six or more depending upon the plan geometry of the structure (Fig. 9.2). Typical cross-sectional geometries are given in Fig. 9.3 for three two-box girder systems. Both single and double bearing arrangements and shallow and deep end-support diaphragms are illustrated. Typical spacing of the

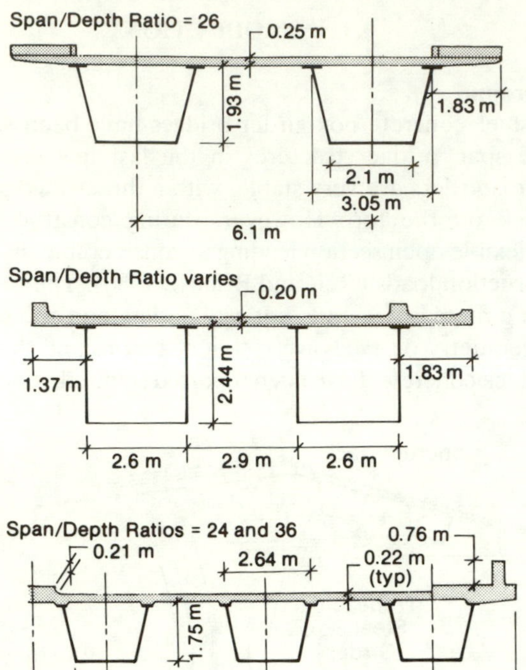


FIG. 9.2. Typical cross-sections.

centreline of the box girders varies from 4·3 to 6·7 m, and the depth of the cast-in-place concrete deck varies from 190 mm to 250 mm. Thus, the ratio of dead load moment to live load moment for box girder structures will depend upon the depth of deck, the cross-section and the span arrangement chosen by the design engineer.

Experience with the construction of box girder systems has indicated that the design specifications (AASHTO, 1983; CSA, 1978; Poellot, 1974) do not provide either design or construction engineers with sufficient guidance regarding the behaviour of thin-walled box girders during the construction phase (United States Steel, 1978). Specifications may not clearly refer to the need for bracing during construction. Interior intermediate diaphragms or cross-frames may be required to retain the cross-sectional geometry of each individual girder during fabrication, handling and placement of the deck concrete. Similarly, diaphragms or cross-

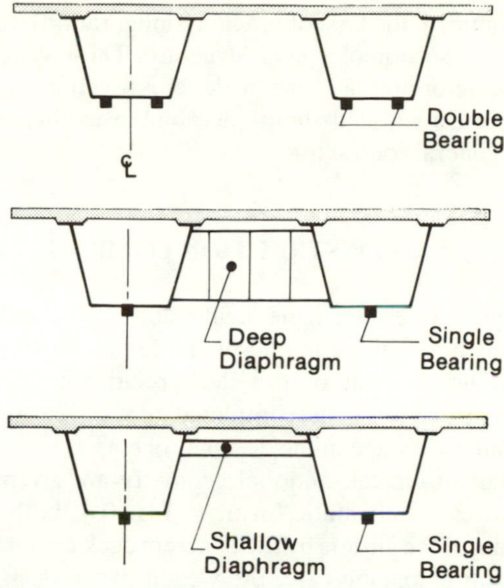


FIG. 9.3. Various support geometries.

frames may be used between girders to ensure the stability and to limit horizontal displacements of the girders during concrete placement.

Responsibility for providing bracing is frequently given to the general contractor or steel fabricator. When the emphasis in design is the strength of the bridge for service conditions, design engineers frequently neglect construction conditions. In these cases the contractor must use expedient procedures to brace box girders. The resulting design of the braces may not prevent lateral movements; or it may be excessive, resulting in waste of material and labour, and conflict between the owner and contractor.

### 9.1.2 Outline

This chapter describes the possible loading configurations present on a thin-walled box girder system during construction, and examines the resulting deformations and stresses induced in typical torsionally open and torsionally quasi-closed box girders. The results are based on tests of model and prototype bridges, supplemented by analyses of individual open and quasi-closed box sections and composite sections. Several bracing systems are examined, the type of bracing configuration being a function of the span and cross-section of a given structure. These systems can be used to minimize the deformation of the girder cross-section and the girder as a whole during construction to limits acceptable to the design engineer, fabricator and general contractor.

## 9.2 CONSTRUCTION LOADING

As part of design, the design engineer will select a cross-sectional geometry for a bridge structure which complies with the functional requirements of the highway. The position of the deck relative to the centreline of individual box girders and the completed box system will generally be established at an early stage in the design process.

Two examples of typical sectional geometry are given, together with various load cases in schematic form, as Fig. 9.4; both concentric and eccentric loading of the girders by the concrete deck are illustrated. For the idealized concentric loading case, the weight of the deck will be equally divided between the four inclined webs. Since the box girders are loaded along the centreline, twist about the shear centre will not occur and the flexural and torsional loadings are easily identified. Bracing of the compression flanges is required to avoid lateral movement or buckling due to the action of a combination of vertical load and horizontal force.

Geometric design constraints may result in the deck concrete being eccentric with respect to the longitudinal centreline of individual girders (Fig. 9.4). The designer may be guided, for example, by specification provisions limiting the cantilever overhang to 60% of the distance between the centres of adjacent top steel flanges of adjacent box girders or less than two metres (AASHTO, 1983; OHBDC, 1983). The case of an eccentric deck concrete results in a uniformly distributed load, $w_c$, located at an eccentricity, $e$, with respect to the centreline acting on the individual box girders. Accordingly, flexural and torsional loading develops, giving rise to vertical deflection and horizontal deflection as a consequence of rotation of the girder about the shear centre.

For both concentric and eccentric loading, end reactions provide a restraint for shear due to the vertical loading. A support restraint should also be provided for torsional loading. A torsional restraint can easily be developed as part of the design of systems 1 and 2 of Fig. 9.3. However, a nominal diaphragm combined with a single bearing (system 3 of Fig. 9.3)

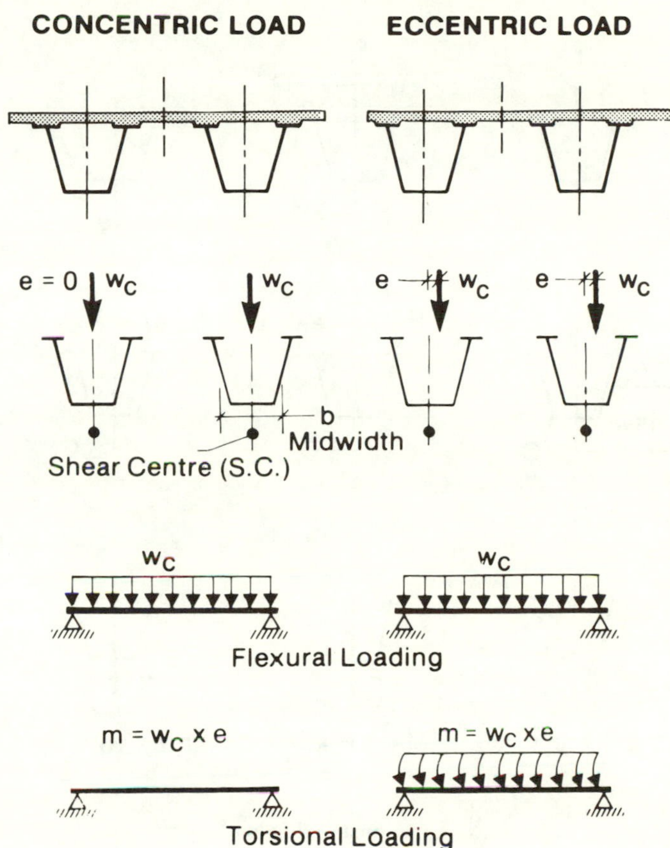


FIG. 9.4. Basic loading.

will require special design and detailing of both the connections and diaphragm member to ensure adequate strength and rotational stiffness. This latter system, with single bearing and nominal diaphragm, is not recommended.

Other loadings can give rise to predictable torsional effects during construction; these include wind loading of the exposed girders, the

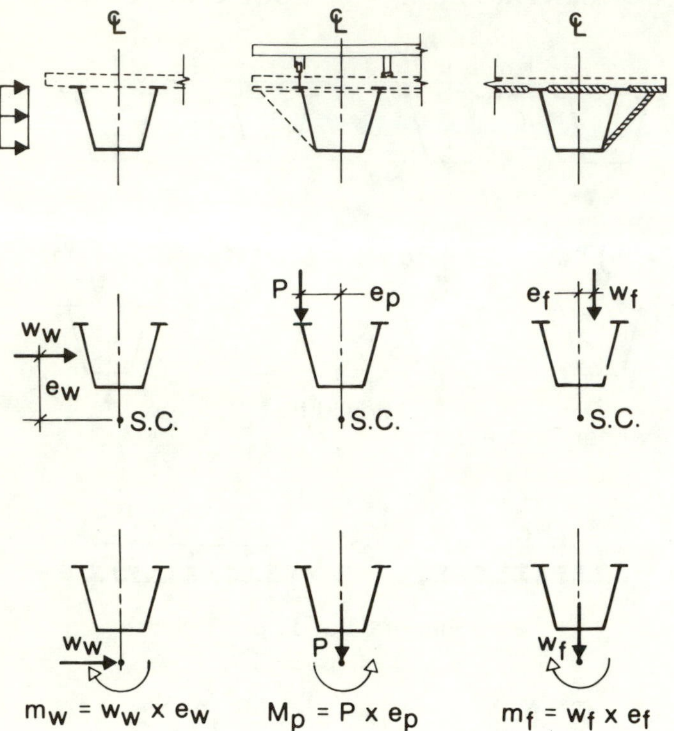


FIG. 9.5. Construction loading.

finishing machine, and formwork (Fig. 9.5). In each case, adequate torsional restraint must be provided, not only to ensure the overall stability of the girders but also to limit the lateral deflections of the individual girders to an acceptable level. Calculation of the torsional loads and movement due to wind, equipment and formwork is possible at the design stage, with the use of a few simplifying assumptions.

There can be other torsional loadings during construction. These may include the concrete handling systems chosen by the general contractor, or the effects of placing the deck concrete in an unsymmetrical fashion. The torsional effects of such loadings can be minimized by careful specification and are generally small relative to those discussed above.

A summary of flexural and torsional loadings, calculated from the design drawings of three representative Canadian box girder bridges, is given in Fig. 9.6. The flexural loading due to the concrete and formwork leads to shear and bending moment diagrams having a form familiar to designers. The concrete, formwork and wind loads combine to give torque and warping moment diagrams similar to those for bending. Values of the uniformly distributed load ($w_c$ plus $w_f$, $w_w$) and the uniformly distributed torque ($m_c$ plus $m_f$, $m_w$) for the vertical and wind loading are given in Fig. 9.6. Values are also given in terms of an eccentricity ratio, $e/b$, where $e$ is the eccentricity of the vertical load relative to the girder centreline and $b$ is the average width of the box. These eccentricity ratios vary from 0·04 to 0·12.

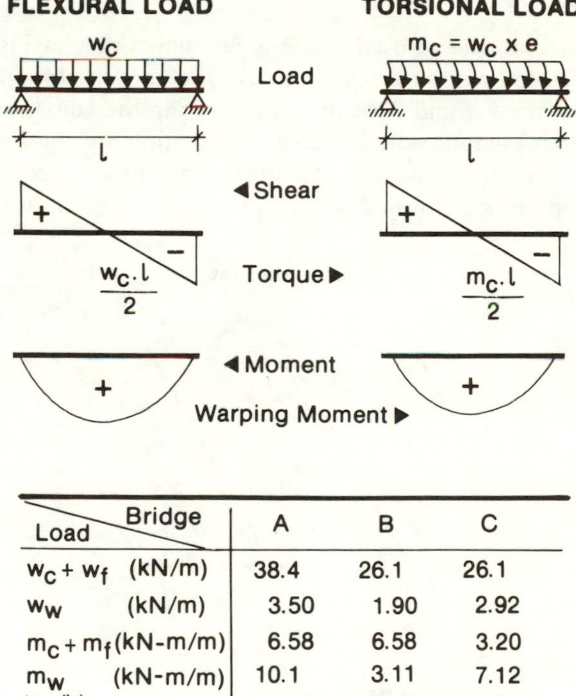


| Load \ Bridge | A | B | C |
|---|---|---|---|
| $w_c + w_f$ (kN/m) | 38.4 | 26.1 | 26.1 |
| $w_w$ (kN/m) | 3.50 | 1.90 | 2.92 |
| $m_c + m_f$ (kN-m/m) | 6.58 | 6.58 | 3.20 |
| $m_w$ (kN-m/m) | 10.1 | 3.11 | 7.12 |
| $(e_x/b)$ | 0.06 | 0.12 | 0.04 |

FIG. 9.6. Values of typical loading.

Vertical and horizontal loads and associated values of torsional loads vary from structure to structure. The values of these loads vary with the choice of cross-sectional geometry made by the design engineer, and the loads can be calculated.

## 9.3 GENERAL BEHAVIOUR

### 9.3.1 Single Member

A trapezoidal box section can be chosen and proportioned to have the following characteristics with respect to torsion:

(a) a torsionally open section;
(b) a torsionally quasi-closed section with a horizontal truss system located near the flange; or
(c) a torsionally closed section with a continuous plate connecting the top flanges of the girder.

The open and quasi-closed sections are illustrated in Fig. 9.7. The torsionally closed section is seldom used for composite bridge structures because of economy and difficulties of ensuring the stability of the thin closure plate in compression. However, calculation procedures frequently refer to a closed section as an idealization of the quasi-closed section.

When a simple span formed from any one of the sections is subjected to

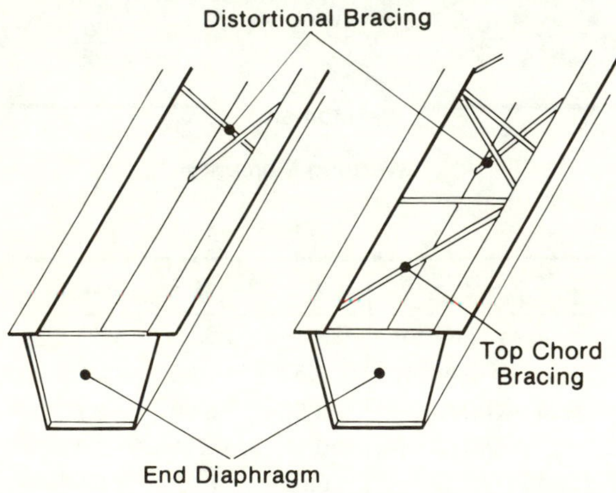


FIG. 9.7. Open and quasi-closed sections.

torsional loading, rotation takes place about the shear centre, the location of which is a function of the geometry of the section used. For typical open sections used in box girder bridges, the shear centre is located approximately one-half of the girder depth below the bottom flange. For a quasi-closed section with a Warren truss type bracing system with members having a slenderness ratio ($l/r$) of approximately 200 (Fig. 9.8), the location of the shear centre is slightly above that of the open section. The shear centre and geometric centroid will be coincident if the girder is a closed box and has two axes of symmetry, vertical webs and equal plate thickness for the top and bottom flanges. Seldom will bridge girders have

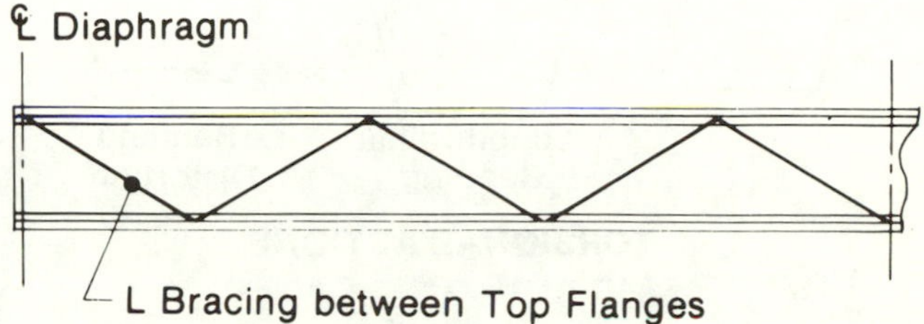


FIG. 9.8. Plan of closed system.

such a geometry. For both torsionally open sections and sections with bracing in the plane of the top flanges, the shear centre and the centre of rotation lie below the bottom flange of the section.

### 9.3.2 Cross-section

An arbitrary line loading on a simple-span box girder (Fig. 9.9a) has gravity and torsional components. Under the gravity load, the section deflects rigidly (longitudinal bending) and deforms (bending distortion) (Fig. 9.9b). Under the torsional loading, the section rotates rigidly (mixed torsion) and deforms (torsional distortion) (Fig. 9.9c) (Branco, 1981; Branco and Green, 1984a).

#### *9.3.2.1 Longitudinal Bending*

When a girder is subjected to transverse loads acting through the shear centre, bending normal stresses arise. Associated shear stresses also occur due to the variation of the bending moment along the beam. These can be calculated from the equilibrium considerations. Such analyses are based on the Navier hypothesis and neglect shear strains due to bending which

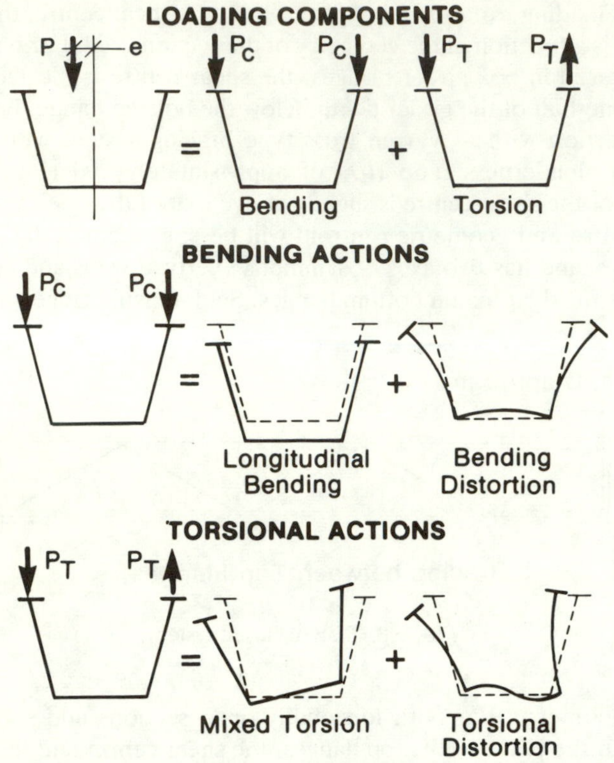


FIG. 9.9. General behaviour of open section.

induce warping displacements in the cross-section. These warping displacements are almost linear in the webs and vary parabolically in the flanges, giving a redistribution of the bending normal stresses, the shear lag effect (Fig. 9.10). This effect is usually small in box girders, except at sections where large concentrated loads or thin, wide girder flanges are present (Trahair, 1977).

### 9.3.2.2 *Mixed Torsion*

When gravity loads do not act through the shear centre, a girder will twist about the longitudinal axis. Uniform torsion and the associated St Venant shear stresses will occur if the rate of change of the angle of twist is constant along the girder and warping is unrestrained. If there is a variation of torque or if warping is prevented along the girder, non-uniform torsion

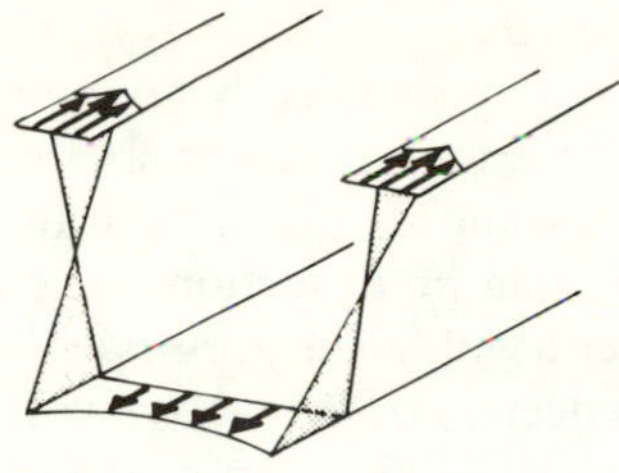

FIG. 9.10. Shear lag.

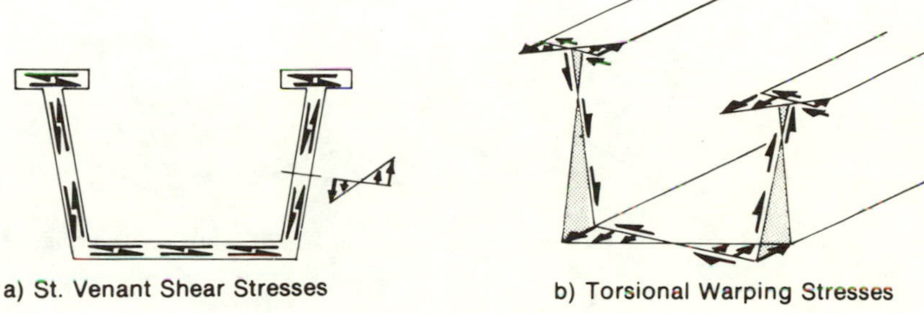


FIG. 9.11. Torsional warping stresses.

(mixed torsion) occurs and longitudinal torsional warping stresses develop (Fig. 9.11). The variation of these stresses induces warping shear stresses in the cross-section which add to the St Venant shear stresses so as to resist the total torque in the section. In open box girders, the torque is mainly resisted by warping torsion because St Venant torsional stiffness is comparatively small.

*9.3.2.3 Bending Distortion*

Due to the transverse bending of the plates forming the girder, the cross-sectional shape changes in the presence of external loads. Bending distortion (spreading) occurs when transverse loads are applied concentrically through the shear centre. In open box girders, this distortion causes outward bending of the webs, upward bending of the bottom flange and in-place bending of the top flanges (Fig. 9.9b). Transverse bending stresses occur in the bottom flange and in the webs. In the top flanges, due to the in-plane bending, longitudinal spreading stresses arise (Fig. 9.12).

*9.3.2.4 Torsional Distortion*

Torsional load tends to deform the cross-section with bending of the walls (Fig. 9.9c). If the load is uniform along a girder having no diaphragms, uniform distortion is restrained only by the transverse stiffness of the plate elements. Under non-uniform distortion, the plate elements also bend in the plane of the element, with shear stresses resisting distortional load. The additional vertical deflection of the webs increases the twist of the cross-section and results in additional distortional warping stresses (Fig. 9.13).

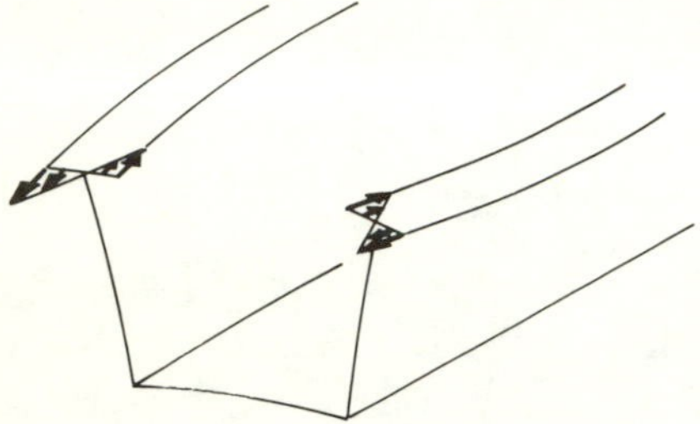

FIG. 9.12. Spreading stresses.

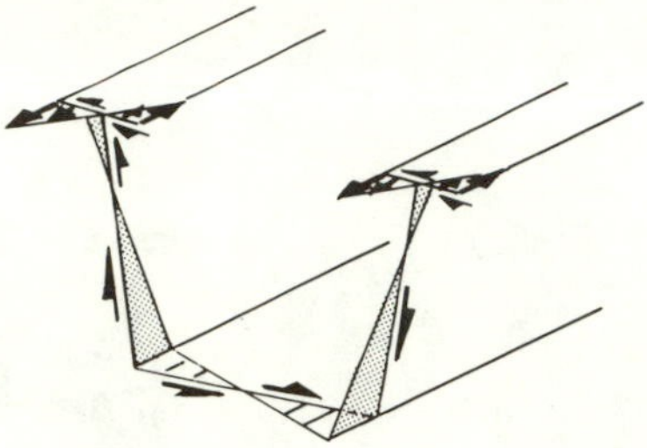

FIG. 9.13. Distortional warping.

### 9.3.3 Bracing Systems

If distortional stiffness is lacking, the cross-section of the open girder can deform, or if the torsional stiffness is small, excessive twist can develop. Bracing systems are usually installed within the box girder to decrease distortion within the section or increase the torsional stiffness, or both, so minimizing these effects (Fig. 9.14).

*Ties* can usually be placed between top flanges to prevent bending distortion, and *distortional braces*, connecting top and bottom flanges, are used if torsional distortion of the cross-section is to be prevented. Neither ties nor distortional braces add to the St Venant torsional stiffness of the girder.

Horizontal bracing can be placed slightly below the top flanges to increase the torsional stiffness or to reduce twist. This bracing can be limited to the girder ends (*torsional boxes*), or be provided throughout the length of the girder (*top chord bracing*). A torsion box (Fig. 9.14c) provides a warping restraint at the supports and effectively changes an open section box from a torsional simple span to a torsional fixed span, so reducing the midspan rotation to one-fifth that of a simple span system. A distortional brace must be included as part of a torsion box. Top chord bracing (Fig.

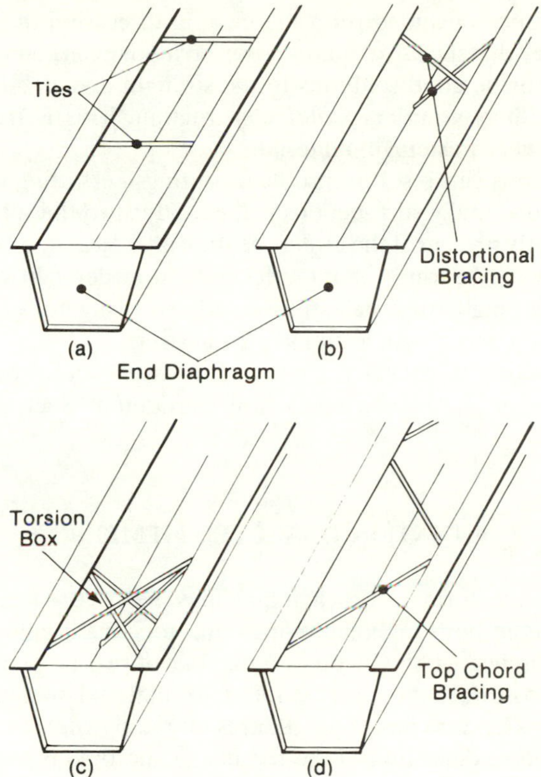


FIG. 9.14. Bracing systems.

9.14d) converts an open box to a quasi-closed box with calculable properties (Kollbruner and Basler, 1969). Depending upon the area and geometry of bracing, the quasi-closed box can be 10 to 20 times stiffer than an open box. Here, stiffness refers to the torque per unit midspan rotation rather than St Venant torsional stiffness.

Stay-in-place metal or wood forms should not be used to close the cross-section instead of top chord bracing, until connection details capable of transferring the shear force between components are developed. Interconnecting bracing systems between adjacent girders, or even shoring, where site conditions are suitable, can also be used to prevent rotation. Stay-in-place forms within a single box girder and interconnecting bracing between girders are not considered in this chapter.

### 9.3.4 Summary

It might well be concluded from the general discussion of bracing for a single member that the use of quasi-closed boxes or even completely closed boxes will provide good solutions to the strength and stability problems associated with box girders under construction. This is true, but such solutions are very wasteful of material.

Such solutions can result in problems during the setting of girders on bearings as torsionally stiff sections will not adjust to the small imperfections frequently present at the girder–bearing interface, due either to small errors in placing bearings or imperfections in girder fabrication. Open sections with small torsional stiffness adjust easily to such errors or imperfections without significant increase in stress.

It is instructive to consider a number of cases where the lateral displacements of a girder as a whole are considered. Such a description follows.

## 9.4 INTERCONNECTED MEMBERS

To illustrate the form of bracing that may be used to restrict lateral displacements in both torsionally open and torsionally quasi-closed box sections, reference will be made to a number of two-box girder systems. Examination of the statics and overall deformational behaviour of these simple systems leads to design procedures for multigirder systems.

The transverse deflection of the top flange and the midspan section of two open-section girders without interconnecting bracing are given in Fig. 9.15. Both girders are subjected to the same value of torsional moment per unit length, but these moments act in opposite senses. The midspan deflection is proportional to the fourth power of the torsion span length (Kollbruner and Basler, 1969; Green, 1978).

If a simple interconnecting tie is introduced at midspan of the two girders, the deflection is reduced (Fig. 9.16). The deflection at midspan is ideally zero. The maximum transverse deflection occurs near the quarter-span point, and is approximately one-twentieth of that of the system without an interconnecting tie. The additional stresses due to the torsional component of loading for the single tie case (Fig. 9.16) are one-quarter of those for the unsupported case (Fig. 9.15).

Girders with identical torsional loadings are not the norm, and consideration of a two-girder case with only one girder torsionally loaded is of

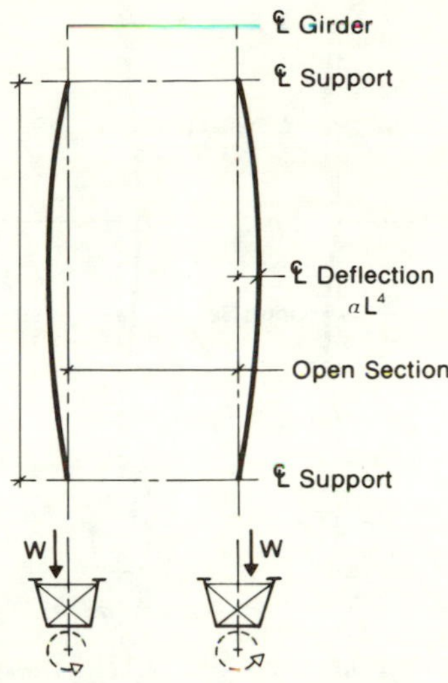


FIG. 9.15. Transverse deflection due to torsional load, open sections, no interconnecting bracing.

greater practical interest. Three arrangements of loading and simple bracing configurations are considered in Figs 9.17, 9.18 and 9.19:

*Case A*—a pair of similar open section girders, with a midspan interconnecting tie.

*Case B*—one open girder and one quasi-closed girder, with a midspan interconnecting tie.

*Case C*—a pair of open girders with a vertical x-brace between girders at midspan.

The torsional loading in case A is shared approximately equally between the two girders (Fig. 9.17). Horizontal displacements are nearly equal for the two girders, and are approximately one-half of those of an unsupported single girder. For case B (Fig. 9.18), the interconnecting tie restrains the midspan horizontal displacement of the open-section girder, and the comparatively stiff quasi-closed section provides torsional resistance. However, this stiff section still rotates and translates at midspan.

The x-brace located at midspan, case C in Fig. 9.19, provides a torsional restraint at midspan, and ideally the rotation of the girders is zero at the

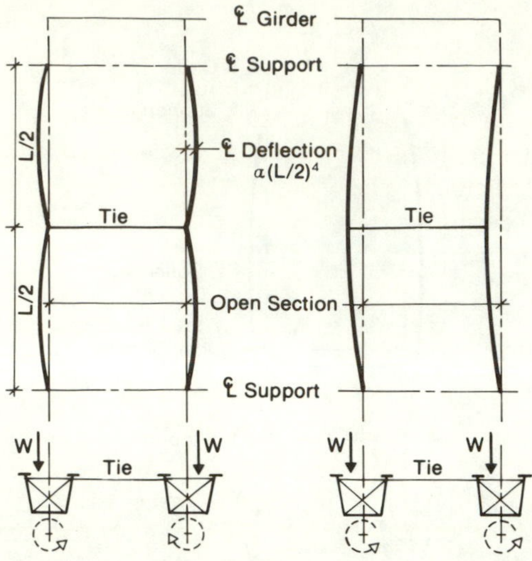


FIG. 9.16. Transverse deflection due to opposing torsional load, open sections, with midspan tie.

FIG. 9.17. Transverse deflection due to complementary torsional load, open sections, with midspan tie.

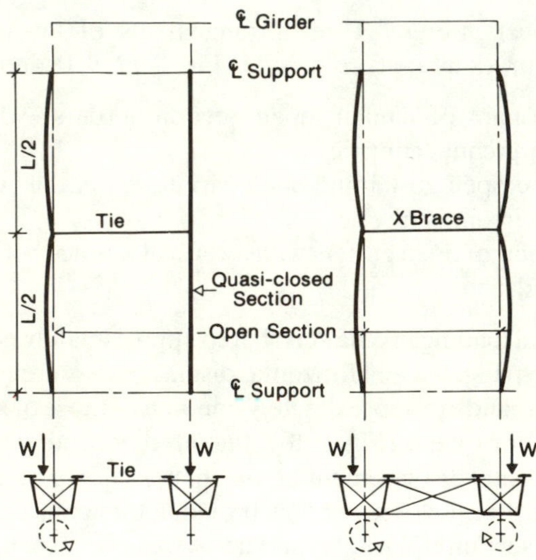


FIG. 9.18. Transverse deflection due to torsional load, open and quasi-closed sections, with midspan tie.

FIG. 9.19. Transverse deflection due to torsional load, open sections, with midspan x-brace.

brace location. The statics of the midspan restraint are illustrated in Fig. 9.20 where a torsional restraint at a support section is shown. Additional girder loads, $V$, developed as a consequence of the restraint can be calculated from simple statics for a known torsional moment, $T$. With $V$ known, the shear in the diagonals can be calculated and thence the forces in the horizontal members. The results of case C provide a basis from which the forces in x-bracing for other torsional loading cases can be found using the principle of superposition.

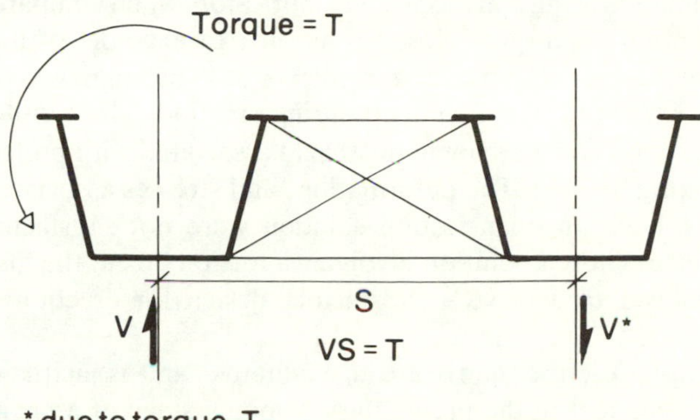


FIG. 9.20. Interconnected girders.

The discussion above assumes that the value of torque provided by the restraining brace is known. An approximate but satisfactory assessment of the torque introduced into the brace system can be found by considering the limiting case where the St Venant torsional rigidity for the open section member is taken as zero. Walker (1977) has shown for this assumption that there is an analogy between flexural and torsional behaviour. Thus, support torque (reactions), torque (shear), bimoment (moment), warping (slope) and rotation (deflection) of a torsionally loaded system can be found from expressions similar to those for bending. Using the flexural analogy, the torsional load to be resisted by the midspan tie or x-brace is five-eighths of the total torque applied to a single torsionally loaded girder. The total midspan reaction for a two-span beam with uniform flexural rigidity and uniform loading and no settlement of the supports is also five-eighths of the total load.

Other simple examples can be given where the lateral movements of single girders can be restricted and where such girders may form members in an interconnecting system later in the construction sequence. A simple span girder subjected to a uniform torsional load may be simply supported for flexure but may incorporate a torsion box (Fig. 9.14) at each extremity, so providing full fixity for warping (Branco, 1981). The lateral displacements for this torsionally restrained case are approximately one-fifth of those without restraint.

Torsion boxes can provide useful and economical solutions to erection problems involving long cantilever girders at sites with restricted access. A torsion box basically requires two additional bracing elements. The use of a small amount of bracing appeals to steel fabricators when compared to the bracing required for a quasi-closed solution. An example of the use of torsion boxes seems appropriate. A cantilever of 45 m length was proposed for a box girder of some 2·5 m deep during erection. The cantilever was subjected to wind and a torsional dead load associated with prefabricated interconnecting bracing. The deformations and stresses associated with a solution involving an open-section solution were not satisfactory. The design engineer chose a solution involving a torsion box at the mid-length of the cantilever to achieve a satisfactory design for erection (Taylor, 1982).

The concepts described above are all qualitative, and as such can be used by a design engineer in the preliminary planning stage of a project. The need exists for simple calculation procedures based on physical evidence of box girder behaviour.

## 9.5 ANALYSIS

### 9.5.1 Experimental Studies

The behaviour of braced girders has been studied by Green (1978) and Branco (1981) in respect of two one-quarter scale model girders. Each girder had a span of 12·2 m, depth of 0·61 m, base width of 0·61 m and top width of 0·91 m (Fig. 9.21a). These proportions are typical of prototype box girder bridges. The girders were supported at the ends on load cells. Girder 1 was loaded with sixteen point loads applied to the top flanges, so simulating a uniformly distributed load on each flange. The girder was extensively gauged at the midspan, three-eighths, quarter and one-eighth points. Girder 2 was used, in conjunction with girder 1, to study the effect of interconnecting bracing between members (Fig. 9.21b).

Angles were used as bracing members. These were connected to the web stiffeners with bolts torqued to a specific value. The angles were gauged on the principal axis to obtain average axial strain, and thence brace forces.

(a)

(b)

FIG. 9.21. Test arrangement: (a) overall view—two-girder test; (b) interconnecting bracing.

Tests were carried out on girder 1 with concentric and eccentric loading and for different types of bracing systems. Tests of the two-girder system included eccentric load on girder 1 and concentric load on girder 2, as well as several types of interconnecting bracing systems. Test results presented in this chapter correspond to a total load level of 90 kN per girder and eccentricity values of either zero or 83 mm ($e/b = 0{\cdot}11$) for girder 1. These values represent the construction load imposed by the concrete deck slab, noting model similitude.

### 9.5.2 Analytical Studies

The finite strip method due to Cheung (1976) was used to study the influence of bracing on the distortion of open girders. The girder was divided into several longitudinal strips connected at their edges (nodal lines), with each strip including longitudinal bending and in-plane deformation. For this analytical model, bracing forces were calculated using the force method. Final displacements and stresses were determined at the nodal lines.

A simplified method—the flange bending analysis—was developed to compute the transverse bending stresses in the flange arising from the bending distortion of the cross-section. The method assumes that the distortion of open box girders is resisted by in-plane bending of the top flanges, and neglects any contribution from the transverse stiffness of the web (Branco, 1981). A continuous beam with elastic supports at the braced sections was used to represent the top flanges. As a result of inclination of the webs, a vertical load acting on the flanges can be decomposed according to Fig. 9.22. The horizontal load components due to concentric or torsional load, $P_S$ or $P_D$, lead to brace forces corresponding to continuous beam reactions. The associated flange bending stresses correspond to longitudinal distortional stresses in the girder top flanges.

Analysis of the open girder, neglecting any distortional effect, was carried out using a torsion–bending analysis (Kollbruner and Basler, 1969; Walker, 1977). Longitudinal stresses arising in bending and torsion are computed from the bending moment and from the bimoment respectively. Shear stresses are caused by shear force or torque (St Venant and warping shear stresses). In an open girder, the St Venant stiffness is small compared with the warping stiffness and can be neglected. This permits a simplified torsional analysis where the differential equation of torsion based on warping stiffness only is similar to that of bending (Branco, 1981). The torsion–bending analysis assumes no distortion in the section and, through a comparison with the finite strip results, the effectiveness of the bracing systems in preventing distortion can be found.

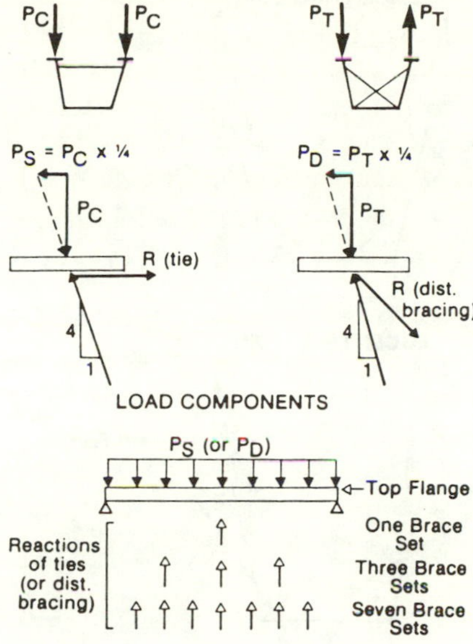


FIG. 9.22. Flange bending model.

Rigid section analysis was combined with the force method to analyse the effects of interconnecting bracing between girders. The influence of an interconnecting tie was studied by considering the compatibility of rotations between girders at the tie section. The effect of an interconnecting truss was found by considering the compatibility of rotation and vertical displacement at the truss section.

### 9.5.3 Bracing to Reduce Distortion, Single Girder

The results of both the finite strip and torsion–bending analyses are compared in Fig. 9.23 with the test results for concentric and eccentric load for a girder with a single tie and distortional brace (Fig. 9.14) at midspan. Close agreement exists between finite strip and test results. The torsion–bending analysis does not predict the stresses in the top flange as bending and torsional distortion of a section is ignored.

The lateral displacements associated with bending distortion (spreading displacements) are shown in Fig. 9.24 for concentric load, and assuming ties at midspan and quarter points. The spreading displacement decreases from 29 mm with no ties to approximately 1 mm with ties. A prototype system would have a deformation which was four times these values under

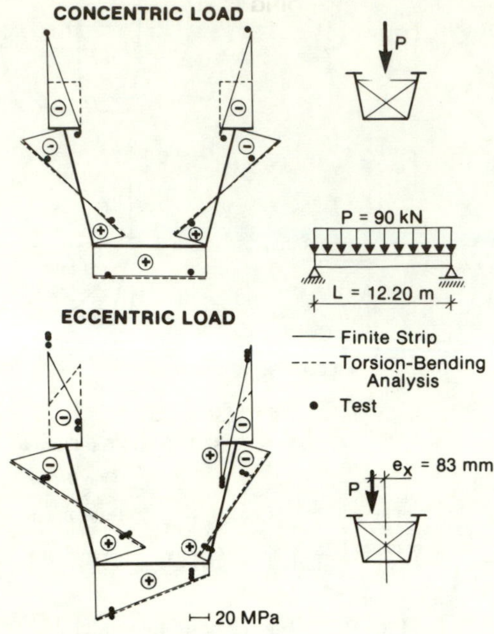


FIG. 9.23. Total midspan stresses, one-brace set.

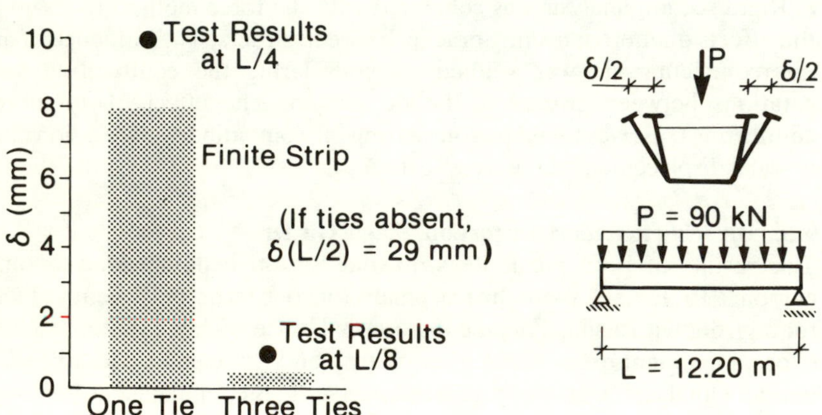


FIG. 9.24. Transverse displacements.

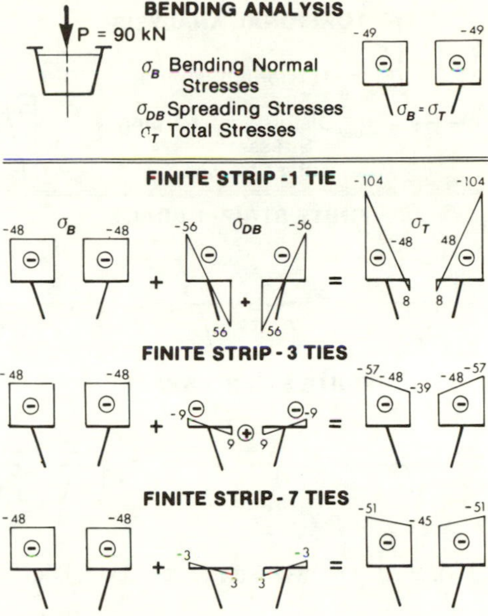


FIG. 9.25. Total midspan stresses, several ties.

prototype loading. A 120 mm lateral displacement would be unacceptable.

The local stresses in the top flange are given in Fig. 9.25 for the concentric load case of Fig. 9.23. The flanges carry stresses due to longitudinal bending, $\sigma_B$, and to distortional bending, $\sigma_{DB}$. The decrease in the maximum total stress values, $\sigma_T$, is clearly evident as the number of ties increases and spacing for the uniformly distributed ties decreases. The spreading effect is almost negligible for seven ties and the total stresses are nearly identical to those due to bending alone. Ties are effective in preventing spreading (bending distortion). The values of $\sigma_{DB}$ are a function of the width (section modulus) of the top flange plates. For very narrow and perhaps somewhat impractical top flange plates, ties may be required at a spacing of less than one-eighth of the span as suggested here.

Torsional distortion occurs when the load is applied eccentrically to a girder. The stress in the top flanges at midspan is shown in Fig. 9.26 for such a load case. Distortional warping stresses ($\sigma_{DW}$) are small even when one bracing set is used. Torsional warping stresses ($\sigma_W$) obtained from the finite strip analysis, for a section without bracing, are similar to the results of the torsion–bending analysis. A rigid body behaviour applies if only distortion is minimized (Branco and Green, 1985).

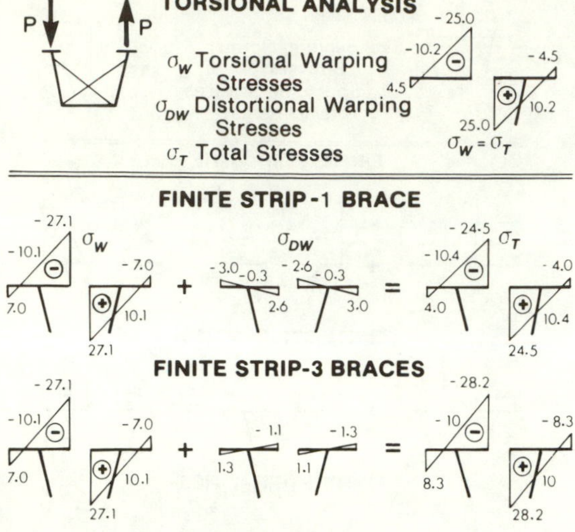


FIG. 9.26. Distortional stresses, several ties.

Branco and Green (1984a) have shown that a stiffened web will further reduce bending and torsional distortion effects. For example, with a stiffness ratio between stiffened web and unstiffened web equal to fifty, distortional stresses are almost negligible and only torsional warping stresses occur. The stiffened web limits torsional distortion. Distortional bracing for the girder and load conditions considered is not required. However, such bracing should probably be included at the span quarter points and midspan as a minimum to maintain the stability of the compression flange and facilitate handling by the fabricator.

The forces occurring in the braces, computed from the finite strip method and flange bending analysis, are shown in Fig. 9.27. Eccentric load, together with one bracing set, three bracing sets and seven ties, is considered. A good agreement is obtained between finite strip and flange bending analysis and test results for one and three bracing sets. Differences are noted for the seven-tie case. These may be due to numerical problems associated with the use of the force method for a highly restrained system, or to experimental difficulties associated with the measurement of small strains in the bracing.

Distortional bracing formed from ties will only restrain bending distortion. Equally spaced angle ties do not contribute to either the St Venant or warping torsional stiffness of an open-section box. Ties do not

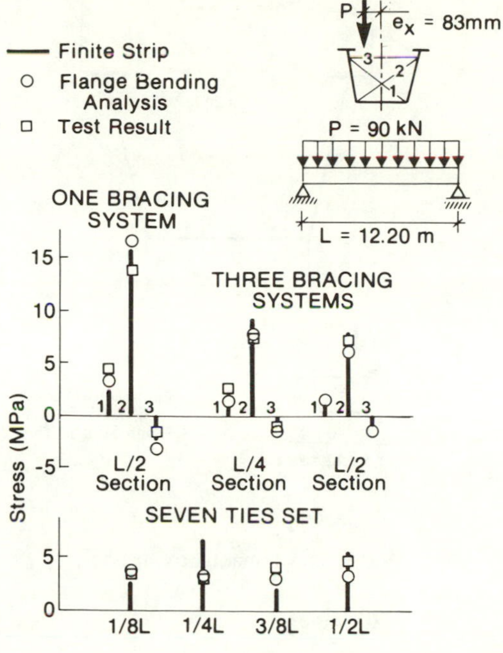


FIG. 9.27. Brace forces.

restrain torsional distortion as both webs of the girder translate an equal amount in the same direction under torsional load (Fig. 9.9c), and there is little or no relative displacement between the ends of the ties. Some design engineers hold the opinion that ties contribute to the torsional resistance of an open section. This is not so.

### 9.5.4 Torsion Boxes

Torsion boxes (Fig. 9.14) provided at the extremities of a girder reduce the longitudinal warping displacements and so decrease the span rotation and associated torsional warping stresses. The magnitude of this reduction of stress is shown in Fig. 9.28 where values at midspan, with and without torsion boxes, are compared. The total stresses occurring at midspan and one-eighth span are presented in Fig. 9.29. The agreement between finite strip and test results is good, as is a torsion–bending analysis considering the girder fixed against warping at the one-eighth and seven-eighth span points. The response of the girder is almost beam-like when compared to the results of Fig. 9.23.

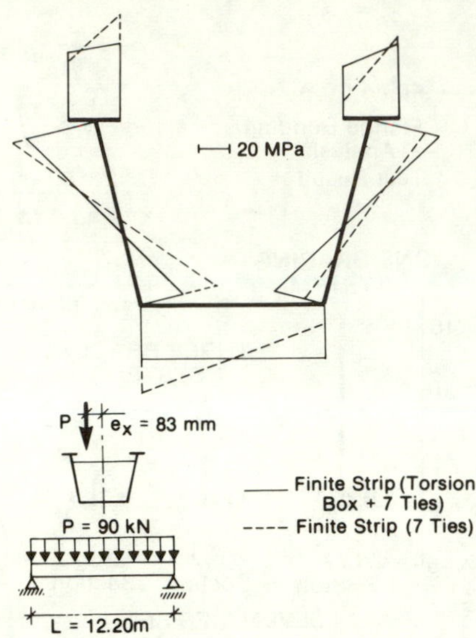


FIG. 9.28. Torsion box stresses.

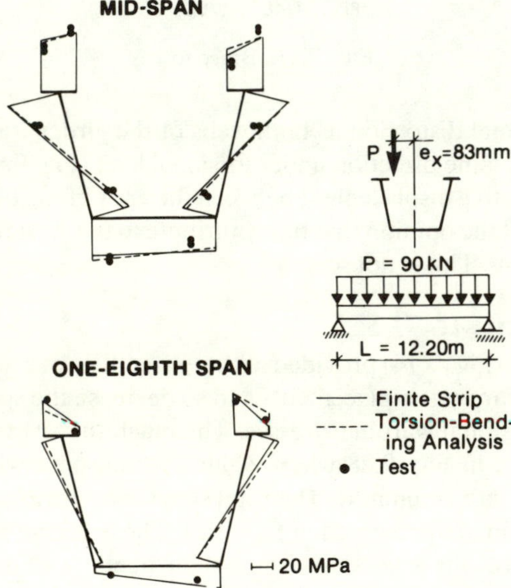


FIG. 9.29. Stresses in torsion box.

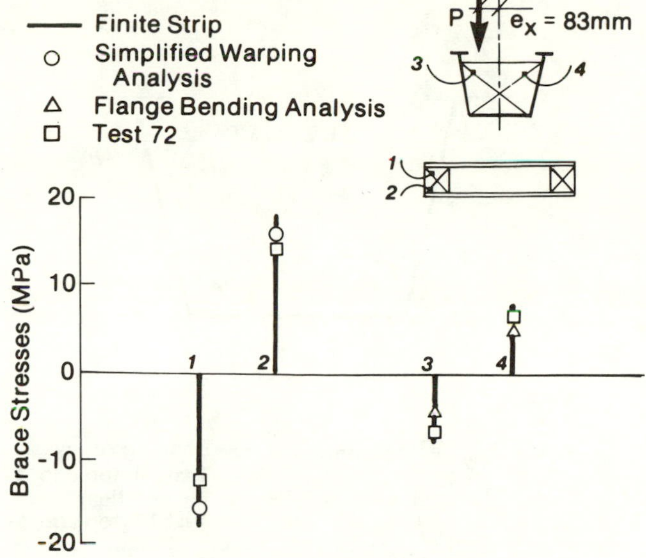


FIG. 9.30. Forces in torsion box bracing.

The forces in the braces of the torsion boxes are given in Fig. 9.30, with good agreement found between finite strip, analysis, and test.

### 9.5.5 Top Chord Bracing

An effective way of reducing warping displacements is to place horizontal top chord bracing near the level of the top flanges. This type of bracing increases the St Venant torsional stiffness, and reduces warping and twist of the girder.

The results obtained at midspan with top chord bracing are presented for eccentric loading in Fig. 9.31. The difference between the results with and without ties is important, since the top chord bracing installed together with ties is more effective in preventing the bending distortion or spreading of the section as the span between unsupported points along the flange becomes one-eighth of the total span rather than one-quarter. The stresses obtained for the test with ties and top chord correspond to pure bending stresses, so indicating the effectiveness of top chord bracing in preventing torsional warping stresses.

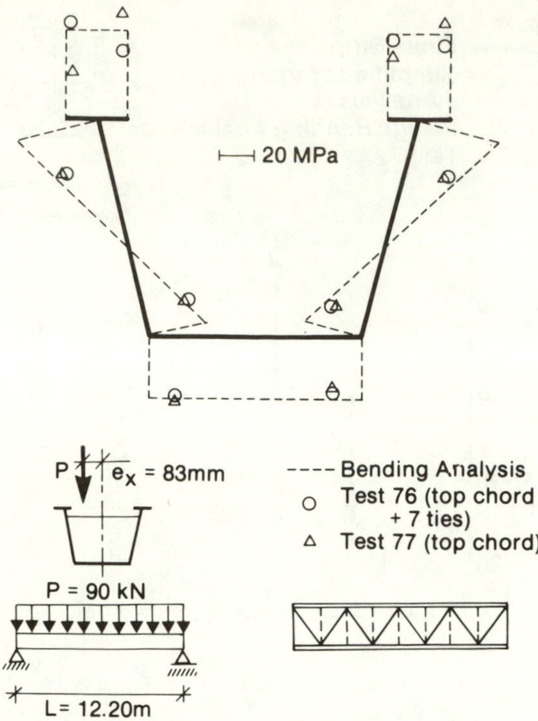


FIG. 9.31. Quasi-closed section, midspan stresses.

## 9.6 COMPLETED GIRDERS

Construction bracing may be left in place. Design engineers should be aware of the influence of such bracing on the service behaviour of a completed bridge. A linear elastic analysis, based on the finite strip method, was used for analysis. Results are compared with experimental values obtained from a model and a prototype bridge (Branco, 1981). The effect of web stiffeners, distortional bracing and interconnecting bracing on the torsional and distortional behaviour of completed box girders is described.

### 9.6.1 Analytical Studies

An analytical method is required which considers both rigid and distortional behaviour of the cross-section, if the influence of the bracing

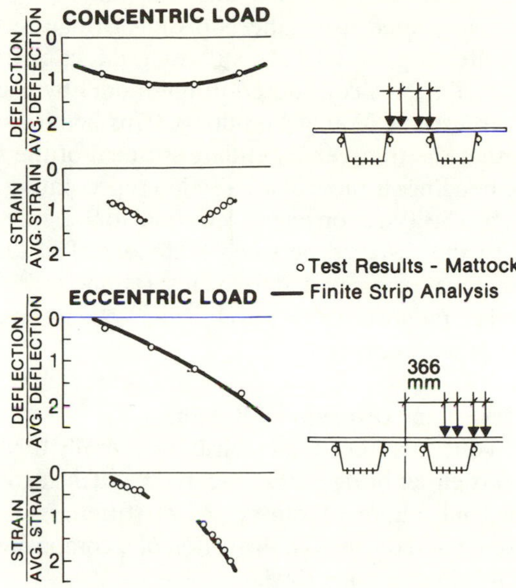


FIG. 9.32. Comparison of test and analysis.

systems on composite box girder bridges is to be studied. The finite strip method was selected over the finite element method because of the geometric properties and boundary conditions of the structures, and the lower computing cost.

The complete structure is idealized as a series of longitudinal strips connected at the edges (nodal lines). Each strip can model shell behaviour with longitudinal bending and in-plane deformation (Cheung, 1976). For the composite box girder bridge, the cross-section was divided into strips to represent the girders and concrete deck. The connection between steel and concrete was simplified using a single nodal line at mid-depth of the deck. Top flanges of the steel box were not considered.

The results from the Mattock bridge model tests were used to verify the finite strip analysis for composite sections (Fountain and Mattock, 1968; Mattock and Johnston, 1968). In Fig. 9.32, the analytical results for concentric and eccentric loading are presented. The good agreement between finite strip and test results for both displacements and strains indicates the effectiveness of the analytical method.

The influence of bracing systems on composite box girder bridge behaviour was investigated using the Nipissing Bridge tests (Holowka, 1979) and the finite strip model. The stiffening effect of transverse web stiffeners on distortion was considered in the model by using web strips with a fictitious transverse Young's modulus. This fictitious modulus was used to ensure that the transverse bending stiffness of the web elements was equal to the bending stiffness of a T-section representing each stiffener and adjacent web. The T-section had a web equal to the stiffener depth and a flange equal to the distance between stiffeners. The load conditions considered in analysis corresponded to the test (Branco, 1981; Branco and Green, 1984b). The influence of various bracing systems on the girder was considered, using the force method.

### 9.6.2 Web Stiffeners and Distortional Bracing

The force effects due to a truck load distribute laterally to all components of a composite box girder bridge. Bending, twist and distortion are present in each box. The contribution of transverse web stiffeners and construction distortional bracing in reducing the distortion of a composite closed box is examined here for the Nipissing Bridge.

Calculated longitudinal midspan strains in the Nipissing Bridge box girders, for both concentric and eccentric truck load, are given in Fig. 9.33 for the case of both stiffened and unstiffened webs. For eccentric loading, larger distortional warping strains are obtained for unstiffened webs than for stiffened webs. Observed results from the bridge with stiffened webs and five bracing systems along the span within each box are also included in the figure. Analysis considering stiffened webs agrees well with the test results. Transverse web stiffeners appear to be effective in preventing both longitudinal and transverse distortional strains, and well-distributed distortional bracing appears to have little effect on these strains.

The influence of the web bending stiffness $D_{xs}$ on the magnitude of the distortional stresses present in a two-girder system with an eccentric load is presented in Fig. 9.34. As the web stiffness ratio $(D_{xs}/D_{xu})$ increases, torsion and distortion stresses $(\sigma_{11}–\sigma_{17})$ (Fig. 9.34) tend to the torsional warping stresses for a rigid section. The distortional warping stresses decrease and are small for a $D_{xs}/D_{xu}$ ratio equal to 43 (Nipissing Bridge).

Fig. 9.35 illustrates the strain values for an eccentric load condition for both unstiffened webs (Fig. 9.35a) and stiffened webs (Fig. 9.35b), including and excluding midspan distortional bracing. Distortional strains have design significance for a box girder with unstiffened webs for both distortional bracing cases. When a single midspan distortional brace is

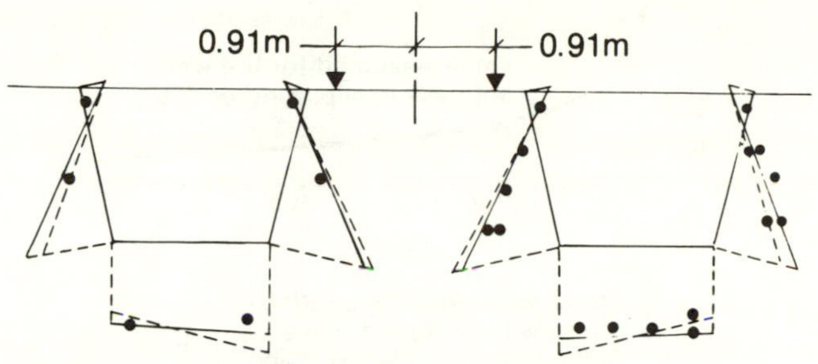


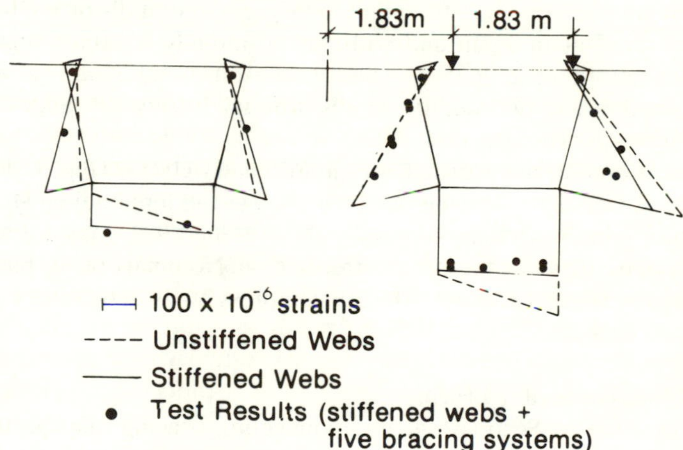


FIG. 9.33. Nipissing Bridge, effect of web stiffness on strain.

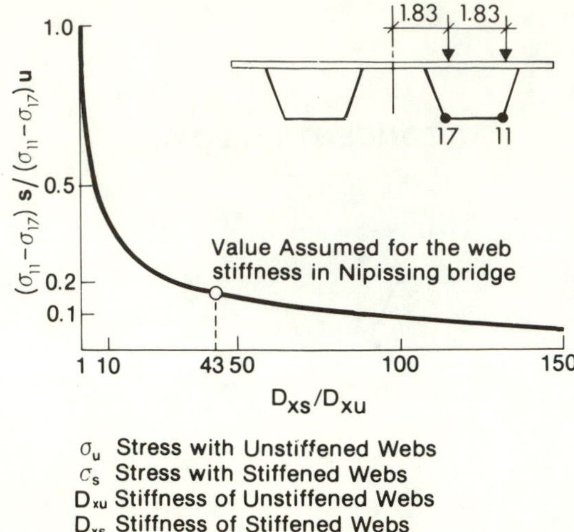


FIG. 9.34. Distortional stresses as influenced by web stiffness.

present, the longitudinal strain due to this restraint of distortion becomes concentrated at midspan and adds to the already existing longitudinal bending strain (Fig. 9.35a). This concentration of strain is virtually eliminated when two additional distortional braces are added at the quarter points (Branco, 1981).

Distortional strains are small for the stiffened web case (Fig. 9.35b). The addition of midspan bracing does not change the longitudinal strain distribution greatly. Bracing forces for the stiffened web case are approximately one-fifth those for the unstiffened case. Nominal bracing based on a slenderness ratio of about 100 for compression is satisfactory for the stiffened web case.

### 9.6.3 Interconnecting Bracing

As suggested in Section 9.4, interconnecting bracing may be installed between box girders during construction to prevent excessive twist. The influence of such bracing on the behaviour of the Nipissing Bridge is described with results for one and three interconnecting bracing sets given, in Fig. 9.36, for an eccentric loading and webs with transverse stiffeners. When bracing is present, there is a small change in distortional strains in the bottom flanges of the box girders as a consequence of distortion due to the interconnecting bracing forces. Agreement with test results for either

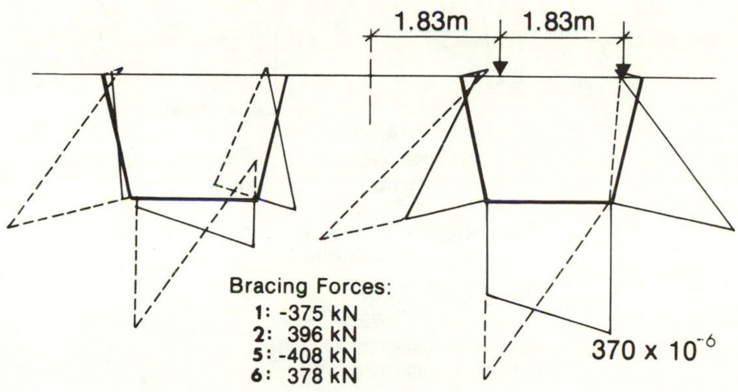


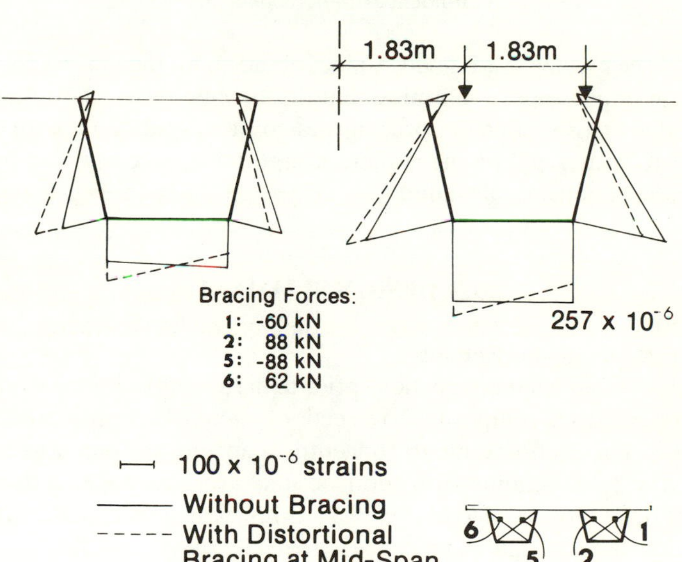


FIG. 9.35. Midspan bracing and distortional stresses.

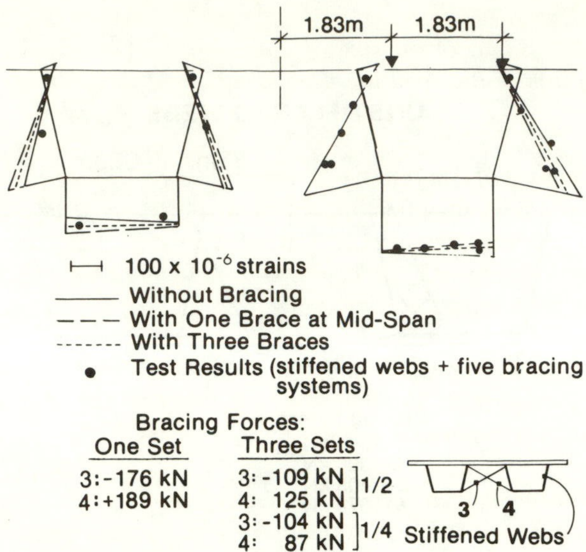


FIG. 9.36. Influence of interconnecting bracing.

one or three bracing sets is good. Values obtained for the bracing forces are of design importance, confirming that this bracing resists deformation of the bridge cross-section. A bracing design based only on the minimum slenderness ratio will be incomplete if used for interconnecting bracing. The forces should be calculated.

## 9.7 DESIGN SUMMARY

### 9.7.1 Distortion—the Actions

Distortion of an open box section, prior to hardening of the concrete deck to form a closed composite box section, should be considered as the superposition of effects due to concentric load and torsional load. Under concentric load bending distortion occurs, while torsional distortion is present due to torsional load (Fig. 9.9). Stresses arising from the effects of distortion have design significance, especially in the top flanges of box girders with inclined webs and torsional load (Figs 9.11–9.13).

#### *9.7.1.1 Bracing for the Actions*

Horizontal ties will limit bending distortion. Vertical distortional bracing can be used to limit torsional distortion effects (Figs. 9.23–9.27). For typical box girder geometries and loads due to construction, ties installed at the eighth points of the span and distortional bracing provided at the quarter points of the span are usually sufficient to minimize distortional effects.

*9.7.1.2 Methods of Analysis*

An approximate flange-bending analysis can be used to determine the transverse bending stresses in the top flanges and the bracing forces arising from distortion (Fig. 9.22). The method should be used only where more than three equally spaced braces are present within the span length. It assumes that the distortion of open box girders is resisted by the in-plane bending of the top flanges, and that the transverse bending stiffness of the web can be neglected. The flange is considered to be a continuous beam with supports at the braced sections.

*9.7.1.3 Transverse Web Stiffeners*

The magnitudes of the distortional stresses are also reduced if a transversely stiffened web is used. The height and spacing of stiffeners should be such that the transverse stiffness of the stiffened web is at least fifty times that of the unstiffened web.

### 9.7.2 Rotation of an Open Box Girder

For typical geometries and constructional loadings, a linear analysis considering only warping stiffness gives acceptable predictions of midspan rotation.

*9.7.2.1 Stress Analysis*

For an open box girder, the torsional warping stresses cannot be ignored in design. Torsional warping stresses can be computed considering the linear analysis based only on the warping stiffness of the section. For typical geometries and constructional loadings, the total stresses computed by the linear analysis have an error of less than 10%.

*9.7.2.2 Bracing to Reduce Rotation*

The rotation of a single girder can be reduced by installing horizontal bracing at the level of the top flanges. This bracing can be placed at the ends of the girder only (torsion boxes) or along the span (top chord bracing) (Fig. 9.14). Top chord bracing is generally not enough to prevent distortion and should be used in conjunction with ties.

For multi-box girders, three equally spaced sets of interconnecting x-bracing installed between girders are sufficient to reduce the warping effects arising from typical constructional load values. This interconnecting bracing must be aligned with distortional bracing within the girders.

## 9.8 CONCLUSIONS

Composite box girder bridges may utilize torsionally flexible open box sections during construction. Excessive rotation or distortion may occur under a variety of constructional loadings. The torsional and distortional stiffness of such open sections can be increased through the use of a variety of bracing systems selected by either the design engineer or the steel fabricator to suit the particular circumstances. The influence of such bracing systems on the behaviour of open and quasi-closed box sections was examined. Horizontal tie and distortional bracing systems were found to be effective in preventing distortion. Stiffened webs were found to be as effective as distortional bracing in minimizing torsional distortion. Bracing forces can be calculated from a flange bending or finite strip analysis.

In a completed box girder, warping stress effects based on rigid section behaviour were found to be small. Distortional warping stresses can be significant if transverse web stiffeners or distortional bracing are not used. At least three distortional braces should be used within each box and each span to avoid concentration of restraint to distortion.

## ACKNOWLEDGEMENTS

The assistance given by the Social Sciences and Humanities Research Council of Canada, the Natural Sciences and Engineering Research Council of Canada, the Technical University of Lisbon, the Ontario Ministry of Transportation and Communications, the Canadian Steel Industries Construction Council and the University of Waterloo during the course of the work described here is gratefully acknowledged.

## REFERENCES

AMERICAN ASSOCIATION OF STATE HIGHWAY AND TRANSPORTATION OFFICIALS (AASHTO) (1983) *Standard specifications for highway bridges*. 12th edn, Washington, DC.

BRANCO, F. A. (1981) Composite box girder bridge behaviour. MASc thesis,

University of Waterloo, Waterloo, ON.

BRANCO, F. A. and GREEN, R. (1984a) Bracing for composite box girder bridges. *Can. J. Civ. Eng.*, **11**, 844–53.

BRANCO, F. A. and GREEN, R. (1984b) Bracing in completed box girder bridges. *Can. J. Civ. Eng.*, **11**, 967–77.

BRANCO, F. A. and GREEN, R. (1985) Composite box girder behavior during construction. *J. Struct. Eng.*, **III**(3), March, 577–93.

CANADIAN STANDARDS ASSOCIATION (1978) Design of highway bridges, CAN3 S6–M78. Rexdale, ON.

CHEUNG, Y. K. (1976) *Finite strip method in structural analysis.* Pergamon Press, Oxford.

FOUNTAIN, R. S. and MATTOCK, A. H. (1968) Composite steel–concrete multi-box girder bridges. *Proc. Can. Struct. Eng. Conf.*, Canadian Steel Industries Construction Council, Toronto, ON, 19–53.

GREEN, R. (1978) Composite box girder bridges—the construction phase. *Proc. Can. Struct. Eng. Conf.*, Canadian Steel Industries Construction Council, Toronto, ON.

HOLOWKA, M. (1979) *Testing of a trapezoidal box girder bridge.* Report RR221, Ministry of Transportation and Communications, Downsview, Ontario.

KOLLBRUNER, C. F. and BASLER, K. (1969) *Torsion in structures.* Springer-Verlag, Berlin.

MATTOCK, A. H. and JOHNSTON, S. (1968) Behavior under load of composite box girder bridges. *ASCE, J. Struct. Div.*, **94**(ST10), 2351–70.

ONTARIO MINISTRY OF TRANSPORTATION AND COMMUNICATIONS (OHBDC) (1983) Ontario Highway Bridge Design Code and Commentary, 2 vols, Downsview, ON.

POELLOT, W. M. (1974) Special design problems associated with box girders. *Proc. ASCE Specialty Conf. on Metal Bridges*, ASCE, St Louis, MO.

REIS, A. and BRANCO, F. A. (1980) The nonlinear behaviour of composite box girder bridges during construction. Conference discussion on the new code for the design of steel bridges, Cardiff. Edited by H. R. Evans.

TAYLOR, P. R. (1982) Bridge erection—The designer's role. *Proc. Canad. Struct. Eng. Conf.*, Canadian Steel Industries Construction Council, Toronto, ON.

TRAHAIR, N. S. (1977) *The behaviour and design of steel structures.* Halsted Press, John Wiley & Son, London.

UNITED STATES STEEL (1978) *Steel/concrete composite box girder bridges—a constructional manual.* Pittsburgh.

WALKER, A. C. (1977) *Design and analysis of cold-formed sections.* Halsted Press, John Wiley & Son, London.

*Chapter 10*

# FERROCEMENT STRUCTURES AND STRUCTURAL ELEMENTS

P. PARAMASIVAM, K. C. G. ONG and S. L. LEE

*Department of Civil Engineering, National University of Singapore, Singapore*

*SUMMARY*

*Ferrocement is defined as a thin-walled reinforced concrete commonly constructed of hydraulic cement mortar reinforced with closely spaced layers of continuous and relatively small diameter wire mesh (ACI Committee 549, 1982). Research into the properties and applications of ferrocement has taken a tremendous stride in recent years and a variety of structures using innovative design and construction techniques have been built and tested. This chapter is intended to provide a review of the recent developments in the design, construction and application of ferrocement.*

*The chapter is divided into seven sections: introduction, materials, mechanical properties, design, construction, applications and conclusion. Wherever possible, results of research are presented in the form of tables and charts. Research studies currently in progress are also included and discussed.*

## NOTATION

| | |
|---|---|
| $a$ | Shear span |
| $b$ | Width |
| $C_c$ | Compressive force in mortar |
| $C_s$ | Compressive force in reinforcement |

| | |
|---|---|
| $c$ | Position of neutral axis |
| $d$ | Position of layer of mesh |
| $d_b$ | Diameter of wire |
| $d_{max}$ | Position of extreme tensile layer of mesh |
| $E_c$ | Elastic modulus of composite |
| $E_m$ | Elastic modulus of mortar |
| $E_R$ | Effective elastic modulus of reinforcement |
| $f'_c$ | Cylinder compressive strength of mortar |
| $f_{cu}$ | Cube strength of mortar |
| $h$ | Total depth |
| $M_n$ | Nominal moment resistance of ferrocement flexural beam |
| $S$ | Mesh size of wire mesh |
| $S_{RL}$ | Specific surface in the direction of loading |
| $S_{Rl}$, $S_{Rt}$ | Specific surface in the longitudinal and transverse direction respectively |
| $T_s$ | Tensile force in reinforcement |
| $V_{cr}$ | Diagonal cracking strength |
| $V_R$ | Volume fraction of reinforcement |
| $V_{RL}$ | Volume fraction in the direction of loading |
| $V_{Rl}$, $V_{Rt}$ | Volume fraction in the longitudinal and transverse direction respectively |
| $\beta$ | Ratio of distance to neutral axis from the extreme tensile fibre and from the outermost layer of steel |
| $\gamma_m$ | Partial safety factor for strength of material |
| $(\Delta l)_{av}$ | Average crack spacing |
| $\epsilon_0$ | Total surface area of bonded reinforcement |
| $\epsilon_R$ | Strain in extreme tensile layer of mesh |
| $\eta$ | Ratio of bond strength to matrix tensile strength |
| $\theta$ | Factor relating average crack spacing to maximum crack spacing |
| $\sigma_{cr}$ | First crack stress of ferrocement |
| $\sigma_{cu}$ | Ultimate strength of composite |
| $\sigma_{mu}$ | Ultimate strength of mortar |
| $\sigma_R$ | Stress in the reinforcement under service load |
| $\sigma'_R$ | Proof stress in reinforcement corresponding to 0·01% strain |
| $\sigma_{Ry}$ | Yield stress of reinforcement |
| $\phi$ | Efficiency factor |
| $\omega_{av}$ | Average crack width |
| $\omega_{max}$ | Maximum crack width |

## 10.1 INTRODUCTION

The idea of reinforcing cement mortar by closely spaced layers of fine wire mesh was originally conceived one and a half centuries ago by Lambot and subsequently promoted by Nervi one century later in the 1940s. Since then the material has been extensively studied and the basic technical information on various aspects of design, construction and application is now available. Experience and expertise gained from its use have shown that ferrocement is a highly versatile construction material with potential for a wide range of practical applications.

Notwithstanding the availability of technical know-how (ACI Committee 549, 1982 and Shah and Balaguru, 1984) and the satisfactory performance record of ferrocement structures already built, large scale applications can be found mainly in Eastern Europe (ISF, 1981; Prawel and Reinhorn, 1983). Isolated cases of field applications, particularly in low-cost rural housing and other minor structures, have been reported in developing countries (Hagenbach, 1972; Castro, 1977; Robles-Austriaco *et al.*, 1981). Recently Matinelli *et al.* (ISF, 1981) and Jennings (1983) reported the completion of several large and successful structures in Brazil and Jordan respectively.

Among the factors hindering wider applications of ferrocement is its expensive mesh system and labour intensive fabrication process. However, there is increasing evidence that even in industrialized countries, ferrocement can be cost competitive through mechanized production and proper choice of mesh reinforcement. Another factor frequently held responsible for its slow growth is the general lack of understanding of the material. But the most important is the lack of design guidelines and specifications, without which building authorities are reluctant to accept this relatively new material.

## 10.2 MATERIALS

Ferrocement is a composite structural material comprising a cement mortar matrix reinforced with layers of small diameter wire mesh uniformly distributed throughout its cross-section. Although meshes of glass or vegetable fibres have been used, the most common form involves steel and it is this type that is covered in this chapter.

The cement mortar paste should be designed for appropriate strength and maximum denseness and impermeability, with sufficient workability

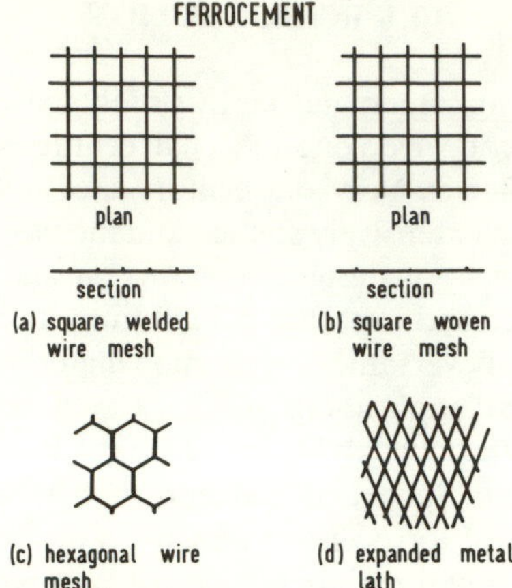


FIG. 10.1. Types of mesh.

to minimize voids. A water/cement ratio of 0·35 to 0·65 combined with an aggregate/cement ratio ranging from 1:1 to 3:1 is generally employed, with the lower values recommended for hand plastering techniques. The aggregate is usually a sharp fine-grade sand and ordinary Portland cements are generally adequate, despite the low covers employed.

The wire mesh is usually fabricated of mild steel and may be of a variety of forms. Figure 10.1 illustrates a number of possibilities including expanded metal. Sometimes regular reinforcing bars in a skeletal form are added to the thin wire meshes in order to achieve a stiff reinforcing cage. The contribution of these extra reinforcements can be considered in the analysis and design. The commonly used composition and properties of reinforcement are summarized in Table 10.1.

## 10.3 MECHANICAL PROPERTIES

### 10.3.1 Reinforcing Parameters

The ACI Committee 549 (1982) listed two important reinforcing parameters, the volume fraction and the specific surface of reinforcement, commonly used to characterize the reinforcement in ferrocement.

$V_R$, the volume fraction of reinforcement, is defined as the total volume of reinforcement divided by the volume of composite (reinforcement and matrix). For a composite reinforced with meshes of square openings,

TABLE 10.1
PROPERTIES AND TYPES OF CONSTITUENT MATERIALS USED IN FERROCEMENT CONSTRUCTION

| *Materials* | *Range* |
|---|---|
| *Wire mesh* | |
| Diameter of wire ($\phi$) | $0{\cdot}5 \leq \phi \leq 1{\cdot}5$ mm |
| Type of mesh | Chicken wire or square woven or welded galvanized mesh or expanded metal |
| Size of mesh opening ($S$) | $6 \leq S \leq 25$ mm |
| Volume fraction ($V_R$) of reinforcement in both directions | $2\% \leq V_R \leq 8\%$ |
| Specific surface ($S_R$) of reinforcement in both directions | $0{\cdot}1 \leq S_R \leq 0{\cdot}4$ mm$^2$/mm$^3$ |
| Elastic modulus ($E_R$) | 140–200 kN/mm$^2$ |
| Yield strength ($\sigma_{Ry}$) | 250–460 N/mm$^2$ |
| Ultimate tensile strength ($\sigma_{Ru}$) | 400–600 N/mm$^2$ |
| *Skeletal steel* | |
| Type | Welded mesh, steel bars, strands |
| Diameter ($d$) | 3 mm $\leq d \leq$ 10 mm |
| Grid size ($G$) | 50 mm $\leq G \leq$ 200 mm |
| Yield strength | 250–460 N/mm$^2$ |
| Ultimate tensile strength | 400–600 N/mm$^2$ |
| *Mortar composition* | |
| Cement | Any type of Portland cement (depending on application) |
| Sand/cement ratio (s/c) | $1 \leq$ s/c $\leq 3$ by weight |
| Water/cement ratio (w/c) | $0{\cdot}35 \leq$ w/c $\leq 0{\cdot}65$ by weight |
| Gradation of sand | 5 mm to dust with not more than 10%* passing 150 $\mu$m BS test sieve |
| Compressive strength (cube) | 30–50 N/mm$^2$ |

*For crushed stone, sand may be increased to 20%.

$V_R$ is divided equally into $V_{Rl}$ and $V_{Rt}$ for the longitudinal and transverse directions respectively.

$S_R$ is the specific surface of the reinforcement. In ferrocement nomenclature it is the total bonded area of reinforcement (interface area) divided by the volume of composite. For a composite reinforced with meshes of square openings, $S_R$ is divided equally into $S_{Rl}$ and $S_{Rt}$ in the longitudinal and transverse directions respectively.

For a ferrocement plate of width $b$ and depth $h$, $S_R$ can be computed from

$$S_R = \epsilon_0 / bh \tag{10.1}$$

where $\epsilon_0$ is the total surface area of bonded reinforcement per unit length.

When wire of diameter $d_b$ is used in square grid meshes the relation between $S_R$ and $V_R$ is

$$S_R = 4V_R / d_b \tag{10.2}$$

### 10.3.2 Modulus of Elasticity

The elastic modulus of elasticity of a steel mesh is not necessarily the same as that of the wire filament. In a woven steel mesh, weaving imparts an undulating profile to the wires. When stressed in tension, the woven mesh made from these wires stretches more than a similar welded mesh made from identical straight wires. A lower elastic modulus is observed during tensile testing. Also, when a woven mesh is embedded in a mortar matrix and tends to straighten in tension, the matrix may resist the straightening, leading to a form of tension stiffening. A similar behaviour is observed for expanded metal and hexagonal mesh. To account for the above effects, the ACI Committee recommends the term $E_R$, effective modulus of the reinforcing system.

### 10.3.3 Behaviour under Compression

Generally, the properties of the matrix control the properties of the ferrocement in compression. This is so in most structural applications where thin ferrocement elements are employed. The nominal resistance of ferrocement sections subjected to uniaxial compression is usually estimated from the load-carrying capacity of the unreinforced mortar matrix. In special circumstances where the reinforcement is very heavy, the contribution of reinforcement may be considered. Accordingly the type, orientation and mode of arrangement of the reinforcement must be considered.

### 10.3.4 Behaviour under Tension

Various investigators have observed that when ferrocement elements are subjected to tensile loading three stages of behaviour are observed (Fig. 10.2a). During the first stage, ferrocement behaves like a homogeneous elastic material till the first crack appears. This is followed by the multiple

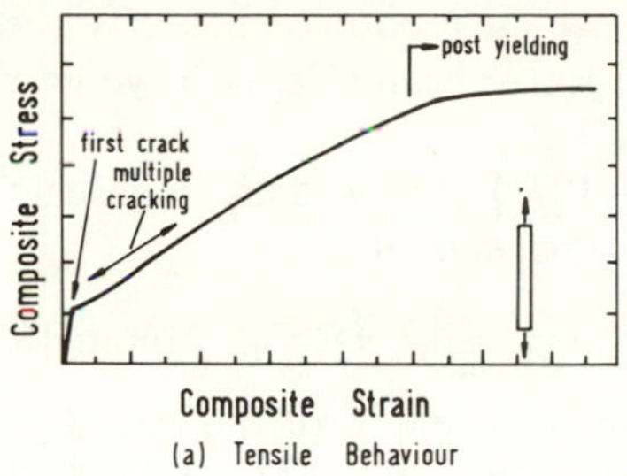


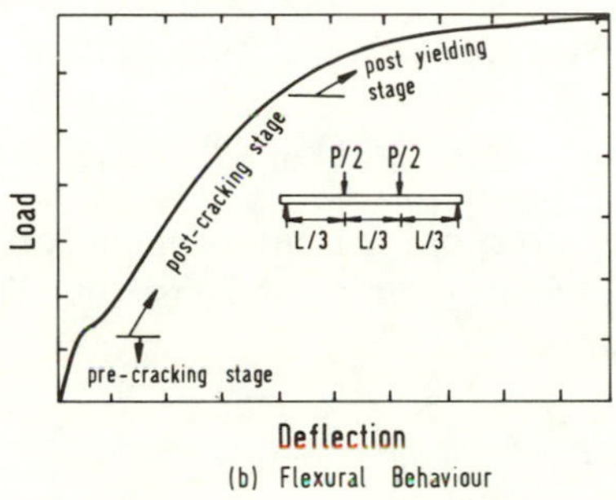


FIG. 10.2. Behaviour of ferrocement.

cracking stage. Theoretically it starts at the occurrence of the first crack and continues up to the point when the reinforcing meshes start to yield. This stage is characterized by an increase in the number of cracks with increasing load, with very little increase in crack widths. The number and size of cracks depend on volume fraction and the dispersion of wire meshes. Generally this stage represents the behaviour of most structural elements under service loads.

In the third stage, called the crack-widening stage, no (or few) additional cracks are formed. This stage begins when the reinforcement starts to yield and continues up to failure. The number of cracks remains essentially constant and crack widths keep increasing.

*10.3.4.1 Elastic Behaviour*

In the elastic stage, the stress–strain relationship is usually defined using a single elastic constant, $E_c$. This is given by

$$E_c = E_m(1 - V_{RL}) + E_R V_{RL} \tag{10.3}$$

where $E_c$, $E_m$ and $E_R$ are the moduli of elasticity of the composite, mortar and reinforcement respectively and $V_{RL}$ is the volume fraction of wire mesh in the direction of loading.

Naaman and Shah (1971) proposed an empirical formula to predict the first crack stress $\sigma_{cr}$ of ferrocement as

$$\sigma_{cr} = \sigma_{mu} + 25S_{RL}(\text{N/mm}^2) \tag{10.4}$$

where $\sigma_{mu}$ is the ultimate strength of the mortar and $S_{RL}$ is the specific surface ($\text{mm}^2/\text{mm}^3$) in the direction of loading. This formula is valid only for $S_{RL}$ less than 0·2 $\text{mm}^2/\text{mm}^3$.

Another method proposed by Nathan and Paramasivam (1974) uses $V_{RL}$ to predict $\sigma_{cr}$. The first crack stress is calculated from

$$\sigma_{cr} = \sigma'_R(V_{RL})^{1\cdot 1} + \sigma_{mu} \tag{10.5}$$

where $\sigma'_R$ is the proof stress of the reinforcement corresponding to 0·01% strain and $\sigma_{mu}$ is the ultimate strength of the mortar. The constant, 1·1, was determined empirically.

*10.3.4.2 Multiple Cracking Stage*

During the multiple cracking stage the contribution of mortar to the stiffness of the composite decreases progressively. A conservative estimate of the Young's modulus of the ferrocement composite is given by

$$E_c = V_{RL}E_R \tag{10.6}$$

Cracking behaviour is mainly dependent on the volume and dispersion of reinforcement. A direct relationship between average crack spacing, $(\Delta l)_{av}$ at crack stabilization, and $S_{RL}$ is proposed by Naaman and Shah (1971) as follows:

$$(\Delta l)_{av} = \frac{\theta}{\eta}\frac{1}{S_{RL}} \tag{10.7}$$

where $\theta$ is a factor relating average crack spacing to maximum crack spacing and $\eta$ is the ratio of bond strength to matrix tensile strength. For square meshes $\theta/\eta$ was found empirically to approximate to unity.

The following empirical design procedure to predict the maximum crack width $\omega_{max}$ in cracked ferrocement tensile elements may be used as

$$\omega_{max} = \frac{3500}{E_R} \qquad \text{for } \sigma_R \leq 345S_{RL} \tag{10.8}$$

and

$$\omega_{max} = \frac{20}{E_R}(175 + 3{\cdot}69(\sigma_R - 345S_{RL})) \text{ for } \sigma_R > 345S_{RL} \qquad (10.9)$$

where $\omega_{max}$ is the maximum crack width in mm, $\sigma_R$ the steel stress under service load in N/mm$^2$, and $S_{RL}$ is the specific surface in the loading direction in cm$^{-1}$.

At ultimate, the load is carried by the mesh reinforcement in the direction of loading. The ultimate strength $\sigma_{cu}$ is usually given by the following

$$\sigma_{cu} = \sigma_{Ry} V_{RL} \qquad (10.10)$$

where $\sigma_{Ry}$ is the yield stress of the reinforcement.

### 10.3.5 Flexural Behaviour

The flexural behaviour of ferrocement also exhibits three stages, viz. the precracking, multiple cracking and post-cracking stages (Fig. 10.2b). The adoption of conventional reinforced concrete theory using either the ACI Code (ACI Standard 318, 1977) or the British Code (CP 110, 1972 or BS 8110, 1985) has been recommended by ACI Committee 549 (1982) to compute the flexural strength of flexural members.

The analysis of ferrocement elements subject to bending (Fig. 10.3) is similar to that of reinforced concrete. The analysis takes into account the effective cross-sectional area and position of the reinforcing layers with

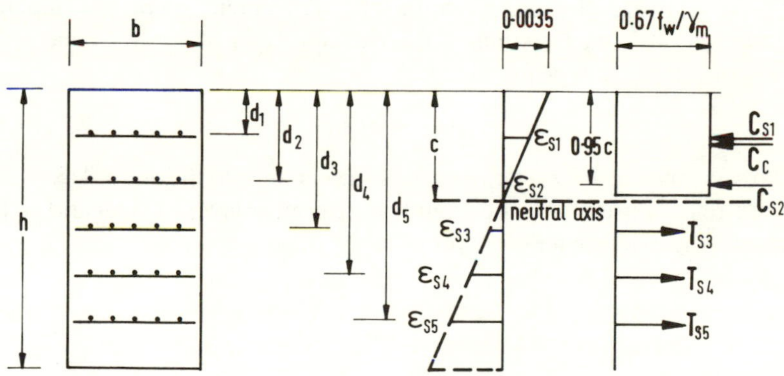


FIG. 10.3. Strain and force distribution at ultimate in ferrocement section under bending (BS 8110, 1985).

respect to the neutral axis. In addition, the type of reinforcement and the orientation, spacing and geometry of the layers of reinforcement, unique to ferrocement, must also be considered.

During the precracking stage, the flexural behaviour may be deduced by using an elastic analysis incorporating the elastic modulus of the reinforcing system and mortar matrix. Subsequently, during the multiple cracking stage, the contribution of the cement mortar in tension may be neglected (Fig. 10.3).

The computation of the nominal moment of resistance of ferrocement sections can be tedious and time consuming. Following an extensive computerized parametric evaluation, Naaman and Homrich (1986) derived the following nondimensionalized equation to predict the nominal moment of resistance $M_n$ of ferrocement flexural beams as

$$\frac{M_n}{f'_c b h^3 \phi} = 0{\cdot}0049 + 0{\cdot}42 \frac{V_{RL} \sigma_{Ry}}{f'_c} - 0{\cdot}077 \left( \frac{V_{RL} \sigma_{Ry}}{f'_c} \right)^2 \qquad (10.11)$$

where $f'_c$ is the cylinder compressive strength of mortar, $b$ and $h$ the size of the beam, $\sigma_{Ry}$ the yield strength of the reinforcement, $V_{RL}$ the volume fraction in the direction of loading, and $\phi$ an efficiency factor.

For investigation of stresses at service loads an elastic analysis similar to that of reinforced concrete may be employed (ACI Committee 549, 1982). Allowable stresses are specified (Section 10.4) for the constituent materials. In the computation of stresses and strains, the cracked transformed section can be used, incorporating the elastic modulus of the reinforcing system and mortar mix.

Under flexure the crack width of ferrocement elements depends primarily on the tensile strain in the extreme layer of mesh. The average crack width, $\omega_{av}$, can be predicted as

$$\omega_{av} = S \beta \epsilon_R \qquad (10.12)$$

where $S$ is the mesh size, $\beta$ the ratio of the distance to the neutral axis from the extreme tensile fibre and from the outermost layer of steel and $\epsilon_R$ the strain in the extreme tensile layer of mesh. Thus

$$\beta = \frac{h - c}{d_{max} - c} \qquad (10.13)$$

and

$$\epsilon_R = \frac{(\sigma_R)_{max}}{E_R} \qquad (10.14)$$

where $h$ is the total depth, $c$ the position of the neutral axis, $d_{max}$ and $(\sigma_R)_{max}$ the position and stress in the extreme tensile layer of mesh and $E_R$ the effective modulus of the wire mesh.

### 10.3.6 Behaviour in Shear

Little or no research is reported in the literature on the behaviour and strength of ferrocement in shear. The writers are aware of only one exploratory study on the shear strength of ferrocement by Collen and Kirwan (1959). Lack of research interest in this area is probably due to the fact that in conventional ferrocement structures the span to effective depth ratio in flexure is large enough to preclude shearing distress. However, with recent developments that broaden the scope of ferrocement applications, transverse shear may become one of the critical design considerations.

A recent study by Mansur and Ong (1987) was undertaken to study the strength of ferrocement in transverse shear. On the basis of flexural tests on simply supported rectangular ferrocement beams, it was found that ferrocement beams are susceptible to shear failure at small shear span to depth ratios when $V_{RL}$ and $f_c'$ are relatively higher. The following empirical expression was proposed to predict the diagonal cracking strength $V_{cr}$ of symmetrically reinforced ferrocement beams:

$$\frac{V_{cr}}{bh} = 3{\cdot}54\left(f_c' V_{RL}\frac{h}{a}\right)^{0{\cdot}76} \tag{10.15}$$

where $b$ and $h$ are the breadth and total depth of the beams, $a$ the shear span and $f_c'$ the cylinder compressive strength.

## 10.4 DESIGN

Ferrocement is characterized by fine diameter mesh reinforcement having a surface area per unit volume of mortar which may be as much as ten times that in conventional reinforced concrete. It is usually recommended that the surface area per unit volume of the mesh reinforcement should not be less than 0.1 $mm^2/mm^3$ and the volume fraction of reinforcement not less than about 1% in each direction. Twice these values are recommended and even more steel may be provided in water-retaining structures.

The compressive strength of the composite is usually directly related to that of the mortar, while the tensile strength is a function of the mesh content and properties. Since the wire mesh is much stronger in tension than the mortar, the mortar's role is to hold the mesh in place properly, transfer stresses by means of adequate bond and give protection to the steel. It should be noted that the cover provided is very low, generally of the order of 3 mm and not greater than 7 mm. This assists in minimizing crack widths due to shrinkage and tensile loading.

The ultimate load of flexural members can be predicted using conventional reinforced concrete theory as given in Section 10.3.5. Stresses in flexural members may also be checked at service loads. The allowable tensile stress in the reinforcement may be generally taken as $0{\cdot}6\,\sigma_{Ry}$, where $\sigma_{Ry}$ is the yield strength measured at 0·0035 strain. For water-retaining structures it may be preferable to limit the tensile stress to 207 MPa.

The accompanying allowable compressive stress in the composite may be taken as $0{\cdot}45\,f'_c$, where $f'_c$ is the specified compressive strength measured from tests on 75 mm × 150 mm cylinders.

All ferrocement members and structures must also satisfy serviceability requirements. The ACI Committee 549 (1982) recommends that the maximum crack width be less than 0·10 mm for non-corrosive environments and 0·05 mm for corrosive environments and/or water-retaining structures. However, no particular deflection limitations are recommended as most ferrocement members and structures are thin and very flexible and their design is very likely to be controlled by criteria other than deflection.

The fatigue stress range in the reinforcement is limited to 207 MPa for structures to sustain a minimum fatigue life of two million cycles. A stress range of 458 MPa may be used for one million cycles. Higher values may be considered if justified by tests.

Shear is seldom a limiting design criterion in ferrocement. The ACI Committee 549 (1982) recommends the design recommendations of the ACI Building Code for punching shear in slabs to be used as a first approximation.

## 10.5 CONSTRUCTION

There are several means of producing ferrocement but all methods seek to achieve the complete infiltration of several layers of reinforcing mesh by a well-compacted mortar matrix with a minimum of entrapped air. The choice of the most appropriate construction technique depends on the nature of the particular ferrocement application, the availability of mixing, handling and placing machinery, and the skill and cost of available labour.

Construction of ferrocement can be divided into four phases:

(1) fabrication of a skeletal framing system;
(2) fixing the mesh reinforcement;
(3) plastering;
(4) curing.

The thinness of the ferrocement elements means that there is little room for poor workmanship, and considerable care is necessary. This applies right the way through the construction process from the placing of the reinforcement through precautions for maintaining the small cover, selection of aggregates, mixing, placing and curing. However, experience has shown that usually only a modest amount of training, production standardization and preparation is required to produce ferrocement of consistent quality.

The skeletal framing system usually involves the fabrication of a stiff reinforcing cage of steel rods or pipes to which the mesh is subsequently tied. It is essential that the mesh reinforcement sections are laid on smoothly and evenly and firmly tied to each other as well as to the skeletal steel. Tie wires used should be properly cut and bent inward.

Mortar plastering and penetration is the greatest problem area in ferrocement construction, and many defects can be attributed to lack of complete infiltration and consolidation. Attempts have been made to mechanize the highly labour-intensive operation of mortar application; wet-mix shotcrete with only air added to the mix at the nozzle to create the spray is the preferred shotcrete method.

Proper curing is essential to develop the desired properties of the mortar. A curing period of 28 days is suggested, although by the first two weeks the mortar will have attained most of its design strength. Curing should commence 24 hours after final application of the mortar. The temperature must be kept well above freezing and the surface not permitted to dry out. Shrinkage cracks may appear on the surface of thin ferrocement structures if curing is improperly carried out. Ferrocement structures are sometimes painted for aesthetic reasons and this may assist in reducing permeability.

## 10.6 APPLICATIONS

One of the earliest applications of ferrocement was the boat built by Lambot in France in 1849. One of his boats, reportedly in good condition, is currently on display in a museum in Brignoles, France (Cassie, 1967).

There was very little application of true ferrocement construction between 1888 and 1942 until Nervi (1951) subsequently promoted its use in civil engineering structures.

Despite evidence that ferrocement was an adequate and economic building material, it gained wide acceptance only in the early 1960s. There has been widespread use of ferrocement throughout the world including Canada, the USA, Australia, New Zealand, the UK, Mexico, Brazil, the Soviet Union, Poland, Czechoslovakia, the Republic of China, Thailand, India, Indonesia and many other developing countries. Applications range from the sophisticated, such as boats and long-span roofing, to the simple, such as sewer linings, dustbins and cattle troughs (Barfoot, 1985). Experience and expertise gained from its use have demonstrated the versatility of ferrocement as a construction material in the building industry.

Ferrocement construction in the early 1970s was labour intensive and particularly suitable for rural applications in developing countries. This does not require heavy plant or machinery and, being a low-level technology, the construction skills can be acquired fairly quickly. Accordingly considerable research effort has been directed in developing low-cost rural housing compatible with local traditions. In urban areas the potential application of ferrocement must be viewed from a different perspective. In order to alleviate the acute shortage and high cost of skilled labour, more so in developed countries, mechanized methods must be employed to expedite the speed of construction.

This section is primarily aimed at identifying the possible areas where ferrocement may be employed judiciously in the building industry. Studies conducted on the application and performance of prototype ferrocement structures are highlighted.

### 10.6.1 Wall Panels

As the traditional method of laying bricks and hollow blocks is a relatively slow process and requires skilled manpower, the application of ferrocement in the construction of wall panels looks promising. Although ferrocement elements are normally slender, the material can easily be formed into suitable shapes like angles and channels (Desai and Joshi, 1976; Sandowicz, 1985) or sandwiched to reduce the slenderness ratio.

Sandwich panels are ideal as non-load-bearing partitions and external walls. A recent study (Lee *et al.*, 1986) was conducted on such panels that contained lightweight polystyrene foam as the core and ferrocement as the facing materials. Six different designs as illustrated in Fig. 10.4 and Table

10.2 were considered. The major parameters of the study were the amount and dispersion of reinforcement and the method of shear transfer between the facings.

Two categories of panels, D and N, are shown in Fig. 10.4. In category D, shear transfer between the facings was achieved by poking straight wires through the core at an angle of about 45° to the plane of the wall, the ends of each wire being spot-welded to the main reinforcement by a mechanized process. Composite action in category N was achieved by providing 25 mm thick mortar ribs at a spacing of 600 mm. Each rib contained a truss made of steel wires. Fig. 10.4 shows the details of each design.

The investigation was divided into the following three parts:

(a) selection of mix proportions for the mortar;
(b) study on panel elements (i.e. facings); and
(c) study on sandwich panels.

In the first part, four different mortar mixes were selected by trial mixing, keeping in view the related cost and the workability required for plastering. Tests were conducted to determine the basic mechanical and shrinkage properties. From the analysis of these test data, the cement/sand ratio chosen for subsequent study on panels was 1:3 by weight, with a water/cement ratio of 0·65.

In the second part of the investigation, direct tension, pure flexure and free shrinkage tests were conducted on specimens representing the facings of each of the six designs shown in Fig. 10.4. The results of these tests were carefully analysed, and on the basis of the cracking behaviour, shrinkage characteristics and ultimate strengths both in tension and flexure, four different designs were selected for further study.

The designs selected for the final part of the investigation were D1, D2, N2 and N4 (Fig. 10.4). Flexural and free shrinkage tests were carried out on prototype panels. In addition, shrinkage and accelerated weathering tests were conducted by simulating actual boundary conditions. All panels were cast in a vertical position, each face being plastered in two layers, separated by an average time interval of about 6 h and then air-cured in the laboratory before testing.

The flexural tests were conducted under third-point loading on panels 2·4 m long and 1·2 m wide. They were simply supported over a span of 2·1 m. The results of these tests, presented in Table 10.3, showed that the cracking of all the specimens occurred at approximately the same load level. However, in the case of D1, cracking was followed by large deflec-

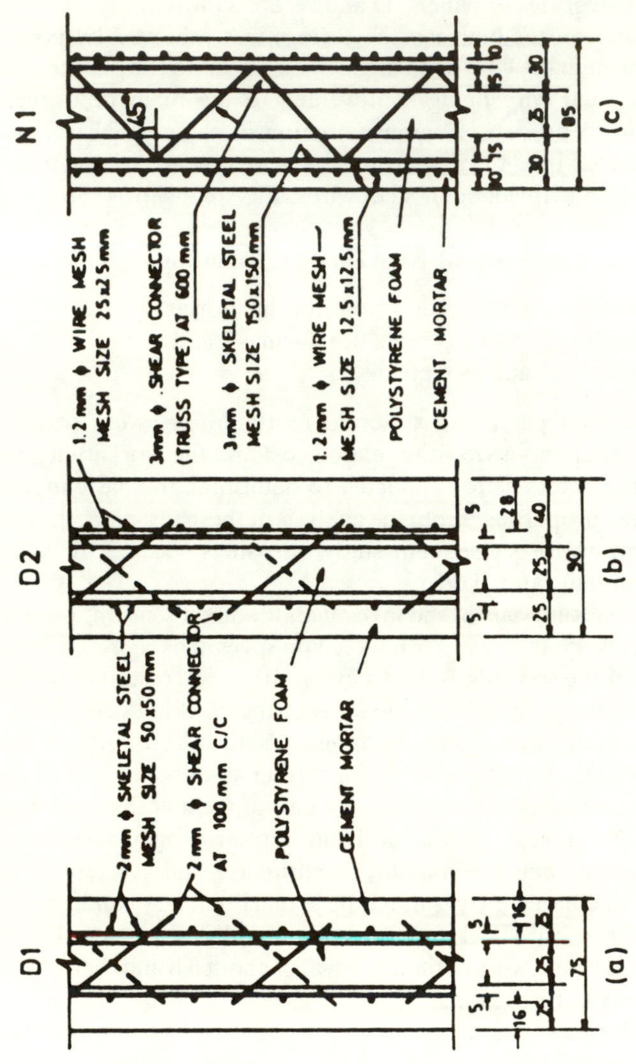
D1
D2
N1
2 mm ϕ SKELETAL STEEL
MESH SIZE 50x50 mm
2 mm ϕ SHEAR CONNECTOR
AT 100 mm C/C
POLYSTYRENE FOAM
CEMENT MORTAR
1.2 mm ϕ WIRE MESH
MESH SIZE 25x25 mm
3mm ϕ SHEAR CONNECTOR
(TRUSS TYPE) AT 600 mm
3 mm ϕ SKELETAL STEEL
MESH SIZE 150x150 mm
1.2 mm ϕ WIRE MESH
MESH SIZE 12.5x12.5 mm
POLYSTYRENE FOAM
CEMENT MORTAR
(a)
(b)
(c)

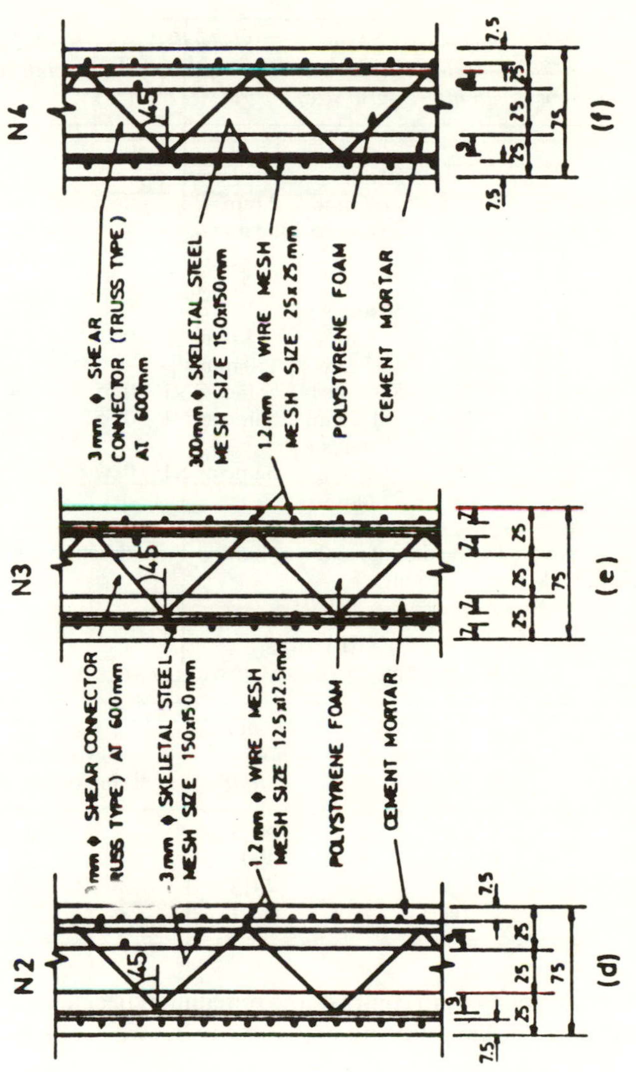


FIG. 10.4. Reinforcement details of sandwich panels.

TABLE 10.2
DETAILS OF SANDWICH PANELS

| *Design* | *Facing* | *Thickness of facing (mm)* | *Reinforcement* Main (wire size grid size) | *Reinforcement* Skeletal (wire size grid size) | $V_f$ (%) | *Total reinforcement*† ($kg/m^2$) |
|---|---|---|---|---|---|---|
| D1 | Each | 25 | 2 mm 50 mm | — | 0·25 | 1·97 |
| D2 | Outer | 40 | 1·2 mm 25 mm sq | 2 mm 50 mm sq | 0·27 | 2·68 |
| | Inner | 25 | 2 mm 50 mm sq | — | 0·25 | |
| N1 | Each | 30 | 1·2 mm 12·5 mm sq | 3 mm 150 mm sq | 0·46 | 4·31 |
| N2 | Each | 25 | Same as N1 | Same as N1 | 0·55 | 4·31 |
| N3 | Each | 25 | *1·2 mm 25 mm sq | Same as N1 | 0·55 | 4·31 |
| N4 | Each | 25 | 1·2 mm 25 mm sq | Same as N1 | 0·37 | 2·89 |

*Two layers; $V_f$ = volume fraction of reinforcement in each direction.
†Web steel is not taken into account

TABLE 10.3
FLEXURAL TEST RESULTS OF SANDWICH PANELS

| *Design* | *First crack moment (N m/m)* | *Ultimate moment (N m/m)* | *No. of cracks within the middle third* |
|---|---|---|---|
| D1 | 1025 | 2013 | 2 |
| D2 | 1040 | 3092 | 3 |
| N2 | 1108 | 4813 | 16 |
| N4 | 1050 | 4521 | 14 |

tion and eventual collapse. In contrast, the remaining specimens exhibited considerable post-cracking strength and ductility before final failure (Fig. 10.5).

Figure 10.6 shows the cracking patterns of the specimens. It can be observed that while numerous cracks appeared in N2 and N4 (c and d), panels D1 and D2 (a and b) exhibited only a few, but major cracks at

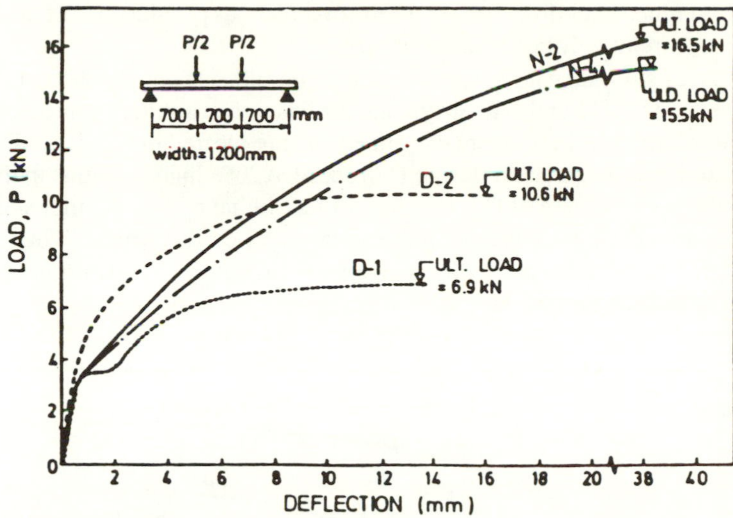


FIG. 10.5. Load–midspan deflection curves of sandwich panels.

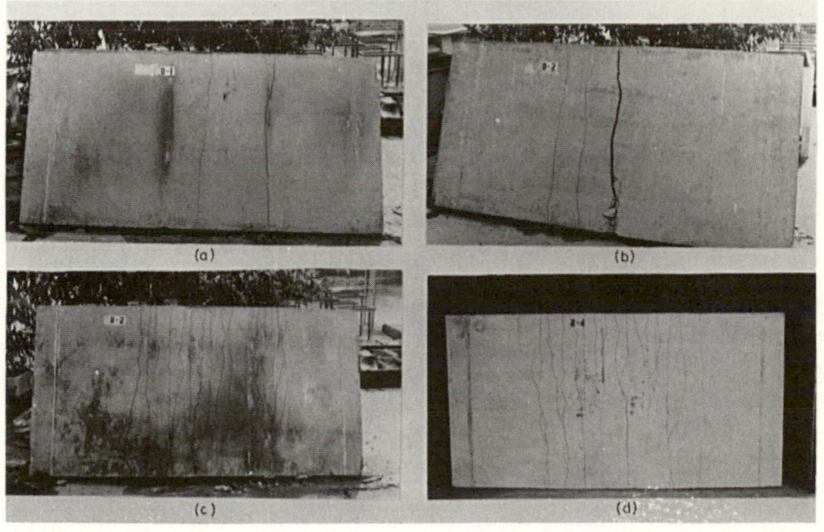


FIG. 10.6. Cracking patterns of sandwich panels in flexure.

failure. Hence finer distribution of reinforcement as provided in N2 and N4 provides a relatively better crack control.

The free shrinkage tests were conducted on full-size panels (2·4 × 1·2 m) of designs D1, D2 and N2. Shrinkage strains under ambient temperature and humidity conditions were measured and are plotted in Fig. 10.7. These indicated that the panel of design D1 recorded 20% higher strains than N2 at 28 days. In the case of D2, the thinner face behaved in a manner similar to N2, but the thicker face exhibited almost twice the strains. Thus, the panel D2 with unsymmetrical facings was subjected to a strain gradient favouring crack formation.

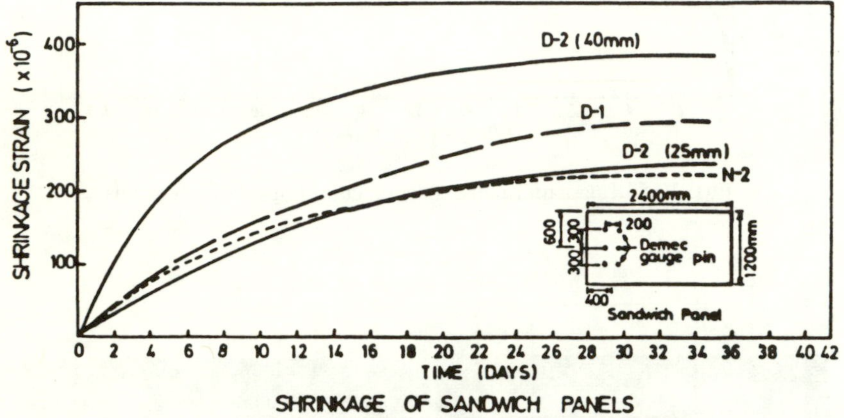


FIG. 10.7. Free shrinkage strains of sandwich panels.

In order to study the effect of end restraint on the shrinkage behaviour of the panels, a simulated test was carried out by imposing tensile strains on the panels. Three full-scale sandwich panels of designs D1, D2 and N2 were cast in an upright position against a 125 mm square steel box section precompressed to a desired strain by using a hydraulic jack. The bases of the panels were bonded to the steel box with epoxy resin. After curing, the prestress in the box section was released in stages. As a result, the box section was elongated, thus putting the faces of the panels in contact with the box beam into equivalent tensile strain similar to shrinkage action.

The cracking patterns of the panels are shown in Fig. 10.8. Cracks in the panel of design D2 appeared at a very early stage of 106 microstrain. The corresponding values for D1 and N2 were 227 and 250 microstrain respectively. Upon full release of prestress, the average crack widths recorded

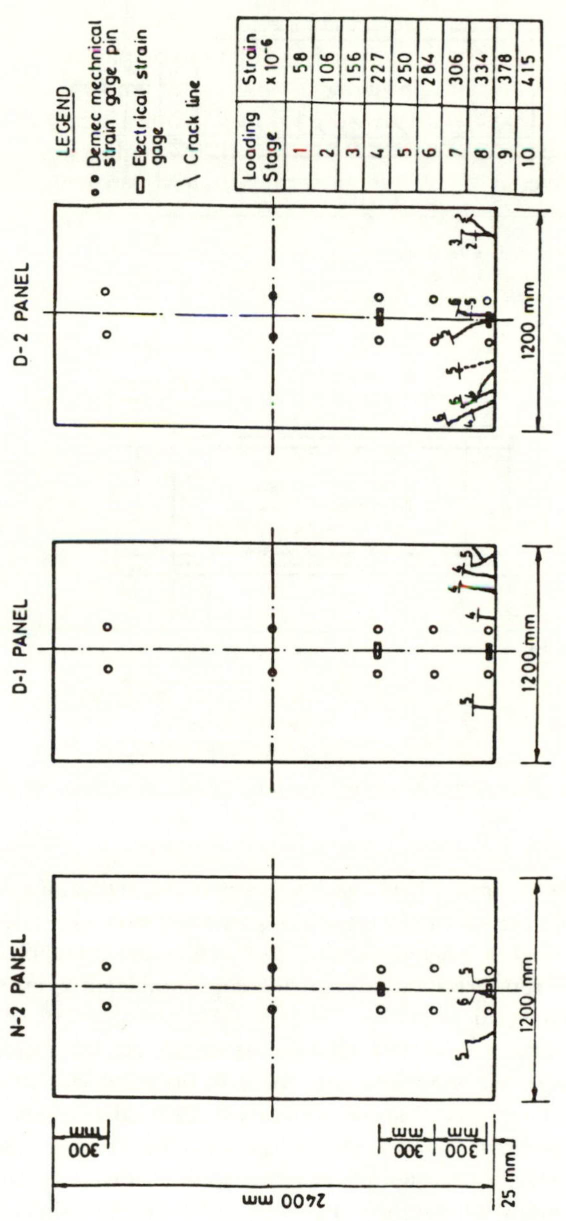


| Loading Stage | Strain x $10^{-6}$ |
|---|---|
| 1 | 58 |
| 2 | 106 |
| 3 | 156 |
| 4 | 227 |
| 5 | 250 |
| 6 | 284 |
| 7 | 306 |
| 8 | 334 |
| 9 | 378 |
| 10 | 415 |

FIG. 10.8. Cracking patterns of panels in simulated shrinkage tests.

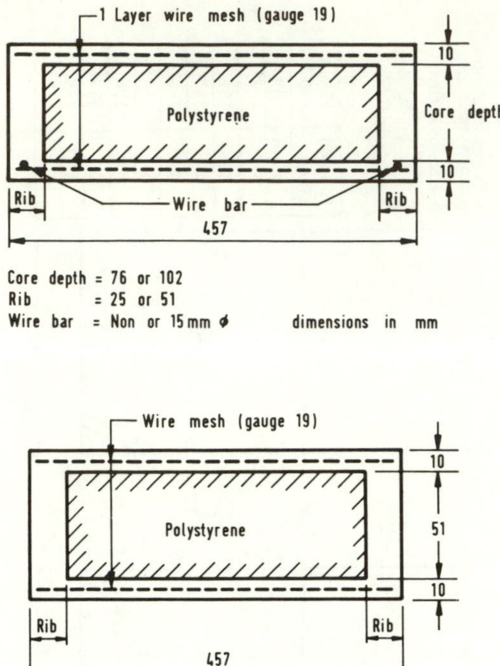


FIG. 10.9. Ferrocement panels tested under bending and in-plane compression.

for the panels of designs D1, D2 and N2 were 0·036, 0·063 and 0·026 mm, and the maximum crack widths were 0·04, 0·10 and 0·04 mm, respectively. Thus, symmetrical arrangement of facings as in D1 and N2 and finer distribution of reinforcement as in N2 can significantly improve the shrinkage characteristics.

Weathering effects on the panels were simulated by subjecting the panels to wetting and drying cycles in a specially prepared weathering tank. Three 300 × 300 mm specimens of designs D1, D2 and N2 with simulated edge restraint were subjected to 100 cycles of such wetting and drying. At the end of this period, no sign of cracking or corrosion of reinforcement was observed, implying that the panels had sufficient durability.

In another study of ferrocement sandwich panels (Fig. 10.9), Nanni and

Chang (1986) tested these panels under bending and in-plane compression. The dimensions of the specimens selected were similar to the ones adopted for the construction of a housing module. They found that the load–deflection behaviour of panels in bending can be modelled using non-linear conventional reinforced concrete analysis. The in-plane compression capacity of the panels can be predicted using the tangent modulus formula for buckling when the facing elements are treated as columns.

Wang (2nd ISF, 1985) described a building system of ferrocement panel elements. The system integrates a prestressed di-ribbed panel of ferrocement (Fig. 10.10) used as a floor or roof slab with a di-ribbed load-bearing external wall panel (Fig. 10.11). Tests for strength, rigidity, sound and thermal insulation of the panels were conducted and found satisfactory. A five-storey building of floor area 1000 $m^3$ was built in 1977. All the elements except for the footings and columns were prefabricated, transported to the site and assembled. The building was subjected to earthquake loading and survived with minimal damage.

### 10.6.2 Sunscreens

Sunscreens for housing are in general of short span, less than 3 m, and designed as one-way slabs, simply supported at the columns. In special cases where longer spans of more than 4 m are required, one-way slab design is no longer economically feasible because of excessive stresses and deflections. In order to overcome this difficulty, prefabricated ferrocement sunshades of folded plate construction may be considered (Lee *et al.*, 1983). With due considerations to such factors as required shade area and visibility, ease of handling during construction, aesthetics and cost, a three-plate design was selected with a simply supported span length of 4·55 m and a thickness of 30 mm.

Three layers of square woven mesh were used to furnish the main reinforcement. Skeletal steel required to provide rigidity during construction consisted of two layers of welded wire fabric. The sunshade was prefabricated. The extended middle plate was then inserted into the slot provided on the sides of the columns, and finally the slots were grouted with cement mortar.

The completed sunscreen shown in Fig. 10.12 was then subjected to uniformly distributed load. The deflections and longitudinal fibre stresses were measured at midspan at various load increments. Fine cracks were observed near the midspan around the upper edge of the top plate when the live load was 3·8 $N/mm^2$. No sign of distress was observed at the end connection throughout the test. The theoretical and experimental load–

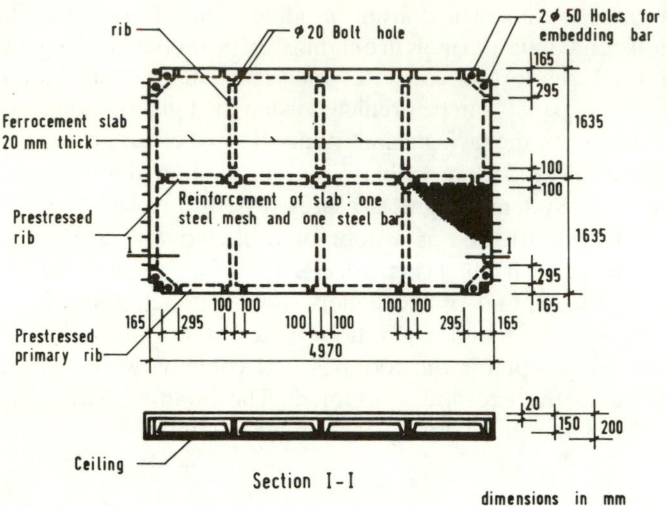


FIG. 10.10. Prestressed di-ribbed panel of ferrocement.

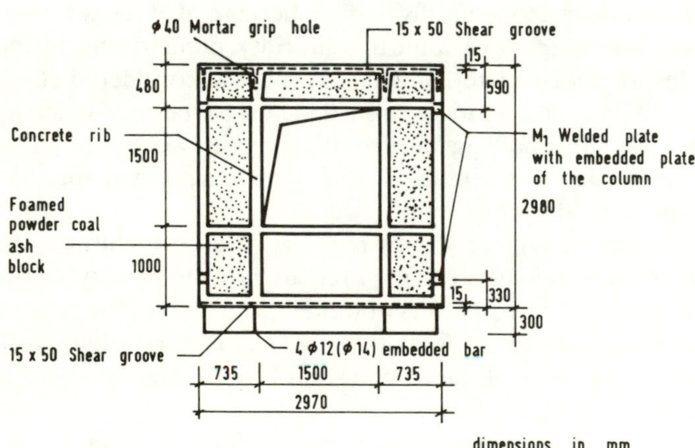


FIG. 10.11. Internal elevation of the complex wall panel.

FIG. 10.12. Three-plate ferrocement sunscreen.

deflection curves were compared, and the load tests indicated that the sunscreen was stronger than the required design strength. The thickness of the section could have been reduced.

L-shape sunscreens were recently installed in several completed multistorey apartment blocks in Singapore. A preliminary cost analysis favoured a precast construction technique. A number of alternative designs using glass fibre reinforced concrete, aluminium, conventional reinforced concrete and ferrocement were carefully analysed. Considerations of the ease of handling and erection of the precast units, connection details, durability, architectural requirements and overall economy led to the use of ferrocement as the most suitable material in this application.

In the design of these sunscreens, due considerations were given to the aesthetic and functional requirements. Bolted connections were used for attachment to the existing building. The top face of the flange was provided with a backward slope to flush out, by rain, the accumulated dust without staining the front face in order to reduce the cost of maintenance. Two layers of welded fine wire mesh with one layer of skeletal steel were used as reinforcement (Fig. 10.13). The sunscreens were cast in steel moulds in a precast concrete factory. After the necessary curing, they were painted and transported to the site. A special lifting device was used during erection.

FIG. 10.13. A view of the reinforcement arrangement.

FIG. 10.14. A view of the building after installation of the sunscreen.

Figure 10.14 shows a view of the building after the installation of the sunscreen. About 300 such sunscreens were precast and erected.

### 10.6.3 Roofing Elements

Ferrocement has a high tensile strength to weight ratio and superior cracking behaviour in comparison to conventional reinforced concrete. Hence it is an attractive material for the construction of thin-wall structures like domes, shells and folded plates in which in-plane forces are dominant. Large obstruction-free areas can be covered to provide modern facilities such as a sports complex, community centre, theatre or opera house. Suitable shell and folded plate elements can be assembled in any desired shape to form a bus shelter, a residential house or a factory building. These structural shapes can also be conveniently used as porch and roof in the construction of a multistorey building.

Three types of roofing elements have been investigated (Lee *et al.*, 1983). These are:

(a) hyperbolic paraboloid shell;
(b) cylindrical shell;
(c) folded plate (Paramasivam *et al.*, 2nd ISF, 1985).

These investigations were primarily aimed at studying the possibility of mass production by using precast techniques and to assess the feasibility of subsequent field installation. Models were made using ferrocement, with special consideration given to possible mass production and subsequent field installation.

#### *10.6.3.1 Hyperbolic Paraboloid Roof*

The use of ferrocement for the construction of shell structures has several advantages such as low dead weight, high tensile strength and excellent resistance to crack propagation compared to other conventional materials. An inverted umbrella is formed by four units of hyperbolic paraboloids, which are joined together by straight horizontal members at the exterior boundary and by straight inclined members at the interior boundary. It has the practical advantage of straight line generatrices. As a result, this type of shell has the particular feature of relatively simple formwork and easy placement of reinforcement.

The shell unit considered for the experimental model is 2·44 m square in plan with a vertical rise of 0·37 m. An overall view of the hyperbolic paraboloid shell constructed is shown in Fig. 10.15. The shell was loaded in five stages up to 4·8 kN/m$^2$. The maximum deflection at this loading was

FIG. 10.15. Hyperbolic paraboloid shell.

1·6 mm, and no distress was observed in any part of the structure. It is feasible to construct a shell of larger plan dimensions using the same shell thickness and reinforcement.

*10.6.3.2 Cylindrical Shell Roof*

A study was also conducted to study the construction of a cylindrical module in ferrocement, each module being supported by a steel column at the centre. The feasibility of mass production, ease in field installation and simplicity in connection details were considered. The module considered for the experimental model was a cylindrical shell, 3 m square in plan with a rise of 0·5 m at the crown. Each unit was prefabricated and the units were then joined together to form a prototype bus shelter as shown in Fig. 10.16. The combined twin vault was tested under a uniformly distributed load to study the response of the structure. At a working load of 1·2 kN/m$^2$, the maximum deflection and principal tensile stress were found to be 7 mm and 1·9 N/mm$^2$ respectively. These values are in close agreement with the theoretical predictions.

FIG. 10.16. Cylindrical shell.

*10.6.3.3 Folded Plate Roof*

Ferrocement is also ideally suitable as a construction material for a folded plate roof. Two prototype units of 3 × 3 m folded plate roof were designed, constructed and connected by simple cast-in-situ connections to form a multiple bay shelter (Fig. 10.17). Steel columns were used to support the structure. No bolts or nuts were used in the connection of the roof structure to the column. Because a rigid joint was desired, the connection was achieved by grouting the gap provided between the column head and the stiffeners. Single and combined units were tested separately under uniform vertical load, and the performance was satisfactory.

*10.6.3.4 Other Roofing Elements*

Mironkov *et al.* (ISF, 1981) reported the use of ferrocement as roofing for halls with spans up to 24 m in the USSR. The roofing element is made up of elements of two types, namely a pyramidal element and the ribbed slab (Fig. 10.18). The system eliminates the need for a suspended ceiling providing good acoustic properties. Openings for natural lighting may be left in the covering. The pyramidal element is a reinforced square-base pyramid of size 1·5 × 1·5 m (Fig. 10.18a). The pyramid is 0·9 m high and is

FIG. 10.17. Folded plate structure.

made of 15 mm thick ferrocement. The ribs at the four corners are 60 mm thick. The apex of the pyramid serves as a support for resting the upper ribbed slabs. The slabs are typically 1·5 × 1·5 m × 15 mm thick. The edges (forming the ribs) are 100 mm thick, integrated together to form a flat roof (Fig. 10.18b).

Barberio (ISF, 1981) described the use of ferrocement as roofing for an incubator for a fish farm in Italy. Six cupolas, paraboloid in the form of a triangular base, were constructed, reinforced with three layers of mesh of grid size 12·5 mm, using wires of diameter 0·9 mm. The matrix consisted of clay foamed up to a maximum pore diameter of 4 mm, local sand and pozzolanic cement. The finished structure is shown in Fig. 10.19.

In Brazil, Martinelli *et al.* (ISF, 1981) reported the use of ferrocement since 1960. Roofs and floor decks composed of 'I', 'V', 'U' and tubular beams were described, ranging from 6 to 35 m span (Fig. 10.20). Use of precast ferrocement caissons for roofs and floor slabs were also described, together with folded plate elements with stiffening central ribs (Figs 10.21 and 10.22). Roofs for swimming pools up to 1000 $m^3$ in size, utilizing an elliptic paraboloid precast shell, have been built using ferrocement elements as shown in Fig. 10.23.

The applications of ferrocement in Czechoslovakia were described by Smola (ISF, 1981). Various structures have been built from 1960 to 1980,

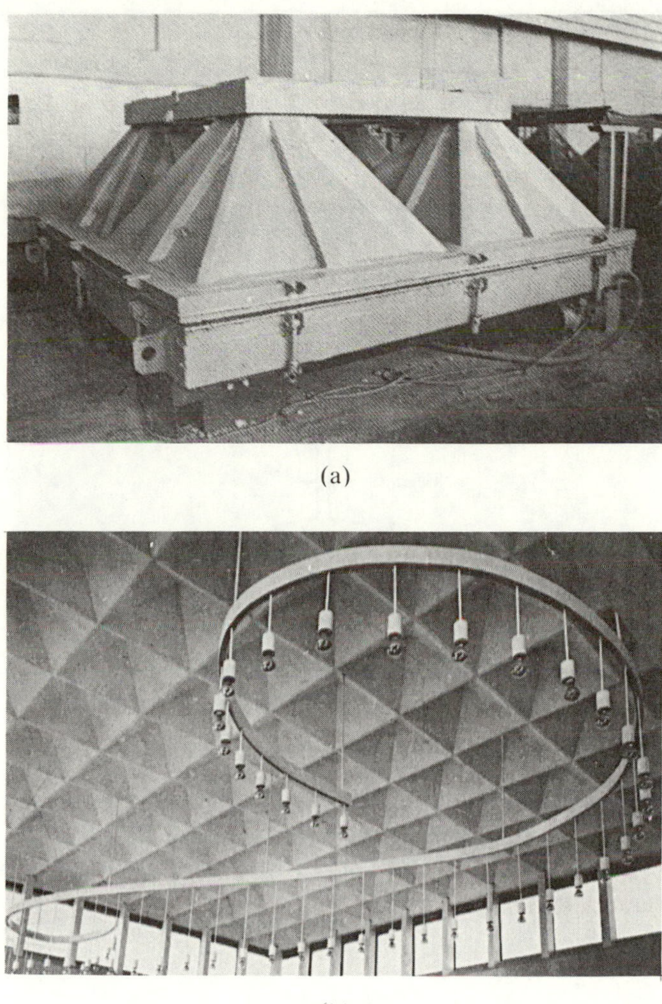

(a)

(b)

FIG. 10.18. (a) Pyramidal ferrocement elements. (b) Interior view of the hall with ferrocement elements.

including shell roofs up to a span of 36 m and suspended roofs for spans up to 30 m (Fig. 10.24). Typically, these roofs can withstand loads of 3 kN/m$^2$.

Jennings (1983) described the construction of mosque domes in Jordan using ferrocement. The central dome of the mosque was 16 m thick at its

FIG. 10.19. A group of ferrocement cupolas for a fish farm.

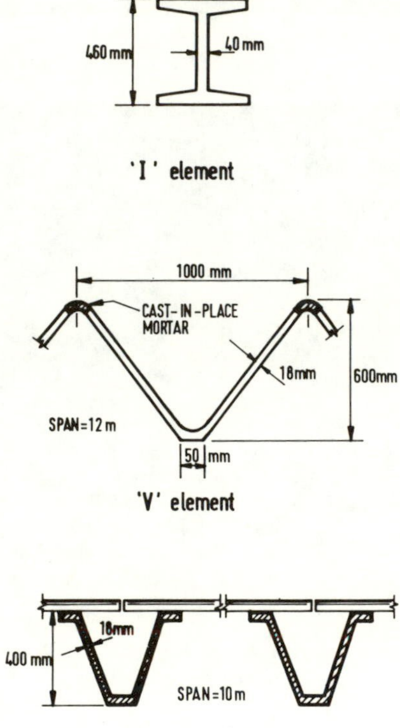


FIG. 10.20. 'I', 'V' and 'U' ferrocement beams for roofs and floor decks.

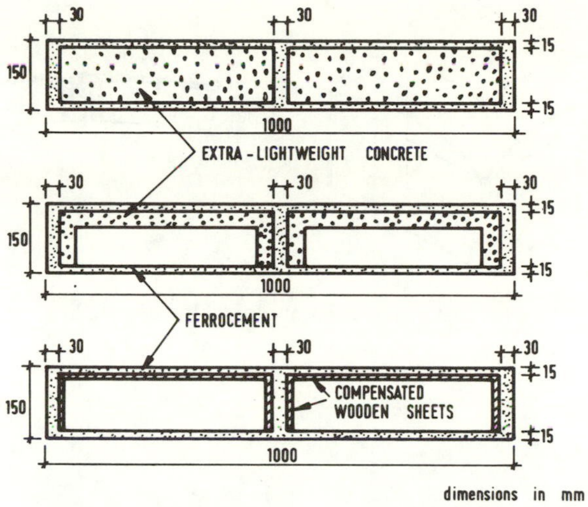


FIG. 10.21. Precast ferrocement caissons for roofs and floor decks.

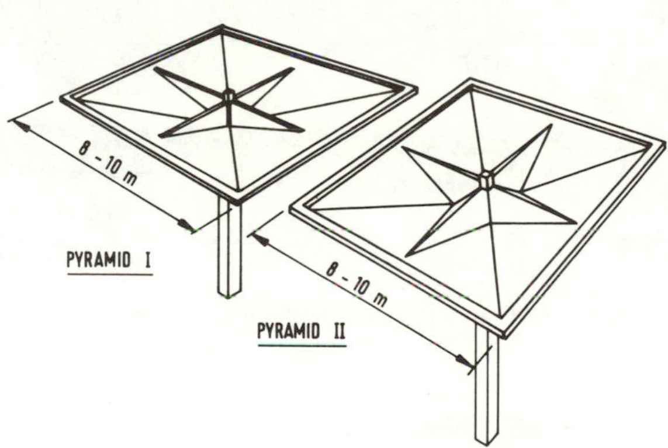


FIG. 10.22. Two different designs for pyramidal ferrocement folded roofing plates.

widest and 10 m high. Five domes were constructed using 25 mm thick ferrocement reinforcement with six layers of mesh of grid size 12 mm square, using wires of diameter 1 mm. Internal ribs reinforced with 8 mm steel rods wrapped with 6 layers of wire mesh were used as stiffeners. A sand : cement : water ratio of 1 : 1 : 0·6 was used for mortar. The shape of the dome was formed using wooden templates (Fig. 10.25).

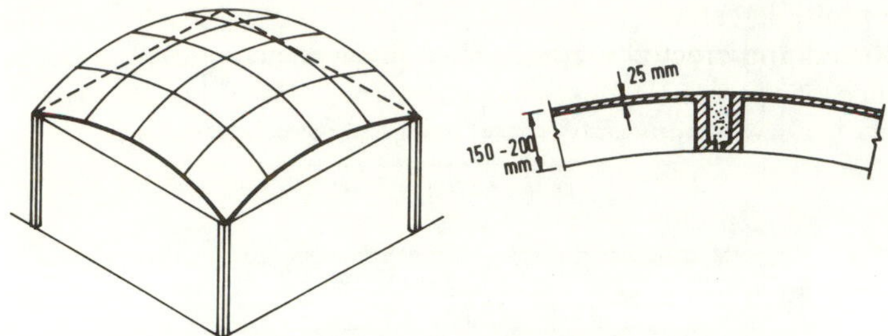


FIG. 10.23. Elliptic Paraboloidal Precast Shell.

FIG. 10.24. Ferrocement dome in Czechoslovakia.

### 10.6.4 Secondary Roofing Slabs

The current practice of using precast concrete units as secondary roofing slabs to provide heat insulation in high-rise buildings in the tropics can well be replaced by the use of ferrocement. In Singapore these slabs consist of 1500 × 600 × 50 mm precast cellular concrete panels containing a centrally placed layer of galvanized welded wire mesh of 50 mm square grids and 3·25 mm diameter. The panels are assembled side by side, each being supported on 150 × 150 × 225 mm precast hollow blocks placed on top of the structural roof to provide an air gap of 225 mm, and the joints are filled with bitumen to complete the secondary roof.

The multiple cracking exhibited by some panels, even before they are transported and erected in place, has caused some concern over the performance of the panels. Although these cracks may not be serious with respect to the ultimate strength requirement, they are potentially dangerous from a durability viewpoint and repair of these cracks after installation of the panels is quite expensive.

A study was conducted at the National University of Singapore to examine the current design, and several possible modifications have been suggested for improved performance of the panels. One of the suggestions

FIG. 10.25. Ferrocement mosque domes in Jordan.

was to use precast ferrocement panels using normal-weight mortar instead of cellular concrete. A thickness of 35 mm with two layers of galvanized fine wire mesh of 12·5 mm square grids and 1·2 mm wire diameter, separated by a layer of skeletal steel (welded wire mesh of 15 mm square grid and 3·25 mm wire diameter) as reinforcement, was found to be adequate. Because of the reduced thickness, the dead weight of the ferrocement panels would remain approximately the same as that of the cellular concrete panels.

Flexural tests were conducted under third-point loading on panels 1·5 m long and 0·6 m wide. They were simply supported on a span of 1·35 m. Figs. 10.26 and 10.27 show the cracking patterns of the cellular concrete and ferrocement panels respectively. Thus the uniform dispersion of re-

FIG. 10.26. Cracking pattern of cellular concrete panels.

FIG. 10.27. Cracking pattern of ferrocement panels.

inforcement in the ferrocement panels significantly improves the cracking behaviour. With regard to the cost, ferrocement panels are more expensive than cellular panels. However, it is expected that the frequency of replacement will be reduced, which may justify the higher initial investment.

Lightweight ferrocement slabs using cellular mortar as the matrix may also be used as secondary roofing slabs. An earlier investigation (Paramasivam *et al.*, 1985) has shown that the desired strength and stiffness can be achieved. The weight of the slab is reduced and the use of cellular mortar provides relatively better heat insulation.

### 10.6.5 Floor Slabs

Ferrocement in its conventional slender form is not suitable for use in floor slabs because of excessive deflection under heavy transverse loads. The problem of large deflection may however be overcome by employing thicker slabs, but the most efficient use of the material can only be achieved by introducing ribs, similar to those of a waffle slab, or by removing the redundant mortar from the tension zone, that is, by providing voids. A study is currently being conducted at the National University of Singapore on the behaviour and strength of various types of ferrocement hollow-core slabs. The results of four such slabs are briefly discussed here.

Fig. 10.28 shows the cross-sectional details of the slabs. All slabs were 2·5 m long, 600 mm wide and 90 mm thick. Slab S1 (not shown in Fig. 10.28) had a solid cross-section. It was included in the test programme to provide a basis for comparison of the behaviour and strength of the hollow-core slabs. Three different core shapes were investigated. Slabs designated as S2, S3 and S4 had longitudinal cores of circular, elliptical and rectangular shape respectively. The top and bottom faces were provided with two layers of fine welded wire mesh (wire diameter 1·2 mm, grid size 12·5 mm square) separated by a layer of skeletal steel (wire diameter 3 mm, grid size 150 mm square) at each face, with a clear cover of 5 mm. Web reinforcement was provided only in slab S3 by a layer of fine welded wire mesh (wire diameter 1·2 mm, grid size 12·5 mm square) in each rib. All slabs were tested in flexure under six equally spaced line loads and at a span length of 2·3 m.

The load versus midspan deflection curves of the slabs are presented in Fig. 10.29. It can be seen that the curves are remarkably similar, which indicates that the removal of concrete core in various amounts does not significantly affect the flexural stiffness of the slab. Similar behaviour was also observed with respect to the maximum crack width. The slabs remain

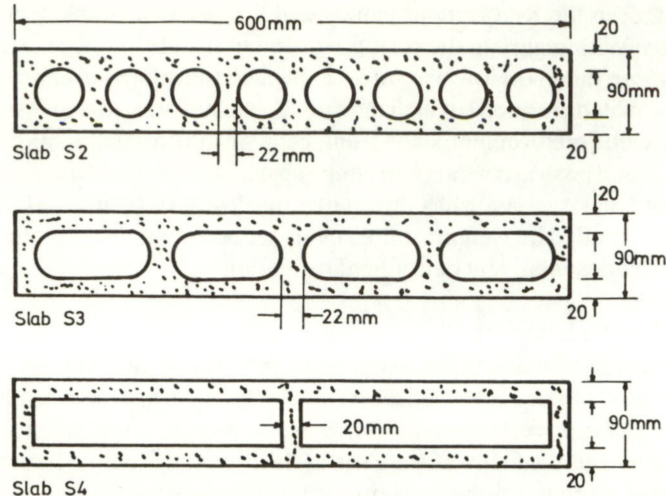


FIG. 10.28. Cross-section of hollow-core slabs.

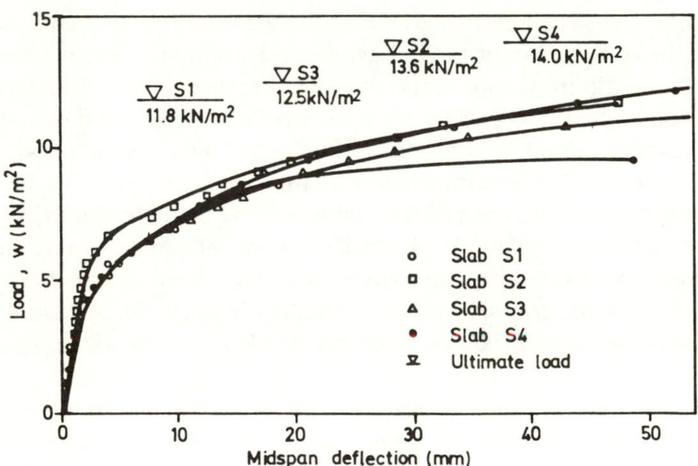


FIG. 10.29. Load–deflection curves of hollow-core slabs.

serviceable at up to 72% of the ultimate load, indicating superior crack controlling characteristics of ferrocement.

All slabs failed in the flexural mode without showing any shearing distress. The ultimate loads are shown in Fig. 10.29. The solid slab S1 failed at the lowest load, which might be attributed to the lower mortar strength ($f_{cu}$ = 22 N/mm$^2$) used as compared with those for the remaining specimens. The cube strengths of the mortar for the slabs S2, S3 and S4 were 28·8, 27·3 and 32·6 N/mm$^2$ respectively. The ultimate strength can be predicted with reasonable accuracy by CP 110 (1972). For the dimensions and the amount of reinforcement used in the present study, the slabs can sustain a minimum uniformly distributed ultimate load of 11·77 kN/m$^2$ at a span length of 2·3 m.

### 10.6.6 Ferrocement Formwork

In recent years, cold-formed steel decks are being increasingly used in the construction of floor slabs. The floor system is constructed by integrating the structural properties of the concrete and formed steel deck. The deck not only serves as a structural formwork supporting the construction and fresh concrete loads, but also furnishes the principal tensile reinforcement for the bottom fibres of the slab. Although formed steel deck offers many advantages, such as ease of construction, simplicity in the installation of building services and savings in construction time and labour, it is difficult to utilize the full potential capacity of the deck in resisting the external loads, due to poor bond between the two materials. The same concept has been extended to ferrocement decking in composite slab construction (Mansur *et al.*, 1984). While retaining most of the advantages of formed steel deck, the bond between the concrete and ferrocement was found to be sufficient to obtain full composite action.

Three types of deck profile, as shown in Fig. 10.30, were investigated. Each deck was 2·5 m long and 610 mm wide. The main reinforcement required for the slab was incorporated within the deck. A typical completed deck is shown in Fig. 10.31.

Composite slabs cast on these decks were then tested under third-point and simulated uniformly distributed loads. Tests have shown full composite action for all three deck profiles up to collapse. For the amount of reinforcement provided and the span (2·3 m) used in this study, the slabs sustained an equivalent uniformly distributed ultimate load of 36·6 kN/m$^2$. The use of ferrocement decks also improves the ductility and cracking behaviour (Fig. 10.32). Although all slabs, irrespective of deck profile,

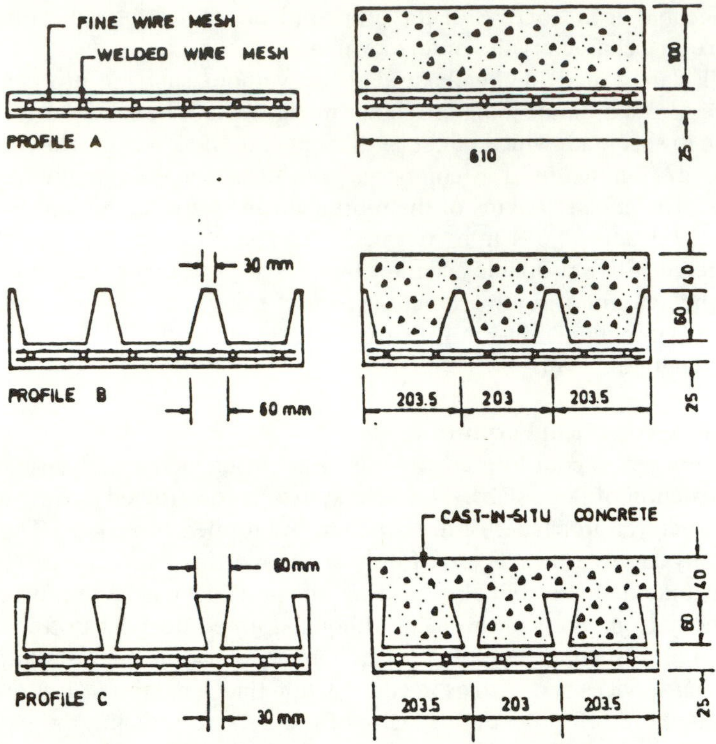


FIG. 10.30. Ferrocement deck profile.

behaved in a similar manner, the provision of ribs eliminates the use of temporary shoring during placement of in-situ concrete.

In a subsequent study, the amount of reinforcement was varied to study the composite action of ferrocement decks. Test results indicated that slabs cast using ferrocement decks can be designed to sustain heavy loads without loss of composite action, using conventional reinforced concrete analysis.

Sandowicz (2nd ISF, 1985) reported the use of precast ferrocement channel units in the construction of houses in Poland. The flange thickness of the channel units is 15 mm, with the rib thickness varying from 20 to 30 mm (Fig. 10.33). These units can be used for several functions within the structural system. The channel units are assembled as shown in Fig. 10.34, serving as formwork for reinforced concrete units such as spandrel beams, columns and floors.

FIG. 10.31. Typical ferrocement deck.

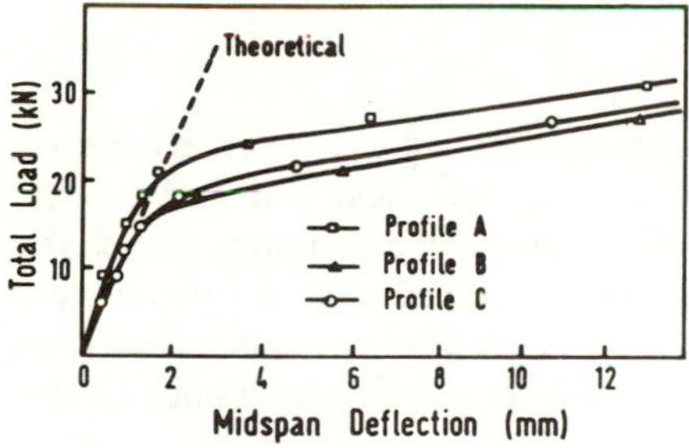


FIG. 10.32. Load–deflection curves of composite slabs.

Rosenthal and Bljuger (1985) investigated the flexural behaviour of rectangular composite beams of low-strength concrete encased in thin-skin elements made of high-strength ferrocement as a prelude in proposing a composite ribbed slab floor system (Fig. 10.35). These ferrocement skins not only serve as forms for reinforced concrete beams and slabs but also provide later concrete cover for the main reinforcement. The composite beams exhibited superior cracking characteristics due to enhanced flexural strength.

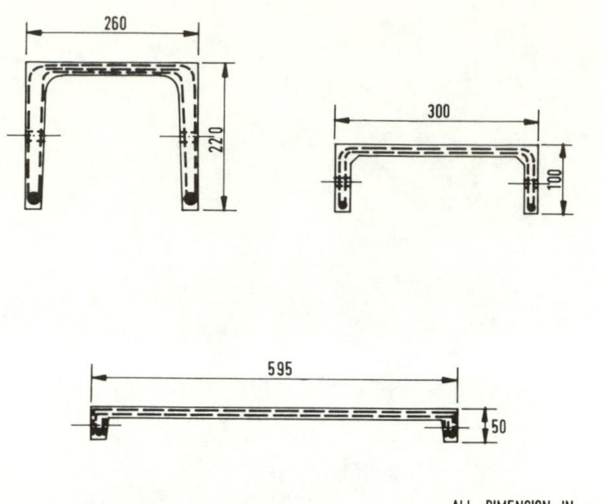


FIG. 10.33. Three types of channel unit.

### 10.6.7 Water-retaining Structures

#### *10.6.7.1 Water Tanks*

Ferrocement construction is labour intensive and suitable for rural applications in developing countries. It does not require heavy plant or machinery and, being a low-level technology, the construction skill can be acquired fairly quickly. Ferrocement is ideally suitable for the construction of thin-walled structures such as water tanks. Ferrocement tanks of 20 $m^3$ capacity have been in use in New Zealand since the late 1960s (NZPCA, 1968).

In high-rise public housing, steel tanks are commonly used to store water for domestic use. This may be due to the availability of steel tanks in prefabricated and modular form. However, the use of such tanks has many disadvantages such as high cost, rusting and consequent deterioration of the quality of stored water, frequent maintenance, and limited life span due to corrosion. A study was carried out to investigate the performance and practical application of ferrocement as an alternative material in the construction of prototype cylindrical and rectangular water tanks by considering cast-in-place construction (Paramasivam *et al.*, 1979). This type of construction poses several problems if it is to be carried out at the top of a building. Therefore research has been directed towards the construction of

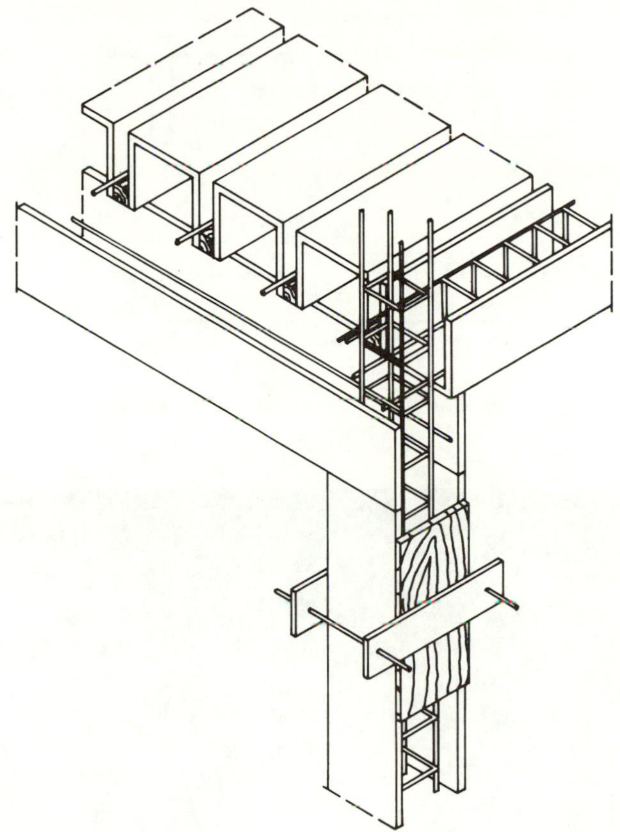

FIG. 10.34. Floor and column assembly using ferrocement units.

these tanks using precast ferrocement panels connected either by bolts and nuts or cast-in-situ joints. With this type of construction, better quality control under factory supervision can be achieved.

The construction of a tank 1·22 m square in plan and 1·22 m in height was undertaken as a prelude to mechanizing the construction of ferrocement water storage tanks in modules, which can be assembled to any required capacity. The reinforcement details for each panel are shown in Fig. 10.36. The precast panels were later joined together by nuts and bolts (Fig. 10.37). The joints were made leakproof by means of rubber gaskets. The performance of the tank has been carefully monitored under tank-full conditions in open exposure (Paramasivam and Nathan, 1984) since its

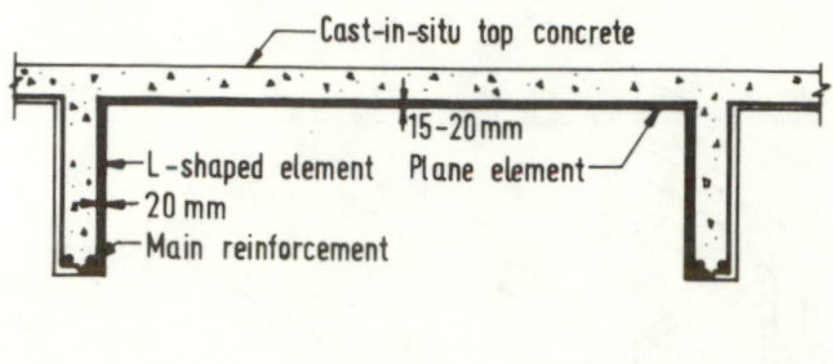


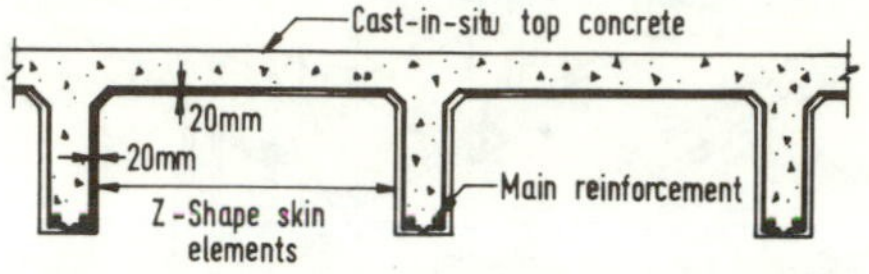


FIG. 10.35. Composite ribbed slab floor systems.

FIG. 10.36. Reinforcement layout for water tank panel.

construction five years ago. The tank is still in good condition. No serious problem of leakage, rusting or staining has so far been observed, except rusting of the mild steel bolts used in the edge connections. It is suggested that these bolts be replaced by stainless steel, galvanized steel or aluminium bolts.

Recently a study was conducted to propose a ferrocement water tank design and attendant construction technique for the collection of rain water in the rural areas of the Philippines. Funded by the International

FIG. 10.37. Assembled ferrocement water tank.

FIG. 10.38. Cylindrical water tank.

Development Research Centre, two prototype cylindrical ferrocement water tanks of the type shown in Fig. 10.38 were analysed, constructed and tested. A construction sequence suitable for rural applications was also presented.

#### *10.6.7.2 Flume*

Flumes for hydraulic model testing are often constructed by assembling steel and glass panels. However, the use of steel has several disadvantages such as high cost and frequent maintenance requirement. Studies conducted in the Hydraulic Engineering Laboratory at the National University of Singapore have shown that ferrocement can be used effectively as an alternative material for such water-retaining structures. A large rectangular flume was required to conduct model tests under wave action. With due regard to the cost of construction and maintenance and the stringent tolerance and alignment requirements, it was decided to use precast ferrocement panels for the base and vertical walls (Paramasivam *et* al., 1984).

The flume was 35 m long, 2 m wide and 1·3 m high, and was supported on steel frames at a spacing of 1·2 m. The base slab was 35 mm thick. Four layers of galvanized welded wire mesh (mesh size 13 × 13 mm, wire diameter 1·2 mm) were used as the main reinforcement, together with three layers of welded wire fabric (mesh size 150 mm × 150 mm; wire diameter 5 mm) as skeletal steel. The wall panels were 30 mm thick and contained three layers of main reinforcement separated by two layers of skeletal steel.

The precast panels were placed on the supporting frames and then connected by in-situ joints. The continuity of reinforcement between the panels was maintained by overlapping the fine wire meshes and connecting the skeletal steel by spot welding. Several bays of glass panels were used for the vertical wall to facilitate observation. Fig. 10.39 shows a complete view of the flume. A preliminary trial run conducted under 1 m deep water showed no leakage of the flume or any distress of the panels. The flume has been under extensive use for the past two years and, so far, there has been no problem of leakage or corrosion.

#### *10.6.7.3 Other Hydraulic Applications*

In China ferrocement hydraulic gates have been used since the 1960s (Zhao and Li, ISF, 1981). Recently the gates have been of gradually longer span and for higher head hydraulic construction. The ferrocement gates (Fig. 10.40) were found to possess several advantages including ease of construction, good crack resistance and lighter weight. Various geometric configurations of gates were used including slab–beam gates, folded plate

FIG. 10.39. An overall view of the flume.

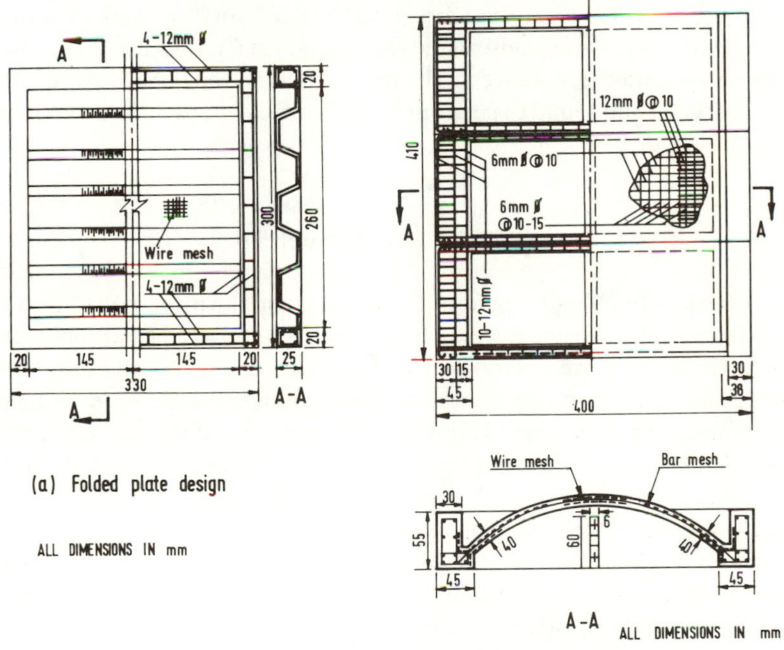


FIG. 10.40. Ferrocement hydraulic gates.

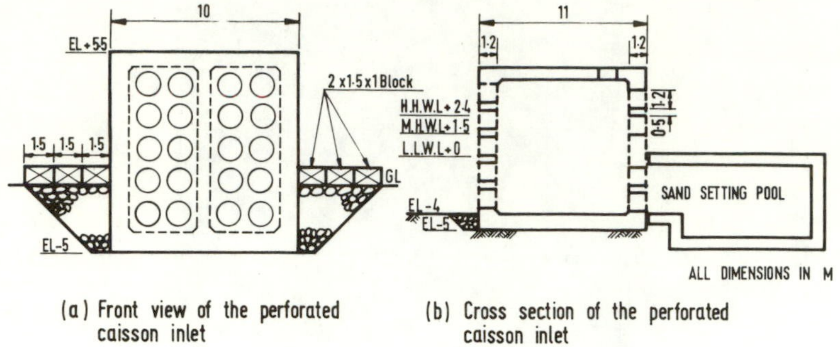


FIG. 10.41. Ferrocement breakwater caisson.

gates (Fig. 10.40a), wave plate gates, arch gates, short cylindrical shell gates (Fig. 10.40b), hyperboloid flat shell gates, etc.

Ferrocement caissons can also be used to reduce the large impact of wave forces on the front part of breakwaters. Su *et al.* (ISF, 1981) reported an experimental study to investigate the effect of ferrocement caissons in reducing the wave forces on a breakwater at the surf zone. In Taiwan two breakwaters had been constructed (Fig. 10.41) at Chia-Lan harbour on the basis of a hydraulic model test. The precast perforated ferrocement caisson was installed and found to perform satisfactorily, with significant economic benefits.

## 10.7 CONCLUSIONS

This chapter highlights some of the possible applications of ferrocement and recent research and development work carried out. These applications of ferrocement have demonstrated the quality and economy that can be achieved by using modern construction techniques such as guniting and prefabrication. Questions of durability and fire resistance of ferrocement have been raised. Fire ratings can be improved by using fireproof coatings for structural components. Durability is a long-term property that can be established only through experience. However, considering the basic constituents of ferrocement, that is, fine galvanized wire mesh and a rich mortar mix, coupled with better crack controlling characteristics, ferrocement is likely to offer more resistance to corrosion than conventional reinforced concrete in spite of its thinner cover. Experiences over the last twenty years have shown that corrosion of reinforcement and hence the durability of ferrocement is not a problem provided that sufficient care is taken during fabrication, that is, preparation of the reinforcement cages and plastering, and that proper maintenance work is carried out at regular intervals. The latter is similarly required for other construction materials. The successful performance of these structural elements confirms the viability, practicality and cost effectiveness of ferrocement in both urban and rural applications.

## REFERENCES

ACI COMMITTEE 549 (1982) State-of-the-art report on ferrocement. *Concrete International: Design and construction*, **4**(8), Aug., 13–38.

AMERICAN CONCRETE INSTITUTE (1977) *Building code requirements for reinforced concrete.* ACI Standard 318–77.

BARFOOT, J. (1985) Fine Forecast for Ferrocement. *Concrete*, **19**(5), May, 25–6.

BRITISH STANDARDS INSTITUTION (1972, amended 1976). *The structural use of concrete.* CP 110: Part 1.

BRITISH STANDARDS INSTITUTION (1985) *Structural use of concrete.* BS 8110: Part 1.

CASSIE, W. F. (1967) Lambot's boats—a personal discovery. *Concrete*, **1**, 380–2.

CASTRO, J. (1977) Ferrocement roofing manufactured on a self-help basis. *J. Ferrocement*, **7**(1), July, 17–27.

COLLEN, L. D. G. and KIRWAN, R. W. (1959) Characteristics of ferrocement. *Civil Eng. & Public Works Review*, **54**(632), Feb., 195–6.

DESAI, P. and JOSHI, A. D. (1976) Ferrocement load-bearing wall elements. *J. Struct. Div., ASCE*, **102**(ST9), Sept., 1903–15.

HAGENBACH, T. M. (1972) Experience of 300 commercially built craft in more than 20 countries, *FAO Seminar on the Design and Construction of Ferrocement Fishing Vessels*, Wellington, Oct., 30 pp.

ISF (1981) *Proceedings of the International Conference on Ferrocement.* Edited by S. P. Shah and E. Ronzoni, Bergamo, Italy, July.

2nd ISF (1985) *Proceedings of the Second International Conference on Ferrocement.* Edited by L. Robles-Austriaco, R. P. Pama, K. Shashi Kumar and E. G. Metha, Bangkok, Thailand, January.

JENNINGS, P. J. (1983) Mosque domes from modern techniques. *Concrete International: Design and Construction*, **5**(11), Nov., 41–4.

LEE, S. L., TAM, C. T., PARAMASIVAM, P., DAS GUPTA, N. C., RAVINDRARAJAH, R. SRI and MANSUR, M. A. (1983) Recent research: ideas tested at the University of Singapore. *Concrete International: Design and Construction*, **5**(11), Nov., 12–16.

LEE, S. L., MANSUR, M. A., PARAMASIVAM, P., ONG, K. C. G. and TAM, C. T. (1986) A study of sandwich wall panel. *J. Ferrocement*, **16**(3), July, 295–313.

MANSUR, M. A. and ONG, K. C. G. (1987) Shear strength of ferrocement beams. *ACI J. Structural J.*, **84**(1), Jan–Feb., 10–17.

MANSUR, M. A., ONG, K. C. G. and TAM, C. T. (1984) One-way concrete slab elements reinforced with ferrocement decking. *J. Ferrocement*, **14**(3), July, 211–20.

NAAMAN, A. E. and SHAH, S. P. (1971) Tensile tests of ferrocement. *ACI J.*, **68**(9), Sept., 693–8.

NAAMAN, A. E. and HOMRICH, J. R. (1986) Flexural design of ferrocement: computerized evaluation and design aids. *J. Ferrocement*, **16**(1), April, 101–16.

NANNI, A. and CHANG, W. F. (1986) Ferrocement sandwich panels under bending and edge-wise compression, *J. Ferrocement*, **16**(2), April, 127–40.

NATHAN, G. K. and PARAMASIVAM, P. (1974) Mechanical properties of ferrocement materials, *Proc. First Australian Conf. Engineering Materials*, Univ. New South Wales, Sydney, 309–32.

NERVI, P. L. (1951) Ferrocement: its characteristics and potentialities. *L'Ingegrnere*, Italy; English translation by Cement and Concrete Association, No. 60, London, 1956, 17 pp.

NEW ZEALAND PORTLAND CEMENT ASSOCIATION (NZPCA) (1968) *Ferrocement tanks and utility buildings*, Bulletin No. CP 10, Wellington.

PARAMASIVAM, P. and NATHAN, G. K. (1984) Prefabricated ferrocement water tank, *ACI J.*, **81**(6), June, 580–86.

PARAMASIVAM, P., NATHAN, G. K., and LEE, S. L. (1979) Analysis, design and construction of water tanks, *J. Ferrocement*, **9**(2), July, 115–28.

PARAMASIVAM, P., DAS GUPTA, N. C. and LEE, S. L. (1984) Design and construction of a prefabricated ferrocement flume, *J. Ferrocement*, **14**(4), Oct. 329–36.

PARAMASIVAM, P., MANSUR, M. A. and ONG, K. C. G. (1985) Flexural behaviour of lightweight ferrocement slabs. *J. Ferrocement*, **15**(1), Jan., 25–33.

PRAWEL, S. P. and REINHORN, A. (1983) Building construction: a competitive building material. *Concrete International: Design and Construction*, **5**(11), Nov., 12–21.

ROBLES-AUSTRIACO, L., PAMA, R. P. and VALLS, J. (1981) Ferrocement an innovative technology for housing, *J. Ferrocement*, **11**(1), Jan., 23–45.

ROSENTHAL, I. and BLJUGER, F. (1985) Bending behaviour of ferrocement-reinforced concrete composite. *J. Ferrocement*, **15**(1), Jan., 15–24.

SANDOWICZ, M. (1985) Application of ferrocement channel elements to housing, *Proceedings of the Second International Conference on Ferrocement, Bangkok, Thailand*, pp. 493–505.

SHAH, S. P. and BALAGURU, P. N. (1984) Ferrocement. In: *Concrete technology and design*, Vol. 2, New reinforced concretes, edited by R. N. Swamy, Surrey University Press, 1–51.

# INDEX

American Association of State Highway and Transportation Officials (AASHTO), bridge deck design, 17, 18, 224, 225, 226, 228
American Concrete Institute (ACI)
  analysis procedure for columns, 204, 207, 209
  Building Code, 197, 297, 300
  committee on ferrocement, 292, 294, 297, 298, 300
Analysis methods
  box girder bridges, 270–1, 285
  girders with deformable connection, 82–107
    finite element method, 84–5
    folded plate analysis method, 85–8
    simplified methods, 88–95
ASCE–AASHTO Task Committee on Flexural Members, on redundant bridge systems, 224, 242
Autostress design method, 3
Axially loaded columns, slender columns, 15–17

Battened columns, 166–8
  advantages of, 168
Bazant–Panula (BP) model, 82
Beam(s)
  analysis, flooring decks, 42–3
  behaviour, deformable connection in, 8–9
  web openings, with, 7–8, 53–76
Beam–column connections, 188–90, 218
Beam–columns, 202
Beam-and-slab bridge decks, 17–18, 221–46
Boat building, ferrocement used, 301, 302
Box girder bridges, 18–19, 249–86
  aesthetics of, 251
  analysis of, 270–1, 285
    analytical studies, 270–1
    bracing effects, 271–5
    experimental studies, 268–70
    top chord bracing effects, 277–8
    torsion boxes effects, 275–7
  bracing, effects of, 19, 280–4
    distortional, 280–2
    interconnecting, 282, 284,
  bracing systems used, 252–3, 262–3
    distortional braces, 262, 263
    stay-in-place forms, 263
    ties, 262, 263
    top chord bracing, 262–3
    torsional boxes, 262, 263
    Warren truss type, 259

Box girder bridges—*contd.*
bracing to reduce
distortion, 271–5, 284–5
rotation of open box girder, 262–3, 285–6
closed top and open top, 19
completed girders, 278–84
analytical studies, 278–80
distortional bracing effects, 280, 282, 283
interconnected bracing effects, 282, 284
web stiffeners, 280, 281
construction loading of, 254–8
concentric loading, 254–5
concrete handling system, loading due to, 256–7
eccentric loading, 254–5
finishing machine, loading due to, 256
formwork, loading due to, 256
typical loadings summarized, 257
wind loading, 255, 256
construction phase
bracing required, 252–3, 262–3
responsibility for bracing, 253
cross-sections used, 251–3
design summary, 284–6
distortion of open box section, 259–62, 284
economies possible, 251
finite strip method used, 270, 271, 272, 273, 274, 275, 276, 277, 279–80
flange bending analysis used, 270, 271, 274, 275, 277
general behaviour of, 258–64
bracing systems, 262–3
cross-sections, 259–62
bending distortion, 261
longitudinal bending, 259–60
mixed torsion, 260–1
torsional distortion, 262
single members, 258–9
interconnected members in, 264–8
ties used, 264–5
torsion boxes used, 268
x-braces used, 265–7

Box girder bridges—*contd.*
Mattock bridge model tests, 279
Nipissing Bridge tests, 280, 281, 282
rigid section analysis used, 271
rotation of open box girders, 285–6
bracing to reduce, 262–3, 285–6
stress analysis for, 285
torsion–bending analysis used, 270, 271, 272, 276
transverse web stiffeners used, 285
Braced frames, 187
Breakwater caissons, ferrocement used, 336
Bridge decks
analysis for load effects, 226–9
construction method effects, 235
current design practice for, 229–35
deflections due to dead load, 229
design codes for, 17–18, 224, 225
fatigue considerations, 232–4, 236–7
flexure analysis of, 18, 229–31
loading analysis of, 17–18, 224–6
dead loads, 225
live loads, 225–6
longitudinal bending moment of, 226–8
longitudinal shear, 228–9
design calculations, 231–2
negative moment section design calculations, 235–6
prestressing effects, 237–9
shear connector design calculations, 232–5
fatigue considerations, 232–4
ultimate strength considerations, 234–5
slab design calculations, 236
temperature differentials in, 246
thermal stresses in, 242–6
transverse bending moment of, 228
welded diaphragms, effect on, 239–42
Bridge girders, testing of, 18
Bridges. *See* Box girder bridges
BS 449: structural steel in building
column design method, 173, 175, 176

BS 5400: steel, concrete and composite bridges column design calculations, 170, 174, 187
  elastic modulus of concrete, 181
  loading analysis in, 4, 18
  source for design rules, 84
BS 5950: structural use of steelwork in building column design, 177, 182
  effective width concept used, 28–9
  masonry panels, on, 10, 11
  testing procedures, 5, 31
BS 8110: structural use of concrete on ferrocement, 327
  stress–strain curve, 181
  theory used for ferrocement, 297
Building Regulations, fire protection required, 190

Capacity reduction factors, 3
Castellated beams, 8
Clawson–Darwin strength model, 63–4, 65
Columns, 11–17, 163–219
  auxiliary reinforcement in, 218–19
  axially loaded, buckling curves for, 175, 176, 177
  Basu and Sommerville approach, 175, 177, 179, 187
  battened columns 166–8
  beam–column connections, 188–90, 218
  biaxial bending behaviour, 205–11
  biaxially loaded, 185–7
  braced frames, 187
  braces/tie-bars, effect on, 187, 212
  cased-strut design method, 173–4
  combination columns, 197
  comparative cost of, 199
  composite columns, definition of, 197
  compressive strengths of concrete in, 14–15
  concrete used, 165
  concrete-encased steel shapes, 166, 167
    load capacities of, 216, 217

Columns—*contd.*
  concrete-encased steel shapes—*contd.*
    longitudinal strength of, 206–7
    shear strength of, 211, 214, 215
    weak axis beam–column strength of, 213–14
  concrete-filled steel tubes, 166, 167, 199
    load capacities of, 216, 217
    longitudinal strength of, 208–9
    shear strength of, 211, 215
  contribution factors used, 13
  cross-sectional strength of, 200–5
  definition of, 165
  design methods, scope of, 11–12
  eccentrically loaded, 202
  effective length concept applied, 187, 216, 217
  end restraints in, 187–8
  failure state for, 203
  fire resistance of, 165, 166, 190, 197, 199
  interaction diagrams for, 14, 183, 185
  limit state for, 202–3
  pin-ended
    axially loaded, 173–82
      composite columns, 177–82
      steel columns, 174–7
    end moments, with, 182–5
  sequence-of-construction effects, 12–13
  short columns, 168–73
    eccentrically loaded, 170, 172–3
    interaction curve for, 173
    squash loads of, 169–70
    ultimate moment of resistance calculated, 170, 171
  slenderness of, 15–17, 176, 179, 214–18
  types of, 166–8, 198–9
  vectorial action effects, 13–14
  Virdi and Dowling approach, 179–80, 187
Compressed steel plates, classification of, 4
Concrete
  creep of, 202, 216
  elastic modulus in columns, 180–2

Concrete—*contd.*
  properties of, 293
  stress-strain curves for, 168, 181, 201
Connections, beam–column 188–90
CONSTRADO design guide, composite beams, 47, 48
Contribution factors, 13
Creep
  concrete in columns, 202, 216
  girders, 82–3
Cupolas, ferrocement used, 318

Deformable connection, girders with, 8–10, 79–112
  analysis methods used, 82–107
    finite element method, 84–5
    folded plate analysis method, 85–8
    simplified methods, 88–95
  effect of extent of concrete area, 111
  practical results using simplified analysis method, 107–11
Design codes
  package nature of, 15, 17
  use of more than one code, 4–5
  *see also* Eurocode . . .
Design compressive strength
  cube strength related to, 15
  cylinder strength related to, 15
Design rules, examples of differences between, 3–4
Detailing, beams with web openings, 75

Earthquake loadings, infilled frames, effect on, 128
Earthquake-resistant structures
  column construction used, 199, 215
Effective length concept; columns, 187, 216, 217
Effective modulus method, 83
Effective width method, flooring deck, 28–9
Elastic modulus
  concrete, definition of, 83
  cylinder strength related to, 15–16

Equivalent frame analogy, infilled frames, 119, 122
Equivalent strut analogy, infilled frames, 119, 120–2
Euler buckling curves, columns, 174, 175
Euler formula, 174
Eurocode 4: composite steel–concrete structures, 5
  column design procedures, 15
  contribution factors used, 13
  shear connections, 4
  source document for, 17
  testing procedures, 5–6
European Convention for Constructional Steelwork (ECCS)
  column design, 170, 174, 177, 184–5, 187
  elastic modulus of concrete, 181–2
  profiled steel sheeting, codes/procedures, 29, 31

Fatigue
  bridge decks, 232–4, 236–7
  ferrocement, 300
Ferrocement
  applications of, 301–36
    boat-building, 301, 302
    breakwater caissons, 336
    floor slabs, 320, 325–7
    formwork, 327–9, 330, 331, 332
    hydraulic gates, 334, 335, 336
    roofing elements, 315–22
      cylindrical shell roof, 316–17
      folded plate roof, 317, 318
      hyperbolic paraboloid roof, 315–16
      pyramidal elements, 317–18, 319, 321
    secondary roofing slabs, 322–5
    sunscreens, 311–15
    wall panels, 302–11
    water flumes, 334, 335
    water tanks, 330–4
  cement mortar used, 291–2, 293
  compressive behaviour of, 294, 299

Ferrocement—*contd.*
construction methods used, 300–1
curing of, 301
definition of, 291
design of structures, 299
disadvantages of, 291, 302
durability of, 336–7
early applications of, 291, 302
elastic behaviour of, 295–6
fatigue aspects of, 300
fire resistance of, 336
first used, 291, 301
flexural behaviour of, 295, 297–9
materials used, 291–2
mechanical properties of, 292–300
mesh used, 292, 293
modulus of elasticity of, 294
multiple cracking stage behaviour of, 295, 296–7, 298
plastering methods used, 301
reinforced concrete theory used, 297, 300
reinforcement required, 299
reinforcing parameters for, 292–4
sandwich panels, 302–10
compression testing of, 310–11
flexure of, 306–7
reinforcement of, 304–5
shrinkage of, 308–10
weathering of, 310
shear, behaviour in, 299
skeletal framing used, 301
tensile behaviour of, 294–7, 299–300
Finite element methods
flooring decks, 51
girders with deformable connection, 84–5
infilled frames
comparison with plasticity theory results, 147, 154
linear finite element analysis, 122–4
nonlinear finite element analysis, 124–5, 132–8
Finite strip analysis method, box girder bridges, 270, 271, 272, 273, 274, 275, 276, 277, 278–9
Fire protection
columns, 165, 166, 190, 197, 199
ferrocement structures, 336
Flange bending analysis, box girders, 270, 271, 274, 275, 277
Flooring
ferrocement used, 320, 325–7
steel sheeting use, 5–7, 21–51
Flooring deck systems
advantages of, 23–4
composite beam behaviour, 23, 42–51
analysis methods used, 42–7
end stud deformation, 49, 50
experimental study of, 47–9, 50
recent analytical developments used, 49, 51
slip variation, 49, 50
composite slab action, 23, 35–42
experimental study of, 36–9
factors affecting failure, 40–1
failure mechanisms in, 40
shear bond capacity evaluated, 35–6, 39
construction deformation of, 25–8
experimental study of, 31–3
prediction in design, 28–31, 34–5
disadvantages of, 24
first used, 22
jointing of sheeting in, 34
profiles used, 22–3
structural action of, 25
Flumes, ferrocement, 334, 335
Folded plate analysis
flooring deck structures, 49–50
stiffness matrix written, 31
girders with deformable connection, 85–8
modelling of connection between deck and girder, 87–8
stiffness matrix written, 85–6
profiled steel sheeting, 30, 34
Folded plate roof structures, ferrocement used, 317, 318
Formwork
ferrocement used, 327–9, 330, 331, 332
steel sheeting as, 22, 23

Highway bridge decks, loading levels for, 3–4, 224–6
Hollow tubes, 10, 95–107
Holorib profiled steel sheeting, 46, 47
Hydraulic gates, ferrocement used, 334, 335, 336

Inelastic behaviour, 83–4
Infilled frames, 10–11, 115–59
  cantilever beam representation of, 119
  composite behaviour of, 117, 118
  dynamic behaviour predicted, 126–8
  early research into, 117, 119
  earthquake loadings affected by cladding, 128
  equivalent frame representation of, 119, 122
  equivalent strut representation of, 119, 120–2
  finite element analysis, 132–8
  integral frames
    collapse mechanism for, 131
    definition of, 117
    design chart for, 156
    design example for, 155–8
    equivalent frame method used, 122
    finite element analysis used, 136–7, 137–8
    general behaviour of, 130
    interface stresses calculated by finite element analysis, 136–7
    load–deflection curves for, 130
    multistorey, plasticity theory used, 142–6
    panel stresses calculated by finite element analysis, 137–8
    single-storey, plasticity theory used, 139–42
    theoretical strength calculated, 147
  linear finite element analysis used, 122–4
  non-integral frames
    collapse mechanism for, 129, 131
Infilled frames—*contd.*
  non-integral frames—*contd.*
    definition of, 117
    design chart for, 157
    design example for, 158–9
    diagonal strut analogy used, 120–2
      finite difference methods used, 120, 121
      strain energy method used, 121
    finite element analysis used, 135, 136, 137
    general behaviour of, 129
    interaction problem encountered, 121, 122
    interface stresses calculated, 135, 136
    load–deflection curves for, 130
    multistorey frames, 152–3
    panel stresses calculated, 137
    single-storey frames, 146, 148–52
    theoretical strength calculated, 154
  nonlinear finite element analysis used, 124–5, 132–8
    frame bending moments calculated, 138
    frame elements used, 134
    interface elements used, 132–3
    interface stresses calculated, 135–7
    iterative procedure used, 135
    panel elements used, 133–4
    panel stresses calculated, 137–8
    representation of interfaces used, 123
  plasticity theory used, 125–6, 139–59
    integral frames
      multistorey frames, 142–6
      single-storey frames, 139–42
    non-integral frames
      multistorey frames, 152–3
      single-storey frames, 146, 148–52
  shear connectors used, 117, 118
Information
  inconsistency of, 2–3
  types of, 2

Interaction curves
beams with web openings, 58, 63, 64, 65, 67, 74
columns, 173, 183, 185, 207, 209
Interaction surfaces
biaxial bending of columns, 210
biaxially loaded column, 186
Italian Standard, composite girders, 109, 110

Lightweight concrete, effect on flooring decks, 48
Limit state design philosophy used, 2
composite columns, 202–3
Linear finite element analysis. *See* Finite element methods
Linear partial interaction design method
flooring decks, 7, 43, 44, 47
*see also* Partial interaction analysis
Location, effect on design approach, 3

Masonry panels, 10–11, 120
Mattock bridge model tests, 279
Metecno profiled steel sheeting, 46, 47
Mosque domes, ferrocement used, 319, 321, 323
Multistorey frames
integral infilled frames, 142–6
non-integral infilled frames, 152–3, 154–5

Nelson (T. R. W.) connector studs, 44, 47
Nipissing Bridge tests, 280, 281, 282
Nonlinear finite element analysis. *See* Finite element methods

Ontario Highway Bridge Design Code (OHBDC), 17, 18, 225, 226–8, 233, 236

Partial shear connection, 4
Partial-interaction analysis, 8, 9, 47
Permissible-stress design methods, 2
Perry–Robertson formula, 175, 176
Pin-ended columns. *See* Columns, pin-ended
Plasticity theory
infilled frames
compared with finite element analysis results, 147, 154
design charts, 126, 156–7
design examples, 153, 155–9
early work, 125–6
integral frames
multistorey frames, 142–6
single-storey frames, 139–42
non-integral frames
multistorey frames, 152–3
single-storey frames, 146, 148–52
Plate girders, 8
Plates. *See* Compressed steel plates
Prasannan–Luttrell formula, 42
Precision Metal Forming Ltd, profiled steel sheeting, 36, 46, 47
Prestressing, bridge decks affected by, 237–9
Prismatic folded plate structure, definition of, 85
Profiled steel sheeting, 5, 22
advantages of, 23–4
buckles formed, 26–8
composite slab formed, 23, 35–42
disadvantages of, 24
dovetail profile used, 22, 35
embossments on, 22, 23, 41, 42
experimental study of, 31–5
folded plate analysis used, 30, 34
joints in, 34
lateral deformation of, 26
matrix stiffness method of analysis used, 31
ponding deformation of, 26
prediction of behaviour in design, 28–31
profiles of, 22–3
structural action of, 23, 25
wet concrete loading of, 25–6, 31
prediction of behaviour, 32

Profiled steel sheeting—*contd.*
wet concrete loading of—*contd.*
simulation of, 31, 33
Push-out tests, 44, 45
modified version, 7, 44–5

Quasi-elastic effective modulus method, 83

Redwood–Poumbouras model, 64, 65
Redwood–Wong model, 64
Regional differences, 3
Rigid beam–column connections, 189–90
Rigid-jointed frames, 187–8
Robertson QL60 profiled steel sheeting, 46, 47
Roik–Bergman design method (for columns), 13–14, 15, 204, 207, 209
Roofing elements
ferrocement used, 315–22
cylindrical shell roof, 316–17
folded plate roof, 317, 318
hyperbolic paraboloid roof, 315–16
pyramidal elements, 317–18, 319, 321
secondary roofing slabs, 322–5

Secant shear stiffness, definition of, 133
Seismic-resistant structures. *See* Earthquake-resistant . . .
Semi-rigid beam–column connections, 189, 190
Sequence (of construction), column resistance affected by, 12–13
Serviceability requirements
beams with web openings, 62–3
ferrocement structures, 300
Shear bond capacity
determination of, 5–6, 35–6
experimental details, 36–9
empirical determination of, 41–2
Shear connection flexibility, definition of, 9, 93
Shear connectors, 4
beams with web openings, 75
bridge decks, 232–5
fatigue considerations, 232–4
ultimate strength considerations, 234–5
infilled frames, 117, 118
spacing on bridge decks, 234
welding of, 44
Shear transference characteristics, determination of, 5–6, 38–9
Shotcrete, ferrocement, 301
Single-storey frames
integral infilled frames, 139–42
non-integral infilled frames, 146–52
Slab bridges, 17–18, 221–46
Squash loads
columns, 202, 203
definition of, 169
Steel
properties of, 293
stress–strain curves for, 168, 201
Structural Specifications Liaison Committee (SSLC), analysis procedure for columns, 204, 207, 209
Stud shear connectors, 224, 232
stiffness and strength of, 6–7
Sunscreens, ferrocement used, 311–15
Swimming pools, ferrocement roofs, 318, 322

Thermal stresses, bridge decks, 242–6
Tower structures
analysis of, 95–107
representation by simple beam, 102–5
Trost–Bazant method, 83
Tubular structures, 10, 95–107
shear stresses at concrete–steel interface, 96, 97
variable cross-section modelled, 99

USA, slab bridge design, 17, 221–46; *see also* American . . . (AASHTO)

Vectorial action effects, columns, 13–14

Wall panels, ferrocement used, 302–11
Water tanks, ferrocement used, 330–4
Web openings, beams with, 7–8, 53–76
  advantages/disadvantages of, 55
  Clawson–Darwin model, 63–4, 65
  design of openings, 63–75
    compared with test results, 74–5
    detailing guidelines, 75
    interaction procedure, 73–4
    maximum moment capacity calculated, 67–9
    maximum shear capacity calculated, 69–73
    procedures used, 65–74
  factors affecting opening capacity, 59, 62

Web openings, beams with—*contd.*
  design of openings—*contd.*
  failure modes of, 57
  forces acting at opening, 55–7
  research into, 55
  response to loading of, 57–62
  ribbed slabs, 58–9, 60–1
  serviceability considerations, 62–3
  solid slabs, moment–shear interaction, 58
  strength at openings, 63–4
Welded diaphragms, bridge decks affected by, 239–42
Welded shear connectors, 44, 47
Wet concrete loading, deformation of steel sheeting under, 25–6, 31
  prediction of behaviour, 32
  simulation of, 31, 33